Structural Integrity

Volume 16

The *Structural Integrity* book series is a high level academic and professional series publishing research on all areas of Structural Integrity. It promotes and expedites the dissemination of new research results and tutorial views in the structural integrity field.

The Series publishes research monographs, professional books, handbooks, edited volumes and textbooks with worldwide distribution to engineers, researchers, educators, professionals and libraries.

Topics of interested include but are not limited to:

- Structural integrity
- Structural durability
- Degradation and conservation of materials and structures
- Dynamic and seismic structural analysis
- Fatigue and fracture of materials and structures
- Risk analysis and safety of materials and structural mechanics
- Fracture Mechanics
- Damage mechanics
- Analytical and numerical simulation of materials and structures
- Computational mechanics
- Structural design methodology
- Experimental methods applied to structural integrity
- Multiaxial fatigue and complex loading effects of materials and structures
- Fatigue corrosion analysis
- Scale effects in the fatigue analysis of materials and structures
- Fatigue structural integrity
- Structural integrity in railway and highway systems
- Sustainable structural design
- Structural loads characterization
- Structural health monitoring
- Adhesives connections integrity
- Rock and soil structural integrity.

**** Indexing: The books of this series are submitted to Web of Science, Scopus, Google Scholar and Springerlink ****

This series is managed by team members of the ESIS/TC12 technical committee.

Springer and the Series Editors welcome book ideas from authors. Potential authors who wish to submit a book proposal should contact Dr. Mayra Castro, Senior Editor, Springer (Heidelberg), e-mail: mayra.castro@springer.com

More information about this series at http://www.springer.com/series/15775

Emmanuel Gdoutos · Maria Konsta-Gdoutos
Editors

Proceedings of the Third International Conference on Theoretical, Applied and Experimental Mechanics

Springer

Editors
Emmanuel Gdoutos
Office of Theoretical and Applied Mechanic
Academy of Athens
Athens, Greece

Maria Konsta-Gdoutos
Civil Engineering Department
The University of Texas at Arlington
Arlington, TX, USA

ISSN 2522-560X ISSN 2522-5618 (electronic)
Structural Integrity
ISBN 978-3-030-47882-7 ISBN 978-3-030-47883-4 (eBook)
https://doi.org/10.1007/978-3-030-47883-4

This Springer imprint is published by the registered company Springer Nature Switzerland AG
The registered company address is: Gewerbestrasse 11, 6330 Cham, Switzerland

Preface

This volume contains sixty-one six-page papers/two-page abstracts presented at the "Third International Conference on Theoretical, Applied and Experimental Mechanics," (ICTAEM_3) held in Athens, Greece, June 14–17, 2020. The papers/abstracts are arranged in three topics and two special symposia. The three topics are: "Materials: Properties, Manufacturing, Modelling" with sixteen contributions, "Fracture" with five contributions and "Miscellaneous (Computational Mechanics, Dynamics, Nanomaterials, Plasticity, Wave Propagation)" with eleven contributions. The two symposia are: "Symposium on Theoretical and Experimental Approaches in Mechanics of Solids with Nonhomogeneities and Defects," by Roman Kushnir with twenty-four contributions and "Fracture Nanomechanics, Fatigue and Fracture at Small Scales (experiments and simulations)" by Pasquale Gallo with five contributions. The papers of the tracks have been contributed from open call, while the papers of the two symposia have been solicited by the organizers, to who we are greatly indebted.

ICTAEM_3 will focus in all aspects of theoretical, applied and experimental mechanics including biomechanics, composite materials, computational mechanics, constitutive modeling of materials, dynamics, elasticity, experimental mechanics, fracture, mechanical properties of materials, micromechanics, nanomechanics, plasticity, stress analysis, structures, wave propagation.

The attendees of ICTAEM_3 will have the opportunity to interact with the most outstanding world leaders and get acquainted with the latest developments in the area of mechanics. ICTAEM_3 will be a forum of university, industry and government interaction and exchange of ideas in an area of utmost scientific and technological importance.

We are sure that besides the superb technical program, the attendees of ICTAEM_3 will enjoy the city of Athens. It is considered as the ancient capital of the world with a long fascinating history dating from the Neolithic Age, the city of the goddess of wisdom Athena, a center for the arts, learning and philosophy, home of Plato's Academy and Aristotle's Lyceum, the birthplace of democracy and of Western civilization, of the most intellectual and artistic ideas, the cradle of democracy, the host city of the first modern-day Olympic Games in 1896, and the Summer Olympics in 2004.

More than a hundred participants attended ICTAEM_3. The participants of ICTAEM_3 came from 20 countries. Roughly speaking 14% came from Europe, 24% from the Far East, 7% from the Americas and 55% from other countries. We are happy and proud to have welcomed in the historic city of Athens well-known experts who came to discuss problems related to the analysis and prevention of failure in structures. Presentation of technical papers alone is not enough for effective scientific communication. It is the healthy exchange of ideas and scientific knowledge, formal and informal discussions, together with the plenary and contributed papers that make a fruitful and successful meeting. Informal discussions, personal acquaintance and friendship play an important role.

We very sincerely thank the authors who have contributed to this volume, the symposium/sessions organizers for their hard work and dedication and the referees who reviewed the quality of the submitted contributions. The tireless effort of the members of the organizing committee as well as of other numerous individuals, and people behind the scenes is appreciated. We are deeply indebted to Dr. Stavros Shiaeles for his hard work and dedication in the organization of the conference. Finally, a special word of thanks goes to Dr. Maria Shiaeles for her continuous collaboration and support.

March 2020

Emmanuel Gdoutos
Maria Konsta-Gdoutos

Contents

Materials: Properties, Manufacturing, Modelling

High-Speed Mode-I Delamination . . . 3
Tianyu Chen, Christopher M. Harvey, Simon Wang, and Vadim V. Silberschmidt

Characterization of Dynamic Fractures for Compressed PMMA Material . . . 9
C. Huang

Tensile Behavior of a Titanium Alloy Additively Manufactured via Selective Electron Beam Melting . . . 14
Yinling Zhang, Shoujiang Qu, Aihan Feng, and Daolun Chen

The Rule of Mixtures on the Elastic Buckling Response of Rectangular GLARE FML Plates Under Shear Stresses . . . 20
Costas D. Kalfountzos, George S. E. Bikakis, and Efstathios E. Theotokoglou

Evaluating Bonding Characteristics of Joggled Lap CFRP Using Acoustic Emission Techniques . . . 26
Claudia Barile, Caterina Casavola, Giovanni Pappalettera, Carmine Pappalettere, and Paramsamy Kannan Vimalathithan

Experimental Study of Low Concentration Diffusible Hydrogen Effect on Mechanical Behaviour of Carbon Steel . . . 32
I. M. Dmytrakh, R. L. Leshchak, and A. M. Syrotyuk

Effects of Temperature on Tensile and Fracture Behavior of Dissimilar Metal Welded Joint for Nuclear Safe-End . . . 38
Lei Wang, Yang Liu, Xiu Song, and Jiahua Liu

A Multi-attribute Information Based Method of Material Strength Distribution Fitting . . . 44
Liyang Xie, Bo Qin, and Ningxiang Wu

Performance of a Carburizing Bearing Steel After Surface Nanosized 52
Yanhui Wang, Fucheng Zhang, Zhinan Yang, Hezong Li, and Qian Yang

Residual Strength and Toughness of Nano and Micro Scale Fiber Reinforced Cementitious Composites 58
Panagiotis A. Danoglidis, Maria S. Konsta-Gdoutos, and Emmanuel E. Gdoutos

Experimental and Numerical Strategy for the Determination of Mechanical Properties Related to Human Cortical Bone Fracture . . . 64
J.-L. Tailhan, Y. Godio-Raboutet, and C. Boulay

Effect of External Field Treatment on the Microstructure and Deformation Behavior of Nickel-Base Superalloy 70
Yang Liu, Lei Wang, Xiu Song, and Yao Wang

Dynamic Speckle Contrast Imaging for Surface Defect Detection 76
Rishikesh Kulkarni, Shanta Patil, and Pramod Rastogi

Experimental Investigation of the Elastic Modulus of Timber by Using Density and Fiber Orientation 82
Wei-Chung Wang and Tzu-Yu Kuo

Effect of pH on Microstructure and Properties of Ultrasonic-Assisted Electroless Ni-P Coatings 86
Jingjing Cao, Xiaoyu Wang, Yifan Zhao, and Guofang Kong

Tailoring the Toughness and Electrochemical Capacitance of CNT Reinforced Mortars 88
Maria S. Konsta-Gdoutos, Panagiotis A. Danoglidis, and Myrsini E. Maglogianni

Fracture

The Creep Resistance of Short-Fibre Reinforced Metal Matrix Composites 93
Vaclav Sklenicka, Kveta Kucharova, Milan Svoboda, Petr Kral, Marie Kvapilova, and Jiri Dvorak

Adhesive Joint Fracture Under Combined Pulsed and Vibrational Loading 100
Yuri Petrov, Andrey Logachev, Nickolay Granichin, and Grigory Volkov

Study on the Fracture Mechanism of the P91 Steel During Small Punch Tensile Testing 106
Hezong Li, Hao Chen, H. K. Al-Abedy, and Wei Sun

Damage Accumulation and Growth Models for the Creep-Fatigue Interaction . . . 112
A. V. Tumanov and V. N. Shlyannikov

Crack Growth and Plastic, Fracture Process and Damage Zones Behavior Under Mixed Mode Loading . . . 117
Valery Shlyannikov and Daria Fedotova

Miscellaneous (Computational Mechanics, Dynamics, Nanomaterials, Plasticity, Wave Propagation)

Parametric Resonance of a Micro-beam with Flexible Support Under Fringing Field Excitation . . . 127
Zhichong Wang

Measurement of Elastic Nonlinearities Using the Fundamental Edge Wave Mode . . . 133
James Martin Hughes, Andrei Kotousov, and Ching-Tai Ng

Unsteady Axisymmetric Electro-Magneto-Elastic Oscillations of a Continuous Cylinder Under the External Displacement Field . . . 140
Vladimir Vestyak and Vasily Scherbakov

Development of Micro-mechanical Models of Fatigue Damage . . . 145
James Vidler, Andrei Kotousov, and Ching-Tai Ng

Research on Springback of 5A02 Aluminum Alloy Considering Thickness Normal Stress in Hydroforming . . . 151
Peicheng Jiang, Lihui Lang, and Sergei Alexandrov

On the Friction Test for Metal Forming Applications . . . 158
Sergei Alexandrov, Marko Vilotic, and Dragisa Vilotic

From the Linear Theory of Elasticity to Bending Equations for Beams with Variable Cross-Section . . . 162
Boris Gusev and Vasily Saurin

On Different Representations of Equations for Beam with Variable Cross-Section . . . 164
Vladimir Poliakov and Vasily Saurin

Shear Residual Stresses Induced by Torsional Loading . . . 166
D. Pérez Gallego, J. Ruiz Hervías, and D. A. Cendón Franco

Elastoplastic Analysis of Re-entrant Cell Formation in Auxetic Foams Under Compression and Heat Treatment . . . 168
Yun-Che Wang, Tsai-Wen Ko, and Hsiang-Wei Lai

Analytical Modeling of the Thermo-Mechanical Behavior of a Friction-Pendulum Seismic Isolator . . . 170
Todor Zhelyazov

Symposium on Theoretical and Experimental Approaches in Mechanics of Solids with Nonhomogeneities and Defects, by Roman Kushnir

Damage Accumulation and Limit State of Welded Pipelines with Corrosion-Erosion Metal Loss Due to Ultra-Low-Cycle Fatigue 175
Alexey Milenin, Elena Velikoivanenko, Galina Rozynka, and Nina Pivtorak

Unsteady Elastic-Diffusion Oscillations of a Simply Supported Kirchhoff Plate Under the Distributed Transverse Load Action 181
O. A. Afanasieva and A. V. Zemskov

A Dynamic Contact Problem of Torsion that Reduces to the Singular Integral Equation with Two Fixed Singularities 187
V. Popov and O. Kyrylova

Some Dynamic Problems for Layered Composites 193
Sergey Pshenichnov

The Influence of Geometric Parameters on the Bearing Capacity of Transformable-Volume Structure 198
L. M. Lobanov, V. S. Volkov, O. V. Makhnenko, S. M. Kandala, and Y. V. Borovyk

The Influence of Mixed Conditions on the Stress Concentration in the Neighborhood of Interfacial Inclusions in an Inhomogeneous Transversely Isotropic Space 204
Oleksandr Kryvyi and Yurii Morozov

On an Approach to the Thermoelastic Analysis of Nonhomogeneous Solids 210
Roman Kushnir and Yuriy Tokovyy

The Stress State Near the Corner Point of the Interface of Piecewise Homogeneous Plane in the Presence of Interfacial Shear Cracks 216
V. M. Nazarenko and A. L. Kipnis

Coupled Non-stationary Longitudinal Vibrations of an Infinite Electromagnetoelastic Rod 222
Thong D. Pham, Dmitry V. Tarlakovskii, and Vladimir A. Vestyak

The Effect of Cracks Interaction in Materials Under Loading Along Cracks 228
Viacheslav Bogdanov and Vladimir Nazarenko

Safe Loading of the Orthotropic Plate with Periodic System of Collinear Cracks 235
Olga Bogdanova

Boundary Element Analysis of Partially Debonded Shell-Like Rigid Inclusions in Anisotropic Medium 242
Heorhiy Sulym, Nataliia Ilchuk, and Iaroslav Pasternak

The Transmission of an Acoustic Wave Through a Composite Plate with Energy Dissipation in a Material Taken into Account Based on Three-Dimensional Equations of Elasticity Theory 248
V. N. Paimushin and R. K. Gazizullin

Vibration of the Plate with Integral Layer Damping: Experimental and Theoretical Studies 254
V. N. Paimushin, V. A. Firsov, V. M. Shishkin, and R. K. Gazizullin

Thermomagnetoelectroelasticity of Bimaterial Solids with High Temperature Conducting Interface and Thin Internal Inhomogeneities 261
Andrii Vasylyshyn, Iaroslav Pasternak, and Heorhiy Sulym

Fracture of Composite Materials Under Compression Along Cracks . . . 268
Mykhailo Dovzhyk, Viacheslav Bogdanov, and Vladimir Nazarenko

Development of Welded Elements of the Railway Freight Car Bogie with Increased Characteristics of Fatigue Resistance and Survivability 273
L. Lobanov, O. Makhnenko, A. Pustovoy, and S. Solovey

Electron Beam Additive Technology Optimization Using Mathematical Modeling 280
Oleh Makhnenko, Nikita Ananchenko, Stepan Kandala, and Andriy Babenko

Non-stationary Contact Problems for Thin Shells and Solids 287
Grigory Fedotenkov and Dmitry Tarlakovskii

Shear Buckling Mode and Failure of Sandwich Specimen Facing Layer Under Four-Point Bending 293
V. N. Paimushin, M. V. Makarov, S. A. Kholmogorov, and M. A. Shishov

Model of Low-Cycle Deformation of Stainless Steel Under Soft Loading 301
Vasilii Gorokhov, Sergei Kapustin, Yuriy Churilov, Dmitriy Kazakov, and Dmitriy Zhegalov

Creep and Long-Term Strength of Structures 305
Ivan Volkov, Leonid Igumnov, and Svetlana Litvinchuk

Mathematical Modeling of Nonlinear Dynamic Deformation and Failure of Metal-Plastic Shells of Revolution 309
Leonid Igumnov, Nikolai Abrosimov, Nadezhda Novoseltseva, and Vasilii Gorokhov

Impact of Transient Pressure on a Half-Space with Membrane Type Coating 312
Elena Mikhailova, Grigory Fedotenkov, and Dmitry Tarlakovskii

Fracture Nanomechanics, Fatigue and Fracture at Small Scales (experiments and simulations) by Pasquale Gallo

Some Considerations on Stress Intensity Factor at Atomic Scale 319
Pasquale Gallo

Initial Intergranular Cracking of Ni-Base Superalloys Due to the Degradation of the Crystallinity of Grain Boundaries Under Creep-Fatigue Loading 325
Wataru Suzuki, Yifan Luo, Kenta Ishihara, Kens Suzuki, and Hideo Miura

Characterization of Microstructurally Small Fatigue Crack Behavior 332
Evgenii Malitckii, Heikki Remes, Pauli Lehto, and Sven Bossuyt

An Atomic-Level Unified Criterion for Brittle Fracture 334
Kai Huang, Takashi Sumigawa, Takahiro Shimada, and Takayuki Kitamura

Fatigue of Micro/Nano Metals 337
Takayuki Kitamura, Takashi Sumigawa, and Kai Huang

Author Index 339

Materials: Properties, Manufacturing, Modelling

High-Speed Mode-I Delamination

Tianyu Chen[1], Christopher M. Harvey[1(✉)], Simon Wang[1,2], and Vadim V. Silberschmidt[1]

[1] Loughborough University, Loughborough, Leicestershire LE11 3TU, UK
c.m.harvey@lboro.ac.uk
[2] Hebei University of Engineering, Handan 056038, China

Abstract. The analytical theory of mode-I delamination propagation in double cantilever beams (DCBs) under high loading rates is developed by accounting for structural vibration and wave dispersion, and by using Euler-Bernoulli beam theory. The developed analytical theory is validated against experimental data and against finite element method (FEM) simulations, showing excellent agreement. It is shown that the developed analytical theory can accurately calculate energy release rate (ERR) for both stationary and propagating delamination, and that structural vibration can have a significant effect on ERR. It is further shown how the theory can be used to post-process experimental results from high-speed delamination tests to determine fracture toughness. Among other examples, the work is therefore expected to be useful to engineers and academic researchers to determine the initiation, arrest and propagation fracture toughness of laminated materials against delamination. The developed theory also provides useful benchmark solutions for the development of numerical codes.

Keywords: Double cantilever beam · Dynamic energy release rate · Dynamic fracture toughness

1 Introduction

Double cantilever beams (DCBs) have been widely used to determine the fracture toughness of materials under mode I loading conditions. A standard method has been established for mode-I interlaminar fracture toughness testing with DCBs [1], but it is only applicable for quasi-static loading. Under dynamic loads, where inertia is significant, there has been comparatively little research carried out. Moreover, what studies there have been did not consider structural vibration and consequently provided a 'smoothed' solution for energy release rate (ERR). Vibration, however, can be significant under high loading rates, which causes ERR to oscillate with a spectrum of frequencies and amplitudes.

For a stationary crack, the authors have presented an analytical solution for the ERR of a DCB under constant high-rate opening [2], which is accurate for the first vibration mode but not for high order vibration modes. Later, the authors developed the analytical solution for the same case, but including higher order vibration modes, for cyclic loading [3], and for cracks on elastic interfaces [4]. The improved representation

E. Gdoutos and M. Konsta-Gdoutos (Eds.): ICTAEM 2020, STIN 16, pp. 3–8, 2020.
https://doi.org/10.1007/978-3-030-47883-4_1

of the uncracked region in the latter work captures the crack tip deformation more accurately, and consequently, the ERR also. The elastic interface in that work is also relevant for the study of crack process zones.

It is worth noting that Ref. [4] uses a global energy balance approach and shows that the dispersive propagation of flexural waves must be considered in order to capture contributions to the ERR from higher-order vibration modes. Qualitatively, this is because not all the energy stored in the vibrating beam is immediately available to the crack tip to drive propagation at a point in time since waves propagate with a finite wave speed. Energy flow in and out of the crack tip region must be considered, which means accounting for the dispersive propagation of flexural waves. By contrast, if the conventional global energy balance approach is used, as in Refs. [5, 6], then when vibration is included, the ERR diverges as higher order vibration modes are added [2].

Whereas the previous work has been for either been stationary cracks [2–6], or for propagating cracks without vibration effects [5, 6], this article will present for the first time the analytical theory for the ERR of a propagating crack with vibration effects at high loading rates. Among other examples, the complete theory is expected to be useful to engineers and academic researchers to determine the dynamic initiation and propagation fracture toughness of laminated materials. It also provides useful benchmark solutions for the development of numerical codes.

2 Analytical Theory for DCBs Under High Rates of Loading

Consider the DCB geometry shown in Fig. 1a, where v is the constant opening rate applied from $t = 0$, and vt is the opening displacement. The Euler-Bernoulli beam assumption applies since $h \ll a$, where h is the thickness of one DCB arm, and a is the crack length. Figure 1b shows the coordinate system and the effective boundary conditions under which the crack tip is assumed to be clamped.

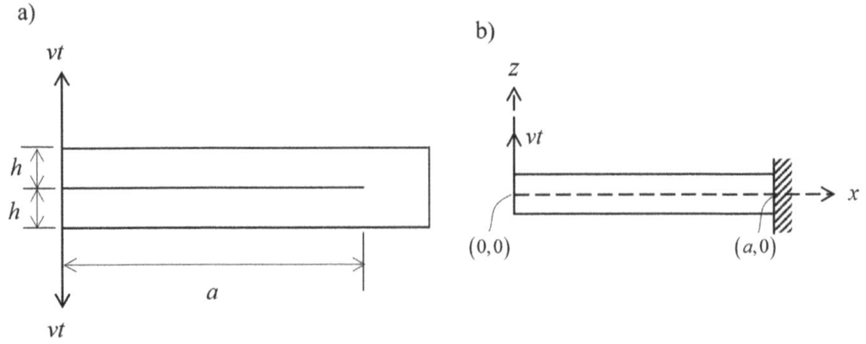

Fig. 1. (a) Symmetric double cantilever beam; (b) effective boundary condition

2.1 ERR Solution for a Stationary Crack

According to Ref. [2], the deflection of the beam in Fig. 1b, which is effectively half a DCB, under constant opening rate is

$$w(x,t) = va^2\sqrt{\frac{\rho A}{EI}}\sum_{i=1}^{\infty}\frac{\Lambda_i}{\lambda_i^3}\phi_i(a-x)\sin(\omega_i t) + \left(\frac{x^3}{2a^3} - \frac{3x}{2a} + 1\right)vt, \tag{1}$$

where for the ith mode, $\phi_i(a-x)$ is the mode shape, ω_i is the angular frequency, λ_i is determined by the frequency equation of $\tanh(\lambda_i) - \tan(\lambda_i) = 0$, and Λ_i is given as $\Lambda_i = (-1)^i\sqrt{\sigma_i^2+1} + \sqrt{\sigma_i^2-1}$ with $\sigma_i = [\cosh(\lambda_i) - \cos(\lambda_i)]/[\sinh(\lambda_i) - \sin(\lambda_i)]$.

According to Freund [7], the dynamic ERR, is

$$G = \frac{M^2(a,t)}{bEI}\left(1 - \frac{\dot{a}^2}{c_0^2}\right), \tag{2}$$

where $c_0^2 = E/\rho$. By combining Eqs. (1) and (2), the dynamic ERR for a stationary crack (that is, $\dot{a} = 0$) is

$$\begin{aligned} G &= \frac{EI}{b}\left[w^{(2)}(a,t)\right]^2 \\ &= \frac{9EIv^2t^2}{ba^4} + \frac{12\sqrt{\rho AEI}v^2t}{ba^2}\sum_{i=1}^{\infty}\frac{\Lambda_i}{\lambda_i}\sin(\omega_i t) + \frac{4\rho Av^2}{b}\left[\sum_{i=1}^{\infty}\frac{\Lambda_i}{\lambda_i}\sin(\omega_i t)\right]^2, \end{aligned} \tag{3}$$

where b is the beam width, and A is the area of the beam's cross-section.

2.2 ERR Solution for a Propagating Crack

Based on Eq. (2), and using the same technique as in Ref. [4] to account for the wave dispersion, the analytical ERR solution for a propagating crack is obtained as

$$G = \frac{9EIv^2t^2}{ba^4}\left(1 - \frac{\dot{a}^2}{c_0^2}\right) + \frac{12\sqrt{\rho AEI}v^2t}{ba^2}\left(1 - \frac{\dot{a}^2}{c_0^2}\right)\sum_{i=1}^{\infty}\frac{\Lambda_i}{\lambda_i}f_i'\sin\left[\left(1 - \frac{\dot{a}}{C_i^{\mathrm{P}}}\right)\omega_i t\right], \tag{4}$$

where f_i' is a reduction factor for wave dispersion and C_i^{P} is the phase speed of the ith mode wave. The derivation of Eq. (4), with full details concerning each term, will be published shortly in a full journal article by the authors.

2.3 Limiting Speed of Crack Propagation

In Eq. (4), the crack propagation speed $\dot{a}$ should be no greater than $c_0 = \sqrt{E/\rho}$ to retain a positive ERR. Another constraint on $\dot{a}$ is that it must be no greater than C_1^{P}, the first mode wave phase speed, since the first mode wave modulates all the other waves and carries the most vibrational energy. That is, if the crack propagates faster than $\dot{a}$, the vibrational energy cannot be supplied to the crack tip to open the crack. This limit in 1D structures is the counterpart to Freund's [7] limiting crack speed of C_R (Rayleigh wave speed) in a semi-infinite space. The ratio of these two limiting speeds is

$$\frac{C_1^p}{C_R} = \frac{\lambda_1}{\sqrt{6}} \frac{1+\nu}{(0.862+1.14\nu)\sqrt{1-\nu}} r , \tag{5}$$

where $r = h/a$ is the aspect ratio of Euler-Bernoulli beam. Equation (5) shows limiting speed in DCBs is not only determined by material properties but also by the geometry. An examination of experimentally measured crack propagating speeds and the theoretical limiting crack speed predicted by Eq. (5) is given in Fig. 2.

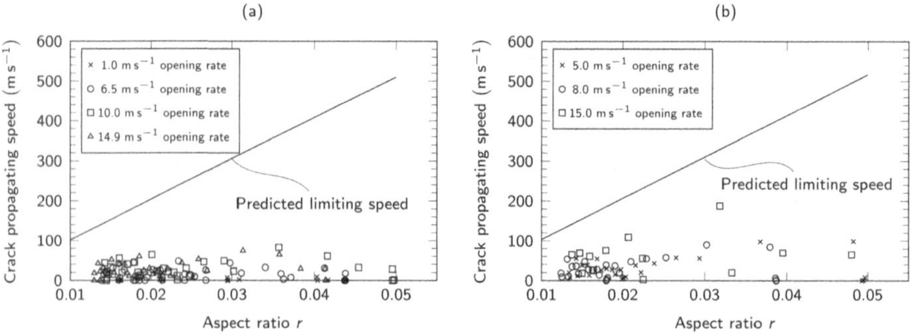

Fig. 2. Comparison of measured crack propagation speeds under various loading rates [6]: (a) PEEK/carbon-fiber composite; (b) epoxy/carbon-fiber composite

3 Experimental and Finite Element Method Verification

There have been relatively few experimental studies of DCBs under high loading rates [8]. Nevertheless, there is a series of experiments reported by Blackman et al. [6] at high loading rates up to 15 ms^{-1}. Considering their results for a loading rate of 10 ms^{-1}, their crack length versus time results can be used to obtain the time-varying crack propagation speed $\dot{a}(t)$, and to calculate ERR using Eq. (3) for the stationary crack and Eq. (4) for the propagating crack. The results are presented in Fig. 3.

In Fig. 3, the fracture toughness for initiation and arrest are adopted from Ref. [9], which numerically simulates the same tests from Blackman et al. [6]. It is seen that the ERR calculated using the developed theory, presented in Sect. 2, can accurately predict the fracture toughness for crack initiation under the dynamic regime. Moreover, the ERR for crack propagation is always above the arrest fracture toughness except for when crack arrest happens at around $t = 2.6$ ms, as expected.

There were not sufficient data points from the experiments to accurately determine the crack propagation speed, and so the developed theory was instead further verified against FEM simulation results from Ref. [9]. The crack length versus time results from the simulation were used to obtain the time-varying crack propagation speed, $\dot{a}(t)$, and to calculate the ERR. Figure 4 compares the analytically predicted and the numerically simulated. Excellent agreement is seen for this case of 10 ms^{-1} opening.

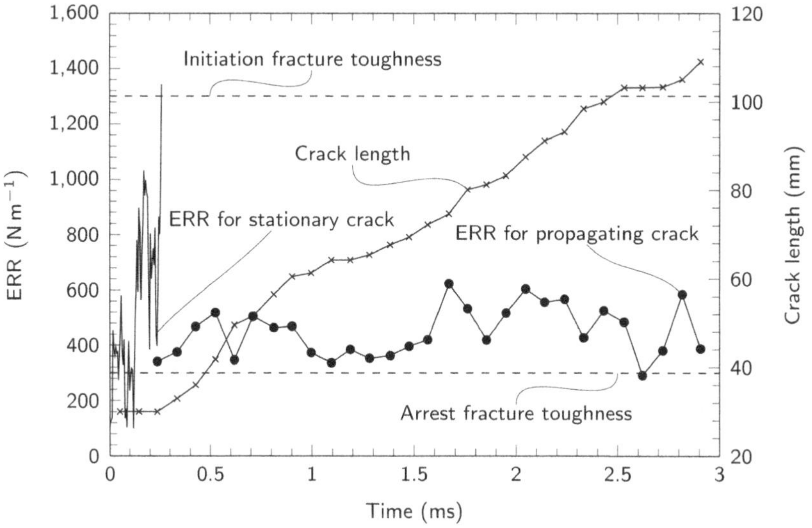

Fig. 3. Dynamic ERR for 10 ms^{-1} opening rate using experiment data

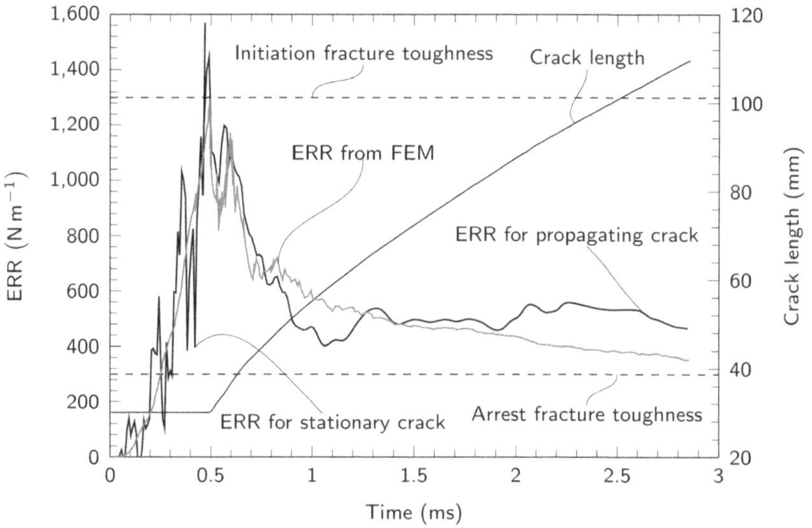

Fig. 4. Dynamic ERR for 10 ms^{-1} opening rate using FEM data

4 Conclusion

Analytical solutions for the dynamic ERR of a DCB resulting from high loading rates, including vibration, are derived for both stationary and propagating cracks. These solutions are verified against experimental results and FEM simulations. It is shown that the developed theory can accurately predict ERR in comparison to numerical simulations; and can accurately determine the fracture toughness for crack initiation and crack arrest in comparison to both experiments and numerical simulations.

To the best of the author's knowledge, the theory presented here for the dynamic ERR of stationary and propagating cracks, including vibration, has not been solved before. These solutions are readily applicable to various engineering applications, and in particular, are expected to be useful to engineers and academic researchers to determine the initiation, arrest and propagation fracture toughness of laminated materials against delamination. The developed theory also provides useful benchmark solutions for the development of numerical codes.

References

1. ASTM D5528-01: Standard test method for mode I interlaminar fracture toughness of unidirectional fiber-reinforced polymer matrix composites (2014)
2. Chen, T., Harvey, C.M., Wang, S., Silberschmidt, V.V.: Dynamic interfacial fracture of a double cantilever beam. Eng. Fract. Mech. (2018). https://doi.org/10.1016/j.engfracmech.2018.11.033
3. Chen, T., Harvey, C.M., Wang, S., Silberschmidt, V.V.: Dynamic interfacial fracture of a thin-layered structure. Procedia Struct. Integrity **13**, 613–618 (2018)
4. Chen, T., Harvey, C.M., Wang, S., Silberschmidt, V.V.: Dynamic delamination on elastic interface. Compos. Struct. **234**, 111670 (2019)
5. Smiley, A.J., Pipes, R.B.: Rate effects on mode I interlaminar fracture toughness in composite materials. J. Compos. Mater. **21**(7), 670–687 (1987)
6. Blackman, B.R.K., Kinloch, A.J., Wang, Y., Williams, J.G.: The failure of fibre composites and adhesively bonded fibre composites under high rates of test. J. Mater. Sci. **31**(17), 4451–4466 (1996)
7. Freund, L.B.: Dynamic Fracture Mechanics. Cambridge University Press, Cambridge (1990)
8. May, M.: Measuring the rate-dependent mode I fracture toughness of composites - a review. Compos. Part A Appl. Sci. Manuf. **81**, 1–12 (2016)
9. Liu, Y., van der Meer, F.P., Sluys, L.J.: Cohesive zone and interfacial thick level set modeling of the dynamic double cantilever beam test of composite laminate. Theor. Appl. Fract. Mech. **96**, 617–630 (2018)

Characterization of Dynamic Fractures for Compressed PMMA Material

C. Huang(✉)

Wuhan Textile University, Wuhan, People's Republic of China
ab6688us@yahoo.com

Abstract. Polymethyl methacrylate (PMMA) is a quasi-brittle material and dynamic behaviors usually present thermoplastic characteristic. But in some cases, PMMA breaks with the complete brittle characteristic. The complex behaviors need to be studied for a further understanding. A series of PMMA specimens are designed to study the behaviors on a Spilt Hopkinson Pressure Bar (SHPB) device. The complex behaviors are found to be dominated by the impulsive energy and there exists a critical energy for the specimens to create the complete brittle fracture.

Keywords: Thermoplastic deformation · Brittle fracture · Impulsive energy

1 Introduction

Dynamic fractures of brittle materials usually present very complex modes and various theories are tried to explain the fracture mechanism from different points of view. The fractures are thought to involve many factors including loads, material properties, microstructures, geometries and so on. The concerning factors have been studied broadly in published literatures, but it is still difficult to determine the factor that can play a dominant role in fractures.

PMMA is generally recognized as a brittle material with the properties of light weight, high strength and transparency. When compressed statically, cylindrical specimen deforms in barrel shape like metal one and presents high ductility with the features of elasto-plastic deformation. However, under SHPB impact loading the primary deformation is characterized by fractures and the material presents strong sensitive characteristics of dynamic behaviors. PMMA is composed of high molecule mass and glass transition occurs at the lower temperature. Under impact loading, thermoplastic deformations usually occur on fragment debris and affect the dynamic behaviors.

In order to study the dynamic fracture characteristics, a series of cylindrical PMMA specimens are designed for SHPB experiments to test size effects on the fractures. The specimens are assigned with two diameters of 8 mm and 15 mm. The lengths of 8, 10, 12 and 14 mm are set for the 8 mm diameter specimens with the aspect ratios (L/D) of 1, 1.25, 1.5 and 1.75, respectively. The 15 mm diameter specimens are designed with the purpose to correlate size effects with the responses, the specimens are designated with the aspect ratios of 1, 1.27, 1.53 and 1.8 that equal to the 8 mm diameter specimens one by one. The specimen lengths are then determined as 15, 19, 23 and 27 mm.

E. Gdoutos and M. Konsta-Gdoutos (Eds.): ICTAEM 2020, STIN 16, pp. 9–13, 2020.
https://doi.org/10.1007/978-3-030-47883-4_2

In the meantime, the 8 length specimens present a monotonic length sequence which can be used to study length effects on dynamic fractures.

2 Size Effects on Dynamic Fractures

To simplify the investigation of size effects, SHPB device sets identical incident loading to all the tested specimens. Under the identical loading, specimen size becomes the only variable in SHPB experiments and dynamic responses can be studied concerning the effects of specimen size. The experiment settings allow the size effects just resulting from the size itself and correlate the fracture modes with the size effects.

Size effects on strengths of the specimens have been discussed and aspect ratio is validated as an important parameter to affect the strengths [1]. This paper focuses on dynamic fractures of the specimens. In the SHPB experiments, the 8 length specimens break in essential different modes. Figure 1 shows the fracture mode of a 27 mm length specimen that has the longest length with the largest volume in the 8 length specimens. In the experiments, impact (or front) end contacting incident bar is colored in black and the rear end is kept in original form to distinguish the fragment debris and make the specimen recovered easily. The fracture shows that large debris comes from the two ends of the specimen and the small and uniform debris primarily breaks from the middle section. The largest debris shows the rear end remained in cylindrical shape with little damages on the circumferential surface, while the impact end is broken into the relatively larger black fragment debris. The fracture difference between the front and rear ends may result from a slightly slowed propagation of stress wave to deliver the stress late to the rear end resulting in the rear stress relatively smaller than the front [2]. Moreover, the end debris is shown apparently larger than the debris from middle section. The fracture modes indicate that transverse deformation is confined by the end friction significantly even though grease is used to reduce the effect of friction. High confinement increases debris size and reduces the amount of debris as the same effect applied by high confinement pressure. The end debris shows a length range of 5–10 mm indicating that the confinement effect has the action range of 5–10 mm. It is worth mentioning that the rear end debris shows lots of adiabatic shear bands on the fracture surface characterizing the significant thermoplastic deformations.

Fig. 1. Fracture of 25 mm specimen

Fig. 2. Fracture of 19 mm specimen

The 19 mm length specimen breaks with the similar modes as the 27 mm one, but presents more characteristics of thermoplastic deformation. The rear end is broken in conical shape that is usually observed in high pressure confinement as the transition of brittle to ductile fracture occurs [3]. Figure 2 shows the rear end debris with the clear adiabatic shear bands on the inclined surface. The surface is formed by a shear fracture indicating that the confinement effect of friction is significant in created complex stress state. The 5–10 mm lengths of the end debris verify the friction confinement to act on the specimens within the length range. Moreover, in comparison with the 27 mm specimen, the 19 mm one shows the reductions of debris sizes as the specimen length decreases to lead to the rise of strain rate. The fractures present the characterized sensitivity of strain rate.

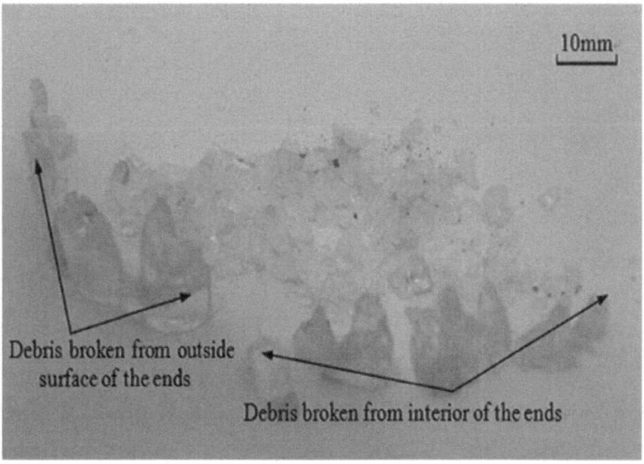

Fig. 3. Fracture of 15 mm specimen

With the further decrease of specimen length, the 15 mm length specimen shows a different fracture mode in comparison with the 19 and 27 mm specimens. As shown in Fig. 3, the specimen is broken completely with relatively uniform debris. In order to minimize the confinement effect of friction, heavy grease is used on the end surfaces. But the end debris still has the 5–10 mm lengths indicating that the friction confinement is changed little for the short specimen. The fracture is primarily under the action of complex stress state. Unlike the 19 and 27 mm specimens, fragment debris of the 15 mm one does not show significant thermoplastic deformations. The specimen breaks with the reduced debris sizes showing the characteristic of brittle fracture. Here, the different fracture modes of the specimens can only be attributed to the length decrease.

Figure 4 shows the fracture of a 14 mm length specimen. The specimen just has 1 mm length difference with the 15 mm specimen. In Fig. 4, the black debris breaks from the impact end with the characteristic of friction confinement effect. The debris shows the lengths around 5 mm that is much less than the 15 mm specimen even though the two specimens break with the similar characteristic of brittle fracture. The debris difference may be concluded resulting from the effect of transverse size as the two just have 1 mm length difference. If the transverse size is the right factor, aspect ratio may be a more desirable geometric parameter to present the effect of transverse size. The aspect ratio has shown the effects on strengths of the specimens [1]. The 14 mm length specimen has the identical aspect ratio with the 27 mm length specimen, but they show essentially different fracture modes. The 14 mm specimen breaks with brittle characteristic of uniform debris as the ends are broken into much smaller debris. If the fracture modes are determined by transverse size, the aspect ratio should present a little correlation with the fractures. There must exist other factor to dominate the fractures.

Fig. 4. Fracture of 14 mm specimen

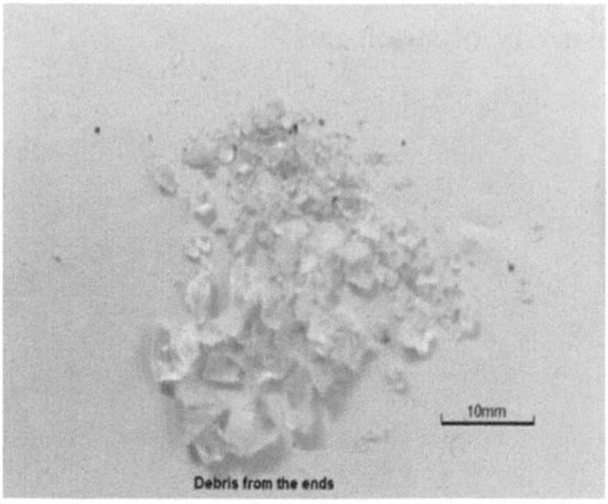

Fig. 5. Fracture of 8 mm specimen

In comparison with the previous specimens, the 8 mm specimen breaks with smaller but more uniform debris as shown in Fig. 5. It can be concluded that under the identical loading condition, with the decrease of specimen sizes the reduced debris becomes more and more uniform and presents the strain rate sensitivity of fractures as the strain rate increases with the specimen length decrease. Moreover, the 15 mm length specimen indicates a transition from the thermoplastic to brittle fractures as the specimens with the length of 15 mm or less than break with the brittle characteristic.

3 Discussion and Conclusion

The transition of fracture modes involves the size changes of both length and transverse diameter, but aspect ratio representing the geometric property of specimen is unable to show the correlation with the fracture modes. Under the identical loading condition, the impulsive energy applying to any given specimen is constant. With the reduction of sizes, specimen volume decreases synchronously and the impulsive energy accumulates the energy much more than the required for breaking the specimen. Impulsive energy appears to be the dominant factor to determine the fracture modes. In the performed SHPB experiments, the identical impulsive energy is applied to the specimens consistently. With the decrease of specimen volume, the impulsive energy exceeds the required fracture energy more and more. The exceeding impulsive energy breaks specimen in a much shorter time and brings the specimen a high deformation strain rate to result in the more uniform fracture. In the case of the fracture, the specimen does not have the time to perform thermoplastic deformation. For the PMMA specimens, the 15 mm one presents the initiation of brittle fracture indicating that the impulsive energy is the critical one to change thermoplastic fracture into brittle one. Under the impulsive energy, the other small size specimens are all broken in brittle fracture regardless of the effect of aspect ratios and with the decrease of specimen volumes, the fracture becomes more uniform with smaller debris. Therefore, impulsive energy is the dominant factor to determine the fracture modes of brittle material.

References

1. Huang, C.: A further exploration on loading strain rate In: Gdoutos, E. (eds.) Structural Integrity, Proceedings of the First International Conference on Theoretical, Applied and Experimental Mechanics, ICTAEM 2018, vol 5, pp. 161–163. Springer, Cham (2018)
2. Wang, Z.G., Meyer, L.W.: On the plastic wave propagation along the specimen length in SHPB test. Exp. Mech. **50**(7), 1061–1074 (2010)
3. Rittel, D., Brill, A.: Dynamic flow and failure of confined polymethylmethacrylate. J. Mech. Phys. Solids **56**(4), 1401–1416 (2008)

Tensile Behavior of a Titanium Alloy Additively Manufactured via Selective Electron Beam Melting

Yinling Zhang[1,2], Shoujiang Qu[1], Aihan Feng[1(✉)], and Daolun Chen[2(✉)]

[1] School of Materials Science and Engineering, Tongji University, Shanghai 201804, China
aihanfeng@tongji.edu.cn
[2] Department of Mechanical and Industrial Engineering, Ryerson University, Toronto, ON M5B 2K3, Canada
dchen@ryerson.ca

Abstract. Additive manufacturing or 3D-printing of titanium alloys with a high strength-to-weight ratio holds significant interest in the aerospace and biomedical industries. The purpose of this investigation was to determine microstructure and tensile deformation behavior of a 3D-printed Ti-6Al-4V alloy via selective electron beam melting (SEBM). Plentiful multi-oriented α-lamellae were present in the prior columnar β grains, which were oriented in the building direction because of the presence of temperature gradients during 3D-printing. The processing parameters selected in this study ensured superior strength and high ductility by controlling the thickness of α-lamellae, with both surpassing the values specified in the ASTM standard. The horizontally orientated 3D-printed alloy demonstrated a certain extent of strain rate sensitivity which decreased with increasing strain, suggesting that dislocation slip was a predominant deformation mode, since the fine and abundant multi-oriented α-lamellae could impede the formation of extension twinning. Fracture surface was observed to be characterized by typical dimples and some entrapped gas pores.

Keywords: Selective electron beam melting · Ti-6Al-4V alloy · Tensile properties

1 Introduction

Ti-6Al-4V alloy has been widely applied to various aircraft components and biomedical industry due to their low density, high strength and excellent fracture toughness [1]. However, the geometrical sophistication of many parts is often difficult to access for the traditional subtractive manufacturing methods along with the inferior machinability of titanium alloys [2].

Recently, additive manufacturing (AM), also known as 3D printing, has become an increasingly popular manufacturing process technique providing a great degree of design freedom and a huge potential for manufacturing optimized and customized components with complicated shapes ideally [3]. Selective electron beam melting

E. Gdoutos and M. Konsta-Gdoutos (Eds.): ICTAEM 2020, STIN 16, pp. 14–19, 2020.
https://doi.org/10.1007/978-3-030-47883-4_3

(SEBM) is one of the most popular AM processes due to the advantages such as the vacuum working environment to avoid impurity elements and relatively slow cooling rate to reduce the process-induced residual stresses [4].

However, studies on the deformation behavior of SEBM-manufactured Ti-6Al-4V alloy are still very limited. Hence, further research is necessary and urgent to evaluate the deformation characteristics of such a 3D-printed titanium alloy, so as to warrant its safe and durable applications in the aerospace and medical industry.

2 Materials and Experimental Procedure

2.1 Additive Manufacturing Processing

Samples fabricated with pre-alloyed Ti-6Al-4V powders, with a chemical composition (wt%) of 6.40 Al, 4.12 V and balance Ti, were processed using SEBM in a vacuum atmosphere. The size of powders used had a size distribution from 45 to 105 μm which meet the specified requirement (45–150 μm). The samples were additively manufactured (or 3D-printed) using an Arcam Q20 system, which was operated at a beam current of 35 mA, a layer thickness of 0.09 mm with a scan speed function of 24 and a line offset of 0.22 mm. The electron beam scanning was set to rotate at an angle of 66° for each layer to improve the bonding between layers with a more uniform distribution of heat combined with a lessened residual stress. The horizontally positioned cylinders with a size of Φ15 × 170 mm were firstly 3D-printed, as shown in Fig. 1. Those cylinders were finally machined and polished into tensile specimens with a gauge length of Φ6 × 32 in the end.

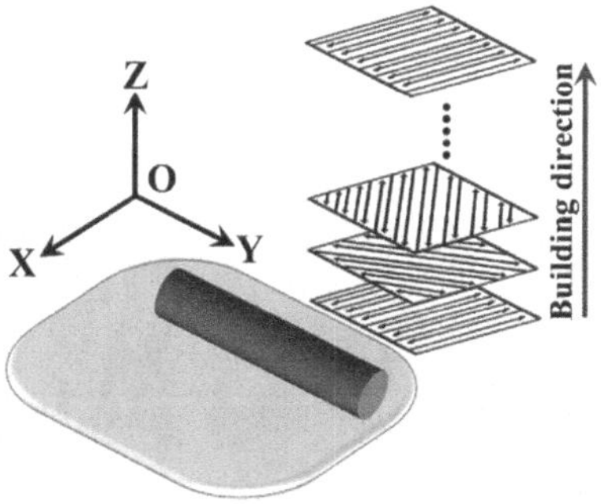

Fig. 1. Schematic diagram of SEBM building strategy and sample positioning.

2.2 Experimental Procedure

Optical microscope (OM) and electron backscatter diffraction (EBSD) were used to analyze the oriented microstructure with an observational surface parallel to the building direction. The sample for the EBSD analysis was prepared first by mechanical grinding, followed by electro-polishing in an electrolyte containing 6% perchloric acid, 34% n-butanol and 60% methanol at 0.8 A and 253 K for ~70 s. Uniaxial tensile tests were carried out under strain control at ambient temperature at varying strain rates from 1×10^{-4} s^{-1} to 1×10^{-2} s^{-1}. Two samples were tested at a given strain rate to determine the yield strength (YS), ultimate tensile strength (UTS) and ductility (%EL). The fracture surfaces were examined using a scanning electron microscope (SEM).

3 Results and Discussion

3.1 Phase Constitution and Microstructure

Figure 2(a) shows the macrostructure of a SEBM-processed Ti-6Al-4V alloy. As seen from the observational surface, columnar grains were basically oriented vertically along the building direction with elongated shapes, which had a morphology of prior-β grains present above the transus temperature of ~950 °C. It is obvious that the highly-oriented columnar morphology was created by the presence of high temperature gradients in the fast heating and cooling process of 3D printing. Almost the entire high-temperature prior-β phase with a body-centered cubic (bcc) crystal structure would unavoidably transform into α phase with a hexagonal close-packed (hcp) crystal structure during the subsequent cooling process (below the transus temperature). Besides, several pores could also be observed, which was due to the trapped gas in the powder bed during 3D printing.

Figure 2(b) shows the EBSD result of inverse pole figure (IPF) orientation map. The black and white lines represent the high-angle (with a misorientation angle of >15°) grain boundaries and the low-angle boundaries (with a misorientation angle in-between 2° to 15°), respectively. The microstructure was characterized by plentiful multi-oriented α-lamellae which were generated following the Burgers orientation relationship (i.e., $\{110\}_{\beta}//(0001)_{\alpha}$ and $<111>_{\beta}//<11-20>_{\alpha}$) during the solid-state phase transformation of titanium alloy [5]. Due to the crystal symmetry, up to six different α variants are possible to be present in each prior-β grain [6].

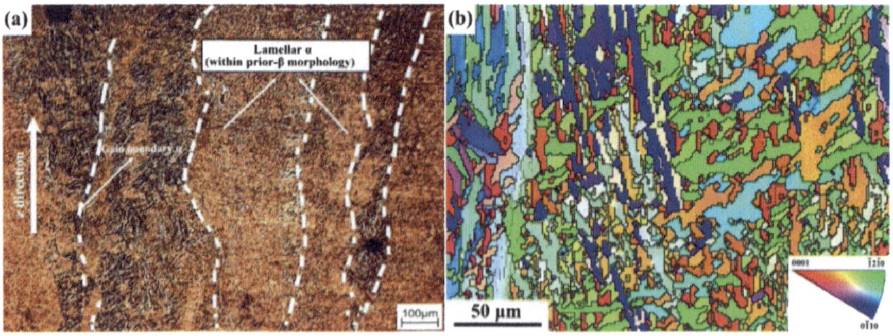

Fig. 2. (a) Optical micrograph and (b) EBSD IPF orientation map of SEBM-manufactured Ti-6Al-4V alloy.

3.2 Tensile Properties

Typical tensile engineering stress-strain curves of the horizontally orientated SEBM-manufactured Ti-6Al-4V samples at varying strain rates are presented in Fig. 3(a). It can be seen that the stress-strain curves shifted higher as strain rate increased. The average value of the YS and UTS of the present SEBM-manufactured Ti-6Al-4V alloy determined at different strain rates is shown in Table 1. The obtained YS (870–961 MPa) and UTS (957–998 MPa) are 5–16% and 7–12%, respectively, higher than the YS (825 MPa) and UTS (895 MPa) specified in the ASTM F2924 standard for the additive manufactured Ti-6Al-4V alloy with powder bed fusion. The obtained ductility of $\sim$13% is also higher than the specified value (6–10%) in this ASTM standard. The hardening capacity (H_c) of a material proposed by Afrin *et al.* [7] can be expressed as follows,

$$H_c = \left(\sigma_{UTS} - \sigma_y\right)/\sigma_y \tag{1}$$

where σ_{UTS} and σ_y represent the UTS and YS, respectively. The hardening capacity values obtained at different strain rates are also tabulated in Table 1, with an average value of $\sim$0.07. Strain hardening exponent (n) can also be used to describe the plastic deformation characteristics of a material, which could be evaluated via the following Hollomon equation [8],

$$\sigma = K\varepsilon^n \tag{2}$$

where σ is true stress, ε is true strain, and K is strength coefficient. The experimental data of σ and ε in the uniform deformation stage from the yield point to the ultimate tensile point are chosen to evaluate the strain hardening exponent. The value of n in the present 3D-printed alloy increased slightly with decreasing strain rate with an average of $\sim$0.06. This is in agreement with that of traditional Ti-6Al-4V ELI alloy with a n value of from $\sim$0.045 to $\sim$0.08 over a wide range of strain rates from $1 \times 10^{-2}\ \text{s}^{-1}$ to $1 \times 10^{-5}\ \text{s}^{-1}$.

Strain-rate sensitivity (SRS) is another important parameter reflecting the deformation behavior of materials. The effect of strain rate on the flow stress of the SEBM-manufactured Ti-6Al-4V alloy can be described using a power law [9],

$$\sigma = C\dot{\varepsilon}^m \tag{3}$$

where σ is flow stress, C is a constant and m is the SRS. A plot based on Eq. (3) at certain true strain values is shown in Fig. 3(b). It can be observed that the m value or the slope in Fig. 3(b) decreased as the true strain increased. It is known that the samples which were prone to extension twinning exhibited an increasing trend of SRS with increasing strain, whereas an opposite trend in the SRS was reported when twinning was not a favorable mode of deformation [10]. This suggests that the slip of dislocations was a predominant mode of deformation, rather than extension twinning in the present 3D-printed Ti-6Al-4V alloy. It is likely as a result of the presence of fine and abundant multi-oriented α-lamellae (Fig. 2(b)), which impede the formation of extension twinning.

Table 1. Tensile properties of the horizontal SEBM-manufactured Ti-6Al-4V alloy determined at different strain rates at room temperature.

Strain rate, s^{-1}	σ_y, MPa	σ_{UTS}, MPa	Elongation, %	H_c	N
1×10^{-2}	961	998	14	0.04	0.04
1×10^{-3}	914	971	15	0.06	0.06
1×10^{-4}	870	957	11	0.1	0.08

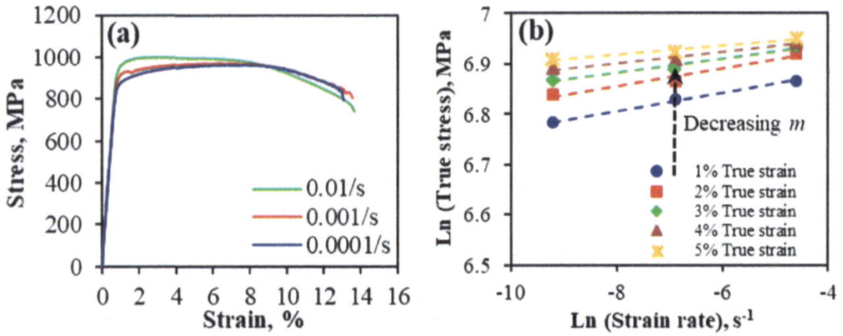

Fig. 3. (a) Typical engineering stress-strain curves of the present SEBM-manufactured horizontal orientated Ti-6Al-4V alloy obtained at three strain rates; (b) strain-rate sensitivity as a function of true strain.

3.3 Fractography

Figure 4 shows a typical example of the fracture surface of the present SEBM-manufactured Ti-6Al-4V alloy tested at a strain rate of $1 \times 10^{-4}\ s^{-1}$. It can be observed that the fracture morphology of this alloy appeared pretty ductile, characterized by the presence of abundant dimples. In addition, a few circular pores also exist on the fracture surface, which appeared during the processing of 3D-printing.

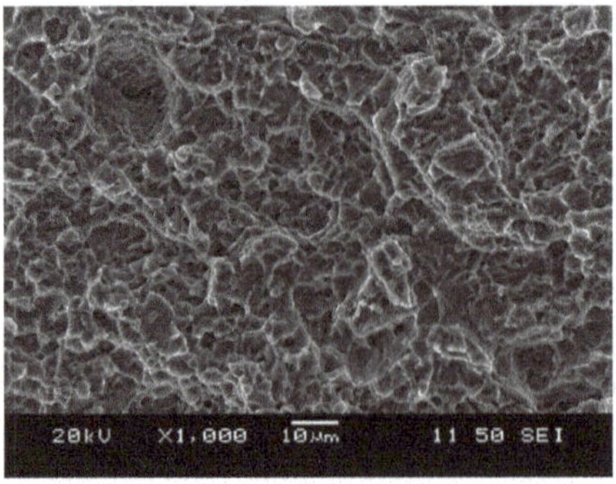

Fig. 4. A typical SEM image of the fracture surface of the present SEBM-manufactured horizontal orientated Ti-6Al-4V alloy tested at a strain rate of $1 \times 10^{-4}\ s^{-1}$.

4 Conclusions

Microstructure and tensile behavior of a horizontally oriented SEBM-manufactured Ti-6Al-4V alloy were evaluated. The following conclusions can be drawn:

(1) The microstructure of as-built SEBM samples was observed to consist of predominantly oriented α-lamellar phase positioned inside the prior columnar β grains along the building direction.
(2) The processing parameters selected in this study were suitable, leading to a superior strength coupled with high ductility of the present 3D-printed alloy, with both values exceeding the specified strength and ductility in the relevant ASTM standard. The alloy demonstrated a certain extent of strain rate sensitivity which decreased with increasing strain.
(3) Fracture surface examination showed a ductile mode of fracture in the tensile tested specimens, as characterized by abundant dimples and some entrapped gas pores.

Acknowledgements. The authors are grateful for the financial support provided by the National Natural Science Foundation of China (NSFC) (Grant No. 51871168) and the Natural Sciences and Engineering Research Council of Canada (NSERC) in the form of international research collaboration. The authors appreciated Q. L, A. M, J. A and R. C for kind assistances.

References

1. Cui, C., Hu, B.M., Zhao, L., Liu, S.: Titanium alloy production technology, market prospects and industry development. Mater. Des. **32**(3), 1684–1691 (2011)
2. de Formanoir, C., Brulard, A., Vivès, S., Martin, G., Prima, F., Michotte, S., Rivière, E., Dolimont, A., Godet, S.: A strategy to improve the work-hardening behavior of Ti-6Al-4V parts produced by additive manufacturing. Mater. Res. Lett. **5**(3), 201–208 (2017)
3. Lavine, M.S.: Fast, continuous, 3D printing. Science **347**(6228), 1325 (2015)
4. Galarraga, H., Lados, D.A., Dehoff, R.R., Kirka, M.M., Nandwana, P.: Effects of the microstructure and porosity on properties of Ti-6Al-4V ELI alloy fabricated by electron beam melting (EBM). Addit. Manuf. **10**, 47–57 (2016)
5. Tan, X., Kok, Y., Tan, Y.J., Descoins, M., Mangelinck, D., Tor, S.B., Leong, K.F., Chua, C. K.: Graded microstructure and mechanical properties of additive manufactured Ti-6Al-4V via electron beam melting. Acta Mater. **97**, 1–16 (2015)
6. Beladi, H., Chao, Q., Rohrer, G.S.: Variant selection and intervariant crystallographic planes distribution in martensite in a Ti-6Al-4V alloy. Acta Mater. **80**, 478–489 (2014)
7. Afrin, N., Chen, D.L., Cao, X., Jahazi, M.: Strain hardening behavior of a friction stir welded magnesium alloy. Scr. Mater. **57**, 1004–1007 (2007)
8. Hollomon, J.H.: Tensile deformation. Trans. AIME **162**, 268–290 (1945)
9. Lin, X.Z., Chen, D.L.: Strain hardening and strain-rate sensitivity of an extruded magnesium alloy. J. Mater. Eng. Perform. **17**, 894–901 (2008)
10. Chun, Y.B., Davies, C.H.J.: Twinning-induced negative strain rate sensitivity in wrought Mg alloy AZ31. Mater. Sci. Eng. A **523**, 5713–5722 (2011)

The Rule of Mixtures on the Elastic Buckling Response of Rectangular GLARE FML Plates Under Shear Stresses

Costas D. Kalfountzos(✉), George S. E. Bikakis, and Efstathios E. Theotokoglou

Department of Mechanics, Laboratory of Strength of Materials, National Technical University of Athens, 9 Iroon Polytechniou, 15773 Zografou, Athens, Greece
koskalf@yahoo.gr

Abstract. Fiber Metal Laminates (FMLs) are hybrid composite materials consisting of alternating metal layers bonded to fiber-reinforced prepreg layers. ARALL, CARALL and GLARE belong to this new family of materials. GLARE is the most successful FML up to now and is currently being used for the construction of primary aerospace structures. In this study, the elastic buckling response of rectangular GLARE FMLs subjected to shearing stresses is investigated using the finite element method and eigenvalue buckling analysis. Simply supported, clamped and mixed boundary conditions are considered. Using validated FEM models, the buckling coefficient-aspect ratio diagrams of seven GLARE grades are obtained along with the diagrams of three UD glass-epoxy composites and monolithic 2024-T3 aluminum. The rule of mixtures is evaluated and found to be a simple method to estimate approximately the elastic buckling stress of the GLARE plates. An approximate formula is derived for the estimation of the critical buckling coefficient of the GLARE plates using the buckling coefficients of their constituents.

Keywords: Elastic buckling · Fiber-metal laminates · Rule of mixtures

1 Introduction

Fiber Metal Laminates (FMLs) are hybrid composite materials consisting of alternating metallayers bonded to fiber-reinforced preimpregnated (prepreg) layers. These materials are developed to fulfill the desire of aircraft manufacturers and airliners for reduction of direct operational costs and structural weight of aircrafts [1]. ARALL (Aramid Reinforced ALuminum Laminates), CARALL (CArbon Reinforced ALuminum Laminates) and GLARE (GLAssREinforced) belong to this new family of materials. GLARE is the most successful FML up to now and is currently being used for the construction of aerospace structures, such as the fuselage of the Airbus A380 air plane, the aircraft cargo floors of Boeing 777, aircraft engine cowlings and aircraft stiffeners with a wide variety of shapes [2–4]. FMLs combine significant advantages of metal and fiber-reinforced composites. Because of the intact bridging of fibers, in the

E. Gdoutos and M. Konsta-Gdoutos (Eds.): ICTAEM 2020, STIN 16, pp. 20–25, 2020.
https://doi.org/10.1007/978-3-030-47883-4_4

wake of the crack, which restrain crack opening, FMLs demonstrate high fatigue resistance [5]. They also have high fracture toughness [6], excellent impact resistance [7], low density, excellent moisture, corrosion and fire resistance [8, 9], which make them ideal for multipurpose aerospace applications.

The elastic buckling of thin plates has great practical importance and it is a classical problem of the strength of materials. Plate buckling must always be considered during the design of many engineering applications where stiffened thin-walled structures are employed. For the purpose of buckling analysis, the wall plates among stiffeners could be analyzed approximately as isolated rectangular plates. FMLs are mainly used for the construction of stiffened thin-walled aerospace structures and, therefore, the buckling strength of FML plates is very important [10, 11].

This study deals with the elastic buckling of rectangular FML plates subjected to shearing stresses, the calculation of their buckling coefficients and the construction of diagrams showing their variation as a function of the plate aspect ratio. More specifically, the main objective of this study is the investigation of applicability of the rule of mixtures for the calculation of the average critical buckling stress of the examined GLARE plates. Furthermore, an approximation formula to predict the buckling coefficients of FMLs is derived using the rule of mixtures for the buckling stresses.

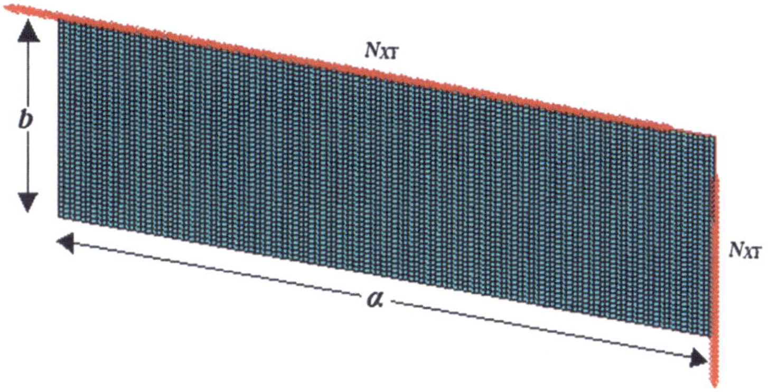

Fig. 1. Finite element mesh of a rectangular 600 × 200 mm plate (*a/b* = *3*) along with the applied shear loading.

2 Problem Definition and Finite Element Modeling Procedure

A thin rectangular plate is considered, under pure shear loading consisting of alternating layers of 2024-T3 aluminum and unidirectional (UD) glass-epoxy bonded to form FML. The plate has length α, width $b \leq \alpha$, as it is illustrated in Fig. 1 and a relatively small thickness t. Since the plate is subjected to pure shear loading (Fig. 1) and each opposite parallel edge is not allowed to move in the direction of N_{xy}, it will buckle for a specific critical value of the applied load N_{xy}. The elastic buckling load will be predicted using the finite element method. ANSYS software [12] is employed for this purpose along with eigenvalue buckling analysis. From the FEM results the shear buckling coefficient k will be determined for variable aspect ratio a/b and (k, a/b) curves of plates consisting of different materials will be constructed.

In order to apply the rule of mixtures, three different material systems are considered, namely FMLs and their constituents, UD composites and monolithic aluminum. Also, three types of bc are considered, namely simply supported, clamped and TSLC. For the FEM modeling we use plane rectangular 4-noded SHELL 181 elements, which are suitable for modeling the thin layered plates studied in this article. Each SHELL element has three translational and three rotational degrees of freedom per node. The material behavior of 2024-T3 aluminum and UD glass-epoxy is idealized using linear elastic material models with isotropic and orthotropic mechanical properties, respectively.

The convergence of the critical buckling load N_{xy} predicted from the eigenvalue buckling analysis is always verified by comparison of results corresponding to fine and very fine mesh density. Furthermore, the ply stress levels corresponding to the critical value of N_{xy} are always below the linear elastic limits. Specifically, the von Mises stresses in aluminum layers and the stress components in prepreg layers are well below the proportional limit of aluminum and the composite ply strengths, respectively.

The shear buckling coefficient of aluminum plates is calculated by [13]:

$$k = \frac{N_{xy} b^2}{\pi^2 D} \tag{1}$$

where D is the bending stiffness of the isotropic plate.

The shear buckling coefficient of the laminates is calculated by [14]:

$$k = \frac{N_{xy} b^2}{\pi^2 \sqrt{D_{11} D_{22}}} \tag{2}$$

where $D_{ii}(i = 1, 2)$ are the bending stiffnesses of the laminate.

The average critical buckling stress is calculated with the relation:

$$\sigma_{xyc} = \frac{N_{xy}}{t} \tag{3}$$

In order to validate our FEM procedure and the suitability of the applied simply supported bc, we have initially compared the FEM results with corresponding theoretical results [13] for monolithic aluminum plates. A very good agreement between our FEM results and the literature predictions has been observed. As a result, the validity of the implemented FEM modeling and the applied bc for the simply supported aluminum plates is ensured. As expected, the clamping of the plates is modeled by imposing zero transverse rotations along their edges in addition to the applied simply supported bc. The TSLC bc along the edges of the plates have been applied as follows: the Transverse edges are Simply supported and the Longitudinal edges are Clamped.

For the validation of our FEM modeling of GLARE and composite laminates, we have compared the obtained FEM results with appropriate literature results [14]. A very good agreement between our FEM results and the literature predictions has been also observed for the analyzed UD composite laminates and the specially orthotropic GLARE grades. Consequently, the FEM modeling of the GLARE and composite laminates is validated.

3 Rule of Mixtures versus FEM Results

We have investigated the applicability of the classical rule of mixtures for the calculation of the average critical buckling stress σ^g_{xyc} of seven different GLARE grades. In ref. [15] other applications of the rule of mixtures can be found. Specifically, the stress σ^g_{xyc} is calculated using the formula:

$$\sigma^g_{xyc} = \sigma^c_{xyc}(1 - MVF) + \sigma^a_{xyc}MVF \tag{4}$$

Where σ^c_{xyc} and σ^a_{xyc} are the average critical buckling stresses of composite and aluminum plates, respectively, with the same dimensions, and MVF is the metal volume fraction of each plate. The composite plates also have the same fiber orientation angles, fiber volume fraction and ply thickness with the composite plies of each GLARE plate. In Fig. 2 the stresses calculated using Eq. (4) are compared with the detailed FEM calculations for GLARE 3 and GLARE 5 respectively. A good agreement between the accurate and the rule of mixtures (σ^g_{xyc}, a/b) curves is found.

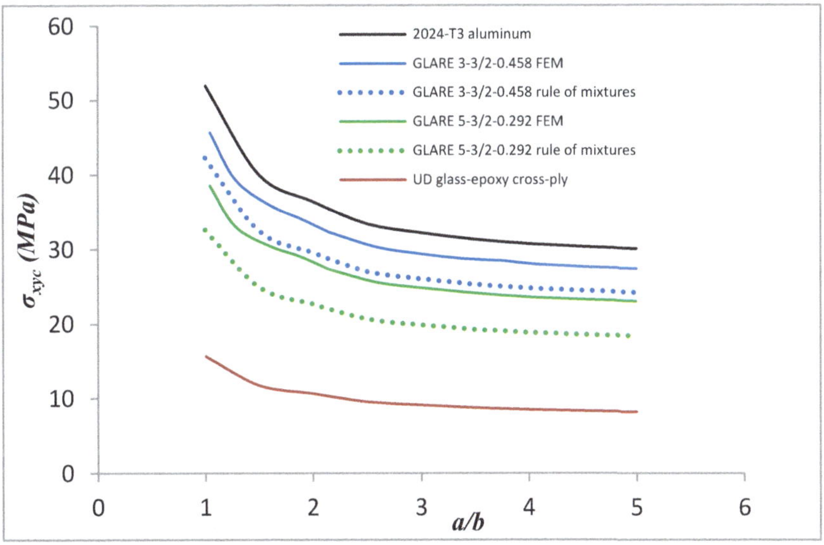

Fig. 2. FEM versus rule of mixtures average critical buckling stresses of simply supported GLARE plates with cross-ply fiber orientation under pure shear.

Using Eqs. (1–3), the average critical buckling stresses of aluminum, composite and GLARE plates can be calculated as a function of their buckling coefficient, and by substitution in Eq. (4) a very useful relation is obtained:

$$k_G = k_C\sqrt{\frac{D_{11C}D_{22C}}{D_{11G}D_{22G}}}(1 - MVF) + k_A\frac{D}{\sqrt{D_{11G}D_{22G}}}MVF \tag{5}$$

where D_{iiC} are the bending stiffnesses of the composite plate, D_{iiG} are the bending stiffnesses of the GLARE plate, k_C is the buckling coefficient of the composite plate, and k_A is the buckling coefficient of the aluminum plate.

Equation (5) can be used in order to estimate approximately the buckling coefficient of a simply supported, clamped or TSLC GLARE plate when k_C and k_A are known. In Fig. 3 the buckling coefficients calculated using Eq. (5) are compared with FEM calculations for GLARE 3 plates with clamped, simply supported and mixed bc. A good agreement between exact and Eq. (5) based (k_G, *a/b*) curves is found. A good agreement has also been found for the other considered GLARE plates of this study. Consequently, the validity of Eq. (5) is demonstrated for the approximate calculation of k_G.

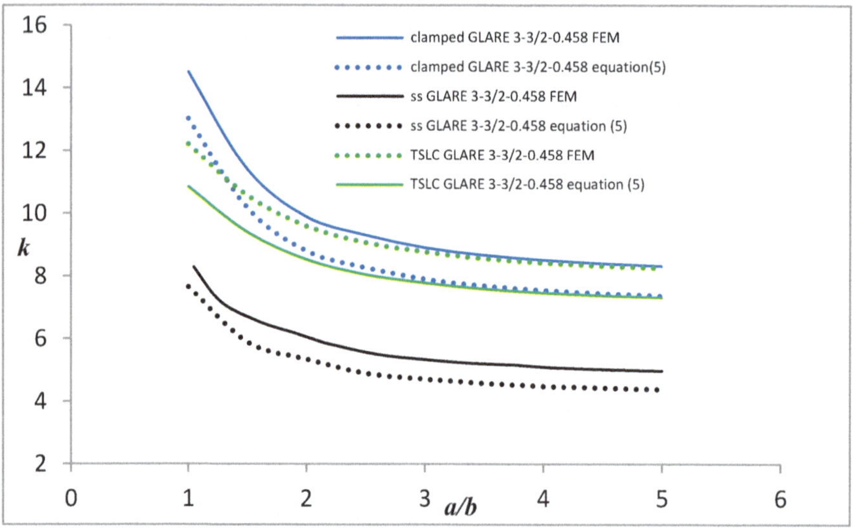

Fig. 3. Exact versus approximate buckling coefficients of GLARE 3 plates under pure shear with clamped, simply supported (ss) and mixed (TSLC) bc.

4 Conclusions

This article deals with the elastic buckling of rectangular simply supported, clamped and TSLC FMLs subjected to pure shear loading. ANSYS FEM software is employed in combination with the eigenvalue analysis in order to predict the buckling coefficient of the laminates. The implemented finite element models have been validated by comparison with appropriate theoretical results concerning simply supported bc.

It is found that the rule of mixtures can be applied in order to estimate approximately the critical buckling stresses of the GLARE plates when the buckling stresses of plates having the same geometry and consisting of one constituent material, aluminum or UD glass-epoxy, are known. The same fiber orientation angles, fiber volume fraction and ply thickness with the composite plies of the GLARE plates are necessary in order to apply the rule of mixtures.

Based on the verified validity of the rule of mixtures, an approximate formula is derived for the estimation of the buckling coefficient of the GLARE plates when the buckling coefficients of plates consisting of the constituent materials are known.

References

1. Verolme, K.: The initial buckling behavior of flat and curved fiber metal laminate panels. Report LR-785, Delft University of Technology (1995)
2. Gunnink, J.W., Vlot, A., De Vries, T.J., et al.: Glare technology development 1997–2000. Appl. Compos. Mater. **9**(4), 201–219 (2002)
3. Woerden, H.J.M., Sinke, J., Hooijmeijer, P.A.: Maintenance of glare structures and glare as riveted or bonded repair material. Appl. Compos. Mater. **10**(4–5), 307–329 (2003)
4. Vermeeren, C.A.J.R.: An historic overview of the development of fibre metal laminates. Appl. Compos. Mater. **10**(4–5), 189–205 (2003)
5. Sinmazçelik, T., Avcu, E., Bora, M.O., Çoban, O.: A review: fibre metal laminates, background, bonding types and applied test methods. Mater.Des. **32**(7), 3671–3685 (2011)
6. Castrodeza, E.M., Bastian, F.L., Perez Ipiña, J.E.: Critical fracture toughness, J_C and δ_{5C}, of unidirectional fibre–metallaminates. Thin-Walled Struct. **41**(12), 1089–1101 (2003)
7. Bikakis, G.S.E.: Simulation of the dynamic response of GLARE plates subjected to low velocity impact using a linearized spring–mass model. Aerosp. Sci. Technol. **64**, 24–30 (2017)
8. Vogelesang, L.B., Vlot, A.: Development of fibre metal laminates for advanced aerospace structures. J. Mater. Process. Technol. **103**(1), 1–5 (2000)
9. Botelho, E.C., Silva, R.A., Pardini, L.C., Rezende, M.C.: A review on the development and properties of continuous fiber/epoxy/aluminum hybrid composites for aircraft structures. Mat. Res. **9**(3), 247–256 (2006)
10. Bikakis, S.E., Kalfountzos, C.D., Theotokoglou, E.E.: Elastic buckling response of rectangular GLARE fiber-metal laminates subjected to shearing stresses. Aerosp. Sci. Technol. **87**, 110–118 (2019)
11. Kounadis, A.N., Sophianopoulos, D.S.: Nonlinear dynamic buckling of a cylindrical shell panel model. AIAA J. **34**, 2421–2428 (1996)
12. ANSYS: Structural Analysis Guide. ANSYS Inc. (2014)
13. Timoshenko, S.P., Gere, J.M.: Theory of Elastic Stability. Mc Graw Hill, Stanford (1961)
14. Gay, D., Hoa, S.V., Tsai, S.W.: Composite Materials Design and Applications. CRC Press LLC, Boca Raton (2003)
15. Moriniere, F.D., Alderliesten, R.C., Sadighi, M., et al.: An integrated study on the low velocity impact response of the GLARE fibre-metal laminate. Comp. Struct. **100**, 89–103 (2013)

Evaluating Bonding Characteristics of Joggled Lap CFRP Using Acoustic Emission Techniques

Claudia Barile(✉), Caterina Casavola, Giovanni Pappalettera, Carmine Pappalettere, and Paramsamy Kannan Vimalathithan

Dipartimento di Meccanica Matematica e Management, Politecnico di Bari, Viale Japgia 182, 70126 Bari, Italy
claudia.barile@poliba.it

Abstract. In this study, the carbon fiber reinforced plastics (CFRP) specimens bonded adhesively in joggled lap configuration are tested for their bonding characteristics. The acoustic emission (AE) technique is used as a characterizing tool and peak amplitude is taken as the primary acoustic descriptor. The peak amplitude distributed in the time domain of the test is clustered by using an unsupervised pattern recognition algorithm (k-means++ algorithm) to differentiate the different damage modes. Furthermore, the waveforms of the acoustic signals recorded were studied using wavelet packet transform (WPT). The frequency band associated with each damage mode is identified using the wavelet packet transform. It is identified that the dominant damage mode responsible for failure is the interfacial debonding and interlaminar crack growth through the thickness of the adhesive layer. Overall, the acoustic emission technique proved to be a powerful tool in evaluating the bonding characteristics of the tested CFRP joggled lap specimens.

Keywords: Acoustic emission · CFRP · Bonding · Wavelet packet transform · Pattern recognition

1 Introduction

The Carbon Fiber Reinforced Plastics (CFRP) has extensively been used in aerospace structures, especially in crucial aerospace structures such as the structures joining lower spar of the fuselage and bulkhead, fuselage and half-wing, etc. Thus, designing and testing a proper joint structure at these crucial parts have become essential.

Acoustic Emission (AE) technique remains a powerful tool in characterizing damage progression in CFRP structures. AE technique records the transient elastic waves developed due to the stressing of a material under loading [1]. The corresponding signals recorded by a transducer (e.g. PZT) are filtered, pre-amplified and stored for further analysis. Since these generated elastic waves are very sensitive even to the grain movement, microcracks or micro-matrix cracking, they are very efficient in analyzing the various features of the damage progression.

E. Gdoutos and M. Konsta-Gdoutos (Eds.): ICTAEM 2020, STIN 16, pp. 26–31, 2020.
https://doi.org/10.1007/978-3-030-47883-4_5

2 Materials and Methods

2.1 Materials

CFRP laminates used in this study comprise high-strength stitched carbon fibers reinforced in epoxy resin and cured by autoclave method. The average resin percentage in each laminate is 43% and the ply thickness is 0.244 mm. The adhesive used for bonding the upper and lower laminates has a shear and peel strength of 25 MPa and 65 MPa, respectively. The upper adherend consists of 8 plies with a thickness of 2 mm and the lower ply with 6 plies and 1.5 mm. The adhesive layer coated to join the upper and lower laminate has a length of 26 mm and a thickness of 3.67 mm. The adhesive was cured at a temperature of 100 °C to 150 °C for about 1 h.

2.2 Testing Methods

The testing was carried out as per ASTM D5868 standard at a crosshead displacement rate of 13 mm/min in 100 kN INSTRON servo-hydraulic machine. To record the acoustic activities during testing, two AE sensors, R30a with an operating frequency of 150 kHz to 450 kHz were fastened to the surface of the specimen. The sensors are placed exactly 40 mm from the middle of the specimens. The recorded AE signals are pre-amplified by 40 dB through 2/4/6 AST pre-amplifier. The waveforms are recorded at a sampling rate of 1 MSPS.

2.3 Unsupervised Pattern Recognition Technique

In this research work, k-means++ unsupervised pattern recognition algorithm is used to cluster the peak amplitude distribution for the entire load history. The k-means++ works on the principle of selecting specific number of centroids; then it assigns the datapoints to each cluster based on the minimum distance between each datapoint and the centroid. Detailed step by step procedure of this technique can be found elsewhere [2]. The peak amplitudes recorded during the testing of all three specimens were classified into three clusters.

2.4 Wavelet Packet Transform

Wavelet Packet Transform (WPT) decomposes the recorded waveform into low and high-frequency components, approximation and detail, respectively, while retaining the orthogonality, smoothness and the localization properties of the parent wavelet. A desired level of decomposition (i) can be assigned to the WPT technique and the total number of WPT components are 2^i. The frequency content of the WPT and the decomposition are carried out in Wavelet Toolbox in MATLAB [3]. The decomposition level for this study is set at i = 3.

3 Results and Discussions

3.1 Mechanical Test Results

The mechanical test results are presented in Table 1. All the specimens have suffered damage in two different stages and accordingly, the peak loads at those junctures are termed as peak load at initial rupture and final rupture.

The mechanical results of all three specimens having same material and geometric configurations show different peaks loads at initial and final rupture. JLS 1 specimen significantly has carried more load before its initial rupture than JLS 2 and JLS 3 specimens. JLS 2, in particular, has a very poor strength both at the initial and final rupture. The differences in these peak loads, despite their similar design configuration is because these three specimens have two adherends each, bonded together using a high-strength adhesive manually. The application of the adhesive, its surface area, thickness, curing conditions could differ between the three specimens. This is the reason why the three specimens have different peak loads at rupture.

Table 1. Mechanical properties of specimens.

Specimen	Peak load *(kN)*	
	Initial rupture	Final rupture
JLS 1	6.02	4.09
JLS 2	2.67	3.89
JLS 3	3.86	5.57
Mean	***4.18***	***4.52***
Std. Dev.	***1.70***	***0.92***

3.2 Acoustic Emission Test Results

The peak amplitudes recorded during the test are clustered using k-means++ unsupervised pattern recognition technique and are presented in Fig. 1. The peak amplitude for all three JLS specimens, clustered into 3 groups, is presented and discussed.

The clustered peak amplitude data shows clearly the differences in damage progressions between the three specimens. In JLS 1 specimen, there is very few events in cluster 1 (higher amplitude cluster) during the initial stages of loading. The peak amplitude accumulates in cluster 1 only at the end of the test around 1.8 s. Similarly, there are very few events in cluster 2 as well until 0.9 s duration. This clearly indicates that no major damage occurred in the specimen until the material reaches 0.9 s. The amplitudes in cluster 3 indicate low amplitude signals, which normally represent microcracking, which is the only major damage mode in JLS 1, until 0.9 s. This occurrence can be connected with the fact that JLS1 shows higher strength compared with JLS 2 and JLS 3.

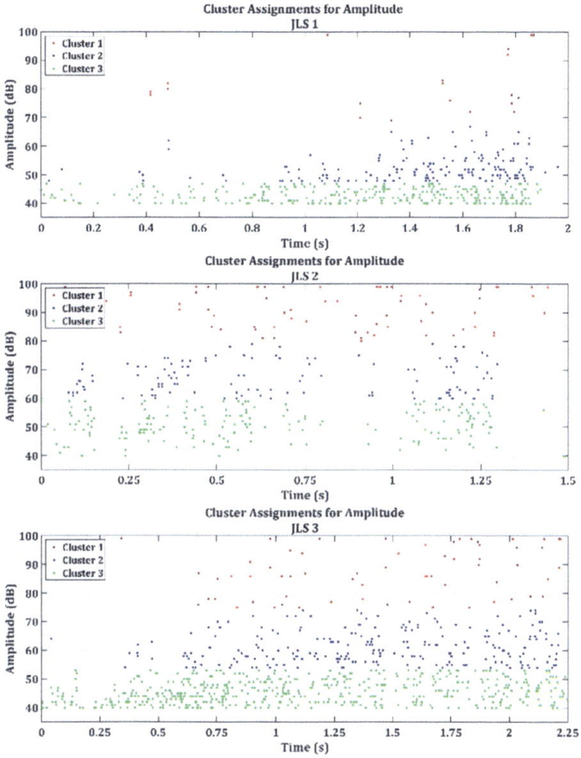

Fig. 1. Peak amplitude clustered by k-means++ algorithm for JLS 1, 2 and 3

Looking at JLS 2, amplitude commenced to accumulate in cluster 1 and 2 at a very early stages, indicating major damages in the specimen at a very early stages of loading. This is the reason for the poor peak load even at the initial rupture. JLS 3, having intermediate strength between JLS 1 and 2 has very few points in cluster 1 at the initial stages of loading. However, events in cluster 2 can be found as early as 0.35 s duration. Thus, major damage as in JLS 2 did not occur in JLS 3 specimens at an early stage, nonetheless, the material suffered significant damage.

Thus, the clustered peak amplitude data of the AE clearly differentiate the damage modes among the three specimens.

3.3 Wavelet Packet Transform Results

Different damage modes in a CFRP emit acoustic signals in different frequency spectra. In that regard, three wavelets each from JLS 1, 2 and 3 specimens were taken for the WPT analysis. One wavelet in each cluster of the three specimens has been taken randomly for the analysis.

Under loading, an adhesively bonded CFRP transfers the load entirely through the adhesive region. The thickness of the adherend does not affect the load in a displacement-controlled condition [4]. Thus, the fiber breakage or fiber pullout is not expected in the fractured specimens. This can be proved evidently also by the WPT results.

In the first WPT results of JLS 1 specimens (Fig. 2(a)), the signal has a very low energy and is centered around 200 kHz and 300 kHz frequency bands. However, in the second and third wavelet, it is centered around 300 kHz and 375 kHz (Fig. 2(b) and (c)). Particularly in the third wavelet, the frequency band remains the same, but it is shifted to the right in time domain (Fig. 2(c)).

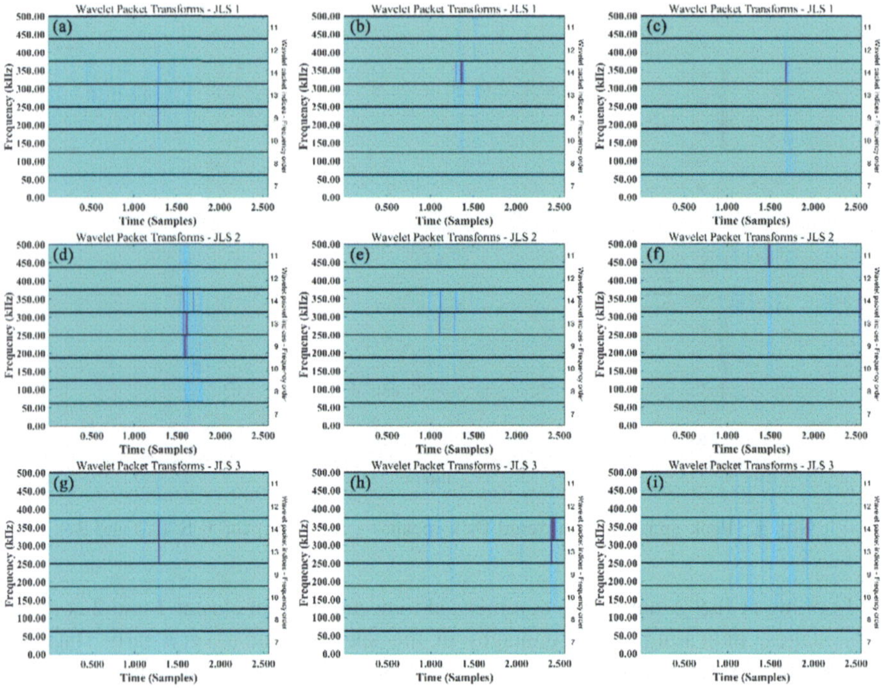

Fig. 2. WPT results of JLS 1, 2 and 3 specimens

In the WPT results of specimen JLS 2, the first wavelet (Fig. 2(d)) has high energy and the frequency band is at 200 kHz to 300 kHz. Unlike Fig. 2(a), this wavelet has higher energy in that region, while comparing the density of the wavelet. In Fig. 2(e), the frequency band has a low energy and is centered around 250 kHz and 375 kHz. The last wavelet is entirely different from the others with two distinct frequency spectrum the time domain (Fig. 2(f)). The first spectrum is at 450 kHz to 500 kHz frequency and the second is shifted in the time domain to the right at 250 kHz and 375 kHz.

The WPT results of JLS 3 specimens are all centered at 250 kHz to 375 kHz, the only difference is being the shift in the time domain (Fig. 2(g), (h) and (i)).

The frequency band between 200 kHz and 300 kHz represents the interfacial debonding. Th frequency band 250 kHz to 375 kHz is mostly misinterpreted as fiber breakage. However, recent studies prove that these bands could represent the interlaminar crack growth [5].

JLS 1 suffered a very slight interfacial debonding under the loading and most of the damage suffered is due to the high frequency signals (250 kHz to 375 kHz). This is why the energy of the 200 kHz to 300 kHz signal in Fig. 2(a) is very low. However, the specimen JLS 2 suffered severe damage at a very early stage. This is indicated by the Fig. 2(d) where the energy is high for the 200 kHz to 300 kHz signal. This means that the specimen suffered the interfacial debonding at a very early stage and failed to carry enough load before rupture. This is the reason why it has higher energies in the frequency bands representing debonding and lower energies in the other frequency domains.

In all the specimens, most of the signals are centered around the frequency band of 250 kHz to 375 kHz. This signifies that the load was carried through the adhesive layer and the interfacial crack growth in the adhesive area are responsible for most of the acoustic activities. Thus, it is safe to conclude that the major damage modes in JLS specimens are interfacial debonding and crack growth.

4 Conclusions

The Joggled Lap Shear specimens are tested under tensile loading. The damage characteristics are studied using AE technique. The k-means++ clustering technique is used to cluster the peak amplitude data of AE to differentiate the damage modes in the specimens. A wavelet from each cluster was taken for WPT analysis to identify the major damage modes in the specimens. The frequency bands indicate that the major damage modes are interfacial debonding and crack growth. The AE proves to be a powerful tool in characterizing the damage modes in the CFRP specimens.

References

1. Grosse, C.: Acoustic Emission Testing, 1st edn. Springer, Heidelberg (2008)
2. Roundi, W., El Mahi, A., El Gharad, A., Rebiere, J.: Acoustic emission monitoring of damage progression in glass/epoxy composites during static and fatigue tensile tests. Appl. Acoust. **132**, 124–134 (2018)
3. Barile, C., Casavola, C., Pappalettera, G., Vimalathithan, P.K.: Damage characterization in composite materials using acoustic emission signal-based and parameter-based data. Compos. B **178**, 107469 (2019)
4. Barile, C., Casavola, C., Pappalettera, G., Vimalathithan, P.K.: Characterization of adhesive bonded CFRP laminates using full-field digital image stereo-correlation and finite element analysis. Compos. Sci. Technol. **169**, 16–25 (2019)
5. Oz, F.E., Ersoy, N., Lomov, S.V.: Do high frequency acoustic emission events always represent fibre failure in CFRP laminates? Compos. A **103**, 230–235 (2017)

Experimental Study of Low Concentration Diffusible Hydrogen Effect on Mechanical Behaviour of Carbon Steel

I. M. Dmytrakh(✉), R. L. Leshchak, and A. M. Syrotyuk

Karpenko Physico-Mechanical Institute of NASU,
5 Naukova Street, Lviv 79060, Ukraine
dmtr@ipm.lviv.ua

Abstract. The presented work is dedicated to the evaluation of the mechanical behaviour of the ferrite-pearlite carbon steel under a low concentration of diffusible hydrogen in the bulk of the metal. Received standard stress-strain diagrams under different hydrogen concentration C_H in the specimens have shown on the existence of some specific value of the hydrogen concentration C_H^* at which the mechanism of hydrogen influence changes, namely: below this value the enhanced plasticity (decreasing of the yield stress) takes places and above – the hydrogen embrittlement occurs. This phenomenon was explained based on the special tests of tensile specimens, which were preliminary charged by hydrogen to concentration C_H^* and then discharged to the concentration $C_H \approx 0$. It has been found that increasing the number of cycles of charging-discharging leads to decreasing the value of yield stress. Thus, even a short-term presence of the hydrogen in the material leads to irreversible changes in the structure of the material, namely: the appearance of defects at the microlevel. This facilitates the deforming ability of steel and reduces the value of the plasticity limit.

Keywords: Carbon steel · Hydrogen concentration · Enhanced plasticity

1 Introduction

Nowadays, this is an indisputable fact that hydrogen can demonstrate various effects on the mechanical properties of structural metallic materials. Although, there is no complete understanding of these effects and the concepts and approaches to the solution of the most important problems of hydrogen materials science are quite contradictory [1]. Furthermore, prevail numbers of investigations in this field have a substantial disadvantage: the true level of hydrogen concentration in the metal was unknown and, hence, the degree of its hydrogenation is taken into account indirectly. However, the volume concentration of hydrogen in the metal and its local concentration in the zones of high mechanical stresses are the critical parameters that define the strength and fracture resistance of materials in given conditions [2, 3].

The presented work removes this disadvantage because the tests were carried out under the known values of the volume concentration of hydrogen in the specimens. The study is dedicated to the evaluation of the mechanical behaviour of the ferrite-pearlitic

E. Gdoutos and M. Konsta-Gdoutos (Eds.): ICTAEM 2020, STIN 16, pp. 32–37, 2020.
https://doi.org/10.1007/978-3-030-47883-4_6

carbon steel under a low concentration of diffusible hydrogen in the bulk of the metal. Here, received dependencies were analyzed as the functions of hydrogen concentration.

2 Experimental Procedure

The object of study was low alloyed pipeline steel (σ_Y = 260 MPa and σ_U = 440 MPa) with nominal chemical composition (in weight %): C = 0.17–0.24; Si = 0.17–0.37; Mn = 0.35–0.65; S < 0.04; remainder Fe. This material consists of grains of ferrite-pearlite, typical for all steel of this class. The standard cylindrical tensile specimens with a diameter of 5 mm were manufactured from the real pipe.

The hydrogen charging of the specimens was made by the electrochemical method under cathodic polarisation at some constant potential $E_{cath} = const.$. With the aim to simulate the hydrogen entry at real operating conditions of the buried pipeline, the following procedure has been applied [4]. The special deoxygenated, near-neutral pH NS4 solution, which is the model of underground water, was chosen as the electrolyte for hydrogen charging of steel. The chemical composition of the NS4 solution is given in Table 1. Taking into account the situation of a freely corroding system that exists for the real pipeline, the potential of polarisation E_{cath} was slightly more negative than the free corrosion potential E_{corr} for given steel.

Table 1. Chemical composition of NS4 solution (gram/litre).

$NaHCO_3$	KCl	$CaCl_2$	$MgCl_2.H_2O$
0.483	0.120	0.137	0.131

The hydrogen concentration in bulk of steel has been determined on the base of the hydrogen discharging process under anodic polarisation with using modified hydrogen electrochemical oxidation method proposed [4, 5].

For the realisation of experimental studies on the hydrogen charging of specimens and determination of the hydrogen concentration in bulk of steel, the special testing stand was developed [6]. This facility is based on the dynamic electrochemical laboratory VoltaLab40 (Radiometer Analytical). As the preliminary stage of the study the experimental dependence "hydrogen concentration C_H in specimen – time of exposure τ" was received. These experimental data were described by power relation:

$$C_H = 0.128 \cdot A \cdot 10^{-6} \cdot \tau^{\mathrm{m}}\ [\mathrm{ppm}], \qquad (1)$$

where A and m are some constants that depend on system "material – environment" and testing conditions. Based on the formula (1) the specimens were hydrogen charged to the assigned level of C_H, before testing.

3 Results and Discussion

Based on the received experimental curves 'applied load – elongation of specimen', the dependencies 'stress σ – strain ε' were constructed for different values of the hydrogen concentration C_H in the specimens. Here it should be noted that calculation of stresses σ was made with taking into account the real values of cross-section area of specimens, i.e. so-called 'true' stress-strain diagrams of were received. These diagrams served as the basis for the determination of the true values of the yield stress σ_Y as the function of the hydrogen concentrations C_H (Fig. 1).

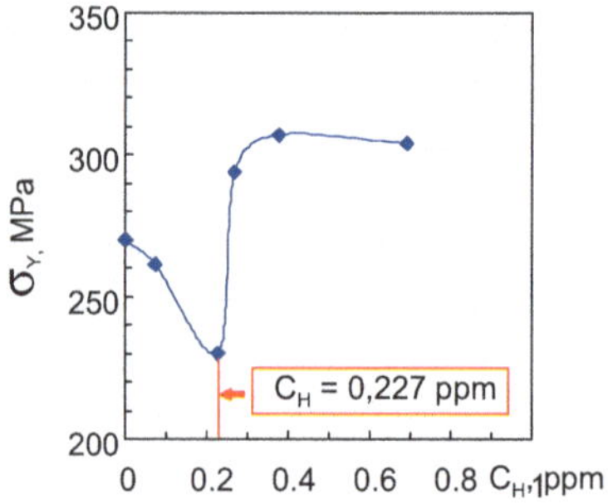

Fig. 1. Dependence of the yield stress σ_Y on the volume concentration of hydrogen C_H in the specimens.

As can be seen in Fig. 1, under increasing of the hydrogen concentration in metal, the decreasing of the yield stress value σ_Y was observed up to $C_H \cong 0.227\ ppm$. This shows the facilitation of a plastic deforming of steel under a given range of C_H in the specimens. After exceeding the value $C_H \cong 0.227\ ppm$, the increase of the yield stress values σ_Y occurs that points to the decrease of plastic deformation under the tensile loading of specimens.

After the tensile test, the fracture surfaces of specimens were examined by the scanning electronic microscopy. The following specific features of the carbon steel fracture on the dependence of the hydrogen concentrations C_H in bulk of material were observed. For non-hydrogenated specimens ($C_H = 0$) fracture surface has the signs of the mixed mechanism of failure. Here, along with 'dimple' relief that showed on the ductile mechanism of fracture, the relief of typical quasicleavage brittle fracture can be found. With increasing of the hydrogen concentrations C_H in material, the increasing of the dimples density and disappearing of quasicleavage elements were found. It points to the increase of the plastic deformation under the fracture process of steel. Such a picture was observed up to the value of hydrogen concentration equal $C_H \cong 0.227\ ppm$. After that, the further increase of the hydrogen concentration in specimen leads to the appearance of some trace of quasicleavage relief on the fracture surface, i.e. the fracture mechanism of steel becomes gradually mixed again.

These data give the basis for the following statement. For this carbon steel, there is some characteristic value of the hydrogen concentration $C_H = C_H^* \cong 0.227\ ppm$ in bulk of material under which the mechanism of hydrogen influences changes, namely: the strengthening (hydrogen embrittlement) occurs at $C_H \geq C_H^*$ and the enhanced plasticity (decreasing of the yield stress value) takes places at $C_H \leq C_H^*$.

For a possible explanation of such phenomenon, a special series of the tests were conducted. Before the tensile test, the specimens were preliminary hydrogen charged to the level $C_H = C_H^*$ and then fully hydrogen discharged ($C_H \approx 0$). This procedure was repeated to obtain the specimens with the different number of cycles of hydrogen charging-discharging N_H, namely: 0, 2 and 5. Such specimens were tested under tensile loading and the nominal and true stress-strain diagrams were constructed for the specimens with the different values of N_H (Fig. 2). It can be seen that the deformation ε, which corresponds to the fracture of the specimen, increases with increasing of N_H. This fact indicates the facilitation of material deforming depending on the value of the parameter N_H.

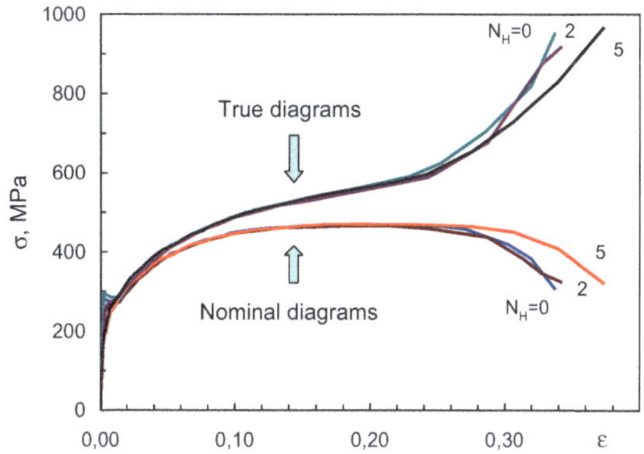

Fig. 2. Nominal and true stress-strain diagrams for specimens with the different number of cycles of hydrogen charging-discharging N_H.

The stress-strain diagrams (Fig. 2) served as a base for determining the yield stress values σ_Y as the function of the parameter N_H (Fig. 3). These results show the monotonic decrease of a yield stress value σ_Y with increasing the charging-discharging cycles N_H. Here, it also may be pointed out on the full coincidence between the data received under conditions of the hydrogen charging-discharging of the specimens and the case of presence of the hydrogen in the specimens (Fig. 1). Thus, it can be suggested that even a short-term presence of the hydrogen in the steel leads to the irreversible changes in the material structure that causes the changes in the mechanical behaviour of the material.

The fractographs of the specimens tested after the different number of cycles of hydrogen charging-discharging N_H showed (Fig. 4) the appearance of defects at the microstructural level. At that, the defectiveness of material increases with increasing of the number of hydrogen charging-discharging cycles N_H (Fig. 5). This facilitates the deforming ability of steel and reduces the value of the plasticity limit.

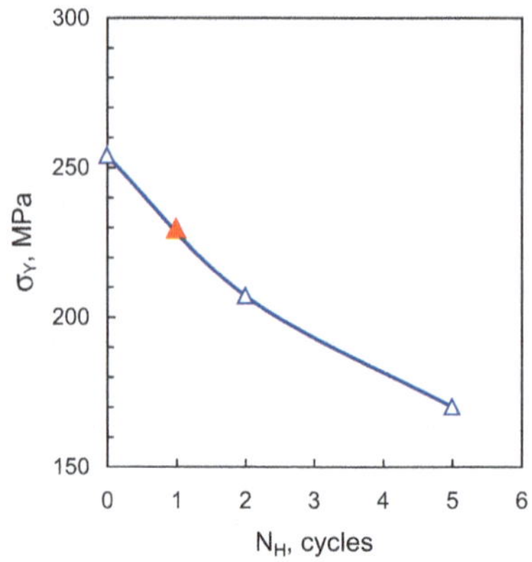

Fig. 3. Dependence of the true values of yield stress σ_Y on the number of cycles of hydrogen charging-discharging N_H. The solid triangle refers to the data presented in Fig. 1.

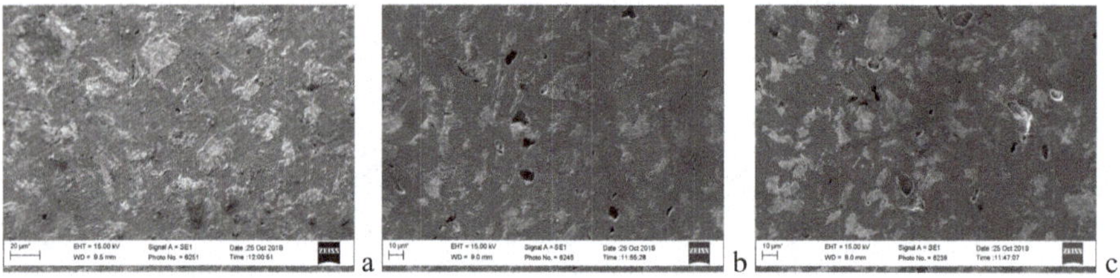

Fig. 4. Fractographs of the specimens tested after the different number of cycles of hydrogen charging-discharging N_H: a – 0; b – 2; c – 5 (x 500). The area of observation was equal to 41300 μm^2.

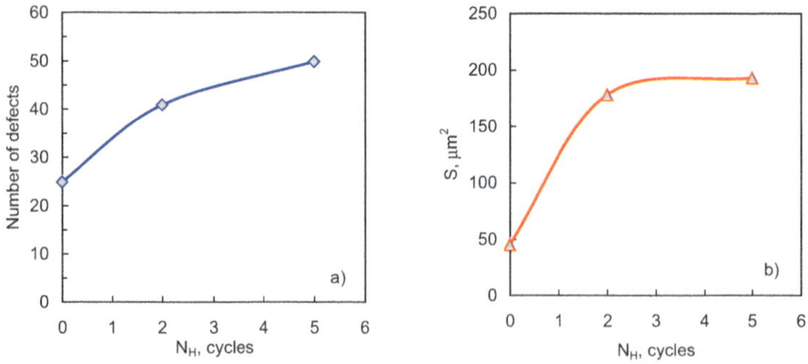

Fig. 5. Number defects (a) and their total square (b) in steel after the tensile test as the function of the number of cycles of hydrogen charging-discharging N_H. The area of observation was equal to 41300 μm^2.

If it may be assumed that the defectiveness of material is strongly related to the dislocations activity, then the received tendencies can be explained by the interaction between hydrogen and dislocations [7]. The hydrogen affects on the dislocations in two ways: promotes their pinning (or dragging) or enhances of their mobility. Competition between these two roles defines whether the resulting phenomenon is positive or negative with the point of view of the fracture resistance of the material.

4 Conclusions

The fact of the yield stress reduction of carbon steel under the presence of the diffusible hydrogen of low concentration has been found. Hydrogenation and de-hydrogenation of the steel do not pass without a trace concerning its microstructure. Even a short-term presence of the diffusible hydrogen of low concentration in the steel leads to the irreversible changes in the material structure that causes the changes in the mechanical behaviour of the material. The obtained results may be useful under the development of technologies for processing steels of this class for the optimization of their service characteristics.

References

1. Somerday, B., Sofronis, P.: Hydrogen-Materials Interaction. ASME Press, New York (2014)
2. Capelle, J., Dmytrakh, I., Gilgert, J., Pluvinage, G.: The effect of hydrogen concentration on fracture of pipeline steels in presence of a notch. Eng. Fract. Mech. **78**, 364–373 (2011)
3. Dmytrakh, I.M., Smiyan, O.D., Syrotyuk, A.M., Bilyy, O.L.: Relationship between fatigue crack growth behaviour and local hydrogen concentration near crack tip in pipeline steel. Int. J. Fatigue **50**, 26–32 (2013)
4. Capelle, J., Dmytrakh, I., Pluvinage, G.: Comparative assessment of electrochemical hydrogen absorption by pipeline steels with different strength. Corros. Sci. **52**, 1554–1559 (2010)
5. Capelle, J., Dmytrakh, I., Gilgert, J., Pluvinage, G.: Sensitivity of pipelines with steel API X52 to hydrogen embrittlement. Int. J. Hydrogen Energy **33**, 7630–7641 (2008)
6. Dmytrakh, I.M., Leshchak, R.L., Syrotyuk, A.M.: Effect of hydrogen concentration on strain behaviour of pipeline steel. Int. J. Hydrogen Energy **40**, 4011–4018 (2015)
7. Murakami, Y., Kanezaki, T., Mine, Y.: Hydrogen effect against hydrogen embrittlement. Metall. Mater. Trans. A **41**, 2548–2562 (2010)

Effects of Temperature on Tensile and Fracture Behavior of Dissimilar Metal Welded Joint for Nuclear Safe-End

Lei Wang(✉), Yang Liu, Xiu Song, and Jiahua Liu

Key Laboratory for Anisotropy and Texture of Materials, Northeastern University, Shenyang 110819, People's Republic of China
wanglei@mail.neu.edu.cn

Abstract. The effects of temperature on tensile and the fracture behavior of dissimilar metal welded joint for nuclear safe-end were studied. The results show that the strength, elongation and fracture toughness of DMWJ significantly decrease with increasing temperature. At room temperature, more deformation twins can make the deflection of the crack propagation direction, so the crack propagation path is extended, and the alloy can absorb more energy during the fracture process, so as to the toughness of the joint is enhanced. While at the evaluated temperature, the second phase particles are more likely to be a crack initiation, which makes the crack propagation easier. In addition, the dislocation density is reduced, resulting in the decreased strength.

Keywords: Dissimilar metal welded joint · Temperature · Deformation · Twining

1 Introduction

The nuclear safety end is located in primary water systems pressure boundary of pressurized water reactors (PWRs) and it is usually connected by dissimilar metal welded joints (DMWJs) so it has been regarded as the key part. The DMWJ is exposed to high temperature and high pressure, as well as complex stress, so that it is the most vulnerable components [1–5]. The safety of the joint can be influenced by the significant microstructure heterogeneity of DMWJ, residual stress [6, 7], weld composition dilution [8], C diffusion [9–13] and others. It is necessary to study the tensile and fracture behavior of the joint, in order to provide a theoretical basis for the safe operation of nuclear power plants. Some literatures have reported the deformation twins had important effect on the toughness of the alloy [14, 15]. However, little attention has been paid to reveal the mechanism of the temperature effect on the tensile deformation of DMWJ. Therefore, in the present study, tensile deformation behavior of a DMWJ was investigated in a temperature range from room RT to 320 °C.

E. Gdoutos and M. Konsta-Gdoutos (Eds.): ICTAEM 2020, STIN 16, pp. 38–43, 2020.
https://doi.org/10.1007/978-3-030-47883-4_7

2 Experimental Procedures

SA508-III-309L/308L-316L DMWJ for the PWRs was used as the experimental material. The tensile and fracture toughness J_Q (with a pre-fatigue crack) tests were conducted from RT (about 23.5 °C) to 320 °C using a SANS CMT5105 universal tensile testing system at a strain rate of 10^{-3} s^{-1}, and the 25 mm extensometer was employed. For optical microscopy (OM) and scanning electron microscopy (SEM) examinations, the specimens were mechanically polished and etched. Thin foils for transmission electron microscopy (TEM) were prepared from the deformed specimen, polished by twin-jet electro polishing machine using an electrolyte consisting of 9% $HClO_4$ and 91% ethanol at a temperature of −30 °C. TEM observation were carried out by a JEOL JEM-2100F operating at 200 kV.

3 Results and Discussions

Table 1 shows the tensile properties of the joints. The yielding strength (YS), ultimate strength (UTS) and the elongation (El) of the joint significantly decrease with increasing tensile deformation temperature. Compared to the joint tested at RT, the YS, UTS and El of that at 320 °C decreased by 63.60%, 28.60% and 49.40%, respectively. Figure 1 shows the J-R curves of the joints at different temperatures, which the specimen was bended with a cross head speed of 0.2 mm/min and the J_Q is determined by off-line of $J = 8\sigma_Y(\Delta a)$. From Fig. 1, it can be seen that the J_Q value of the joint decreases with the increasing of temperature, 246.12 kJ/m^2 at RT, 155.96 kJ/m^2 at

Table 1. Tensile properties of DMWJ at different temperatures

Temp (°C)	YS (MPa)	UTS (MPa)	EI (%)
RT	332	561	27
150	306	460	17
240	251	427	15
320	121	400	13

320 °C, the J_Q loss ratio reaches 36.63%.

The microstructures of the tested joints (near the fracture position) at RT and 320 ° C are shown in Fig. 2(a) and (b), respectively. It can be noticed that the grain has been

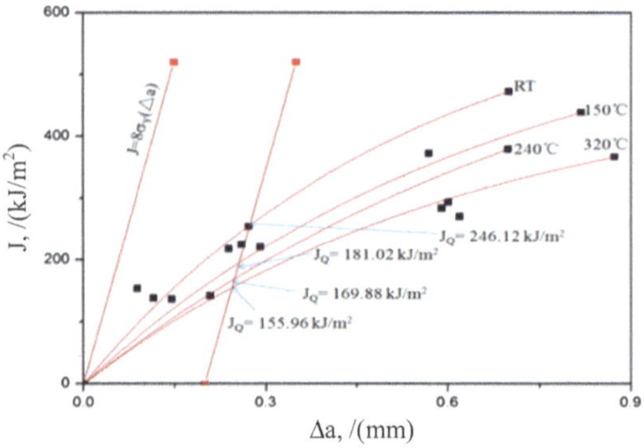

Fig. 1. *J-R* curves of DMWJ at different temperatures

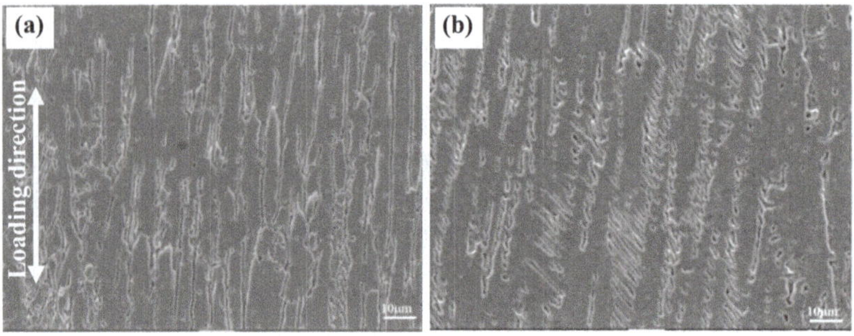

Fig. 2. SEM micrographs of longitudinal section of specimens tensile tested at (a) RT, (b) 320 °C

elongated along the tensile direction, which means the fracture positions of the joints are consistent at different temperatures.

The fracture surface morphologies of the joints tested at different temperatures are shown in Fig. 3. The fracture surfaces illustrate typical ductile fracture characteristics

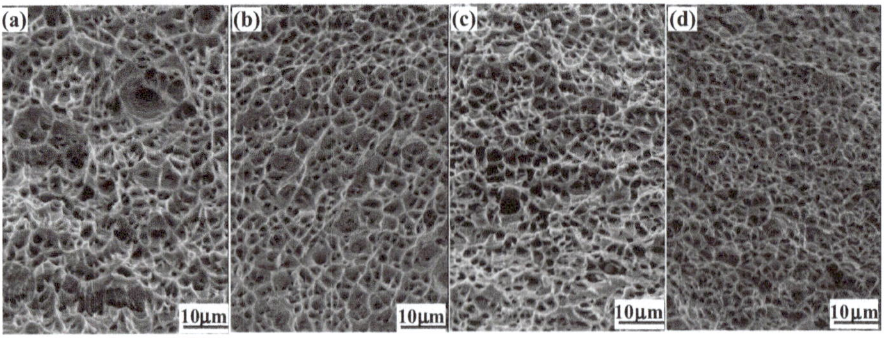

Fig. 3. Fracture surface morphologies of joint tensile tested at: (a) RT, (b) 150 °C, (c) 240 °C, (d) 320 °C

with dimples at different temperatures. However, with increasing temperature, the dimple size decreases, and the toughness of the joint decreases.

Figure 4 shows the back scattering morphologies of the fracture surface after fracture toughness test at different temperatures. The main cracks expand from the left to right. The gray porphyritic in backscattering is ferrite. It is clear that the secondary cracks can be observed from the specimens at different temperatures except RT. It can be seen that crack propagated with transcrystalline fracture, and some-times propagated with second phase particle to another (Fig. 4(d)). That means the second phase particles is the main crack source at high temperature. As well known, the dislocations will start to move under a low stress, and then pile up around the second phase particles, when the stress is large enough, the micro-voids will form along the interface between the

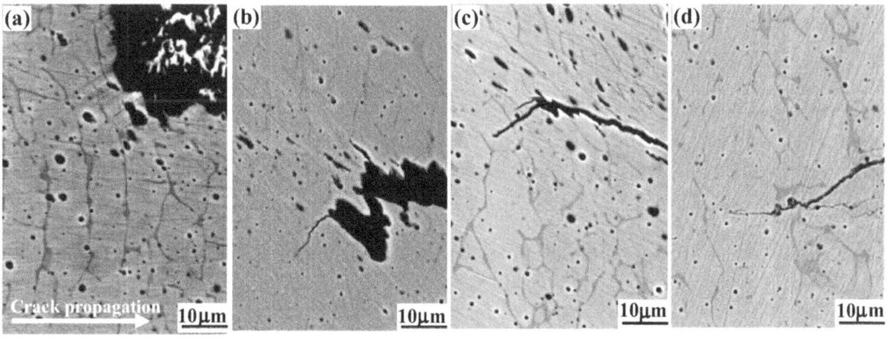

Fig. 4. Cross-section surface morphologies of joint fracture toughness tested at: (a) RT, (b) 150 °C, (c) 240 °C, (d) 320 °C

second phase particles and matrix by the stress concentration with dislocation pole up. And finally the combination of microvoids became a cracking and lead fracture.

Figure 5 shows the dislocation configuration of the tensile tested joints from RT to 320 °C. After tensile testing at RT, the number of twins is higher, and a lot of dislocation tangles can be observed in twins. While, both number of twins and dislocations decrease obviously for the joint tested at 320 °C. Since the thermal vibration of atoms increases with the increasing of testing temperature, more slip systems can be activated at higher temperature, so that the dislocation density will be decreased. The joints can also be coordinately deformed by twins because of relatively low stacking fault. But with the increasing temperature, the number of twins decreases, so the elongation of the joint will decrease.

It is well known that fracture toughness of a material is depended on the ability of coordinated deformation at crack tip. If the plasticity of a material is very good, more absorbed energy is needed for the crack tip enlarging, resulting in higher fracture toughness. In addition, when deformation twining occurs, the stress concentration can be released, so the plasticity of the joint increases. It has been reported that crack growth direction is easy deflected by the deformation twining [16, 17]. On the other hand, the dislocation slip can be hindered with the deformation twining during

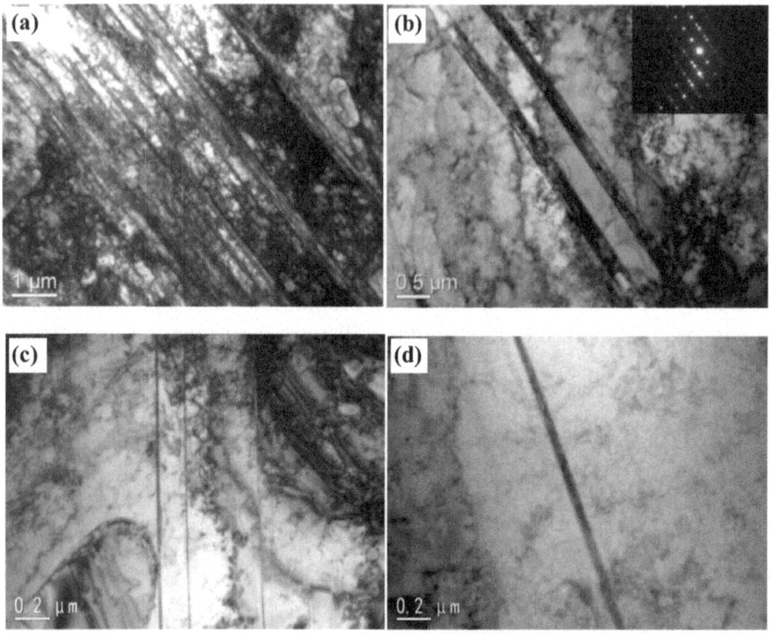

Fig. 5. Dislocation configuration of joint tensile tested at: (a) RT, (b)150 °C, (c) 240 °C, (d) 320 °C

deformation, then dislocations pile up on the twin boundaries, and this leads to the stress concentration. When a crack propagates to a deformation twin, the stress concentration occurs at the twin boundary. When the stress concentration is difficult to release by the slip of dislocations at the twin boundaries at crack tip, crack will propagate along the twin boundary, therefore the direction of crack propagation will be changed. With such direction changing, the crack propagation route will become longer and longer. So that, more absorb energy is needed for the fracture processing of a joint at RT, comparing with those at elevated temperatures.

4 Conclusions

(1) The main crack source of a joint is the second phase particles. With the increasing of testing temperature, the atom vibration increases, and more slip system can be started to motion. It makes the decreasing of dislocation density, as well as the strength of the joint. Because the stacking fault energy is low, the deformation can be coordinated by twinning during tensile deformation at the lower temperatures. With increasing of testing temperature, while a number of deformation twining decreases, the elongation of the joint decreases.

(2) Cracks easy propagate along the twin boundary, thus the crack propagation direction will be changed and the crack propagation route becomes longer. The

fracture toughness of the joint is higher at RT than those at elevated temperatures, because the longer the route is, the more the absorbed energy needs.

References

1. Seifert, H.P., Ritter, S., Shoji, T., et al.: Environmentally-assisted cracking behavior in the transition region of an Alloy182/SA 508 Cl.2 dissimilar metal weld joint in simulated boiling water reactor normal water chemistry environment. J. Nucl. Mater. **378**(2), 197–210 (2008)
2. Jang, C., Lee, J., Kim, J.S., et al.: Mechanical property variation within Inconel 82/182 dissimilar metal weld between low alloy steel and 316 stainless steel. Inter. J. Press. Vessels Pip. **85**(9), 635–646 (2008)
3. Kim, J.W., Lee, K., Kim, J.S., et al.: Local mechanical properties of Alloy 82/182 dissimilar weld joint between SA508 Gr.1a and F316 SS at RT and 320 °C. J. Nucl. Mater. **384**(3), 212–221 (2009)
4. Qin, R.Y., Duan, Z.L., He, G.: Microstructure and ductility-dip cracking susceptibility of circumferential multipass dissimilar weld between 20MND5 and Z2CND18-12NS with Ni-base filler metal 52. Metall. Mater. Trans. **44A**(10), 4661–4670 (2013)
5. Wang, H.T., Wang, G.Z., Xuan, F.Z., et al.: Local mechanical properties of a dissimilar metal welded joint in nuclear power systems. Mater. Sci. Eng., A **568**, 108–117 (2013)
6. Qiao, D., Zhang, W., Pan, T.Y., et al.: Evaluation of residual plastic strain distribution in dissimilar metal weld by hardness mapping. Sci. Tech. Weld. Joining **18**(7), 624–630 (2013)
7. Joseph, A., Rai, S.K., Jayakumar, T., et al.: Evaluation of residual stresses in dissimilar weld joints. Inter. J. Press. Vessels Pip. **82**(9), 700–705 (2005)
8. Chen, Z.-B.: Failure analysis of dissimilar steel joints of A/P. Central China power (4), 26–30 (1996)
9. Pan, C.: Microstructure and transformation mechanism of dissimilar metal welded joints of composite parts. Wuhan Trans. Tech. Univ. (1998)
10. Pavlinov, L.V.: Iron and carbon diffusion in tetra component iron Chromium Nickel Molybdenum alloys in region of gamma solid solutions. Fiz. Met. Metalloved. **41**(2), 344–350 (1976)
11. Yang, H., Shi, G., Zhang, Y., et al.: Finite difference of carbon transfer at high temperature for dissimilar steel welded joints. Weld. J. **4**, 77–80 (2001)
12. Zhang, Y., Yang, H., Cao, S., et al.: Analysis on the phenomenon of carbon transfer in dissimilar steel welded joints. J. Weld. **02**, 89–92 (2001)
13. Breedis, J.F.: Influence of dislocation substructure on the martensitic transformation in stainless steel. Acta Metal. **13**(3), 239–250 (1965)
14. Murr, L.E.: Stacking-fault anomalies and the measurement of stacking-fault free energy in f.c.c. thin films. Thin Solid Films **4**(6), 389–412 (1969)
15. Michiuchi, M., Kokawa, H., Wang, Z.J., et al.: Twin-induced grain boundary engineering for 316 austenitic stainless steel. Acta Mater. **54**(19), 5179–5184 (2006)
16. Chuaiphan, W., Chandra-Ambhorn, S., Sornil, B., et al.: Microstructure, mechanical and corrosion behaviour of dissimilar weldments. Sheet Metal **410–411**, 533–541 (2009)
17. Chung, W.C., Huang, J.Y., Tsay, L.W., et al.: Microstructure and stress corrosion cracking behavior of the weld metal in Alloy 52-A508 dissimilar welds. Mater. Trans. **52**(1), 12–19 (2011)

A Multi-attribute Information Based Method of Material Strength Distribution Fitting

Liyang Xie(✉), Bo Qin, and Ningxiang Wu

Laboratory of Vibration and Control of Aero-Propulsion System, Northeastern University, Shenyang, China
{lyxie, bqin, nxwu}@me.neu.edu.cn

Abstract. Information fusion technique has been widely applied to a variety of subjects such as fault diagnosis and image identification. Bayes estimation is a special type of information fusion technique applied to parameter estimation for probability distribution of random variable. The present paper presents a new type of information fusion technique for material strength distribution estimation in the situation of small size sample. To precisely describe material strength, three-parameter Weibull distribution is used. To find out a reasonable location parameter in the situation that only a few experimental observations are available, the knowledge and information from different aspects are utilized. First, an empirical shape parameter is chosen with reference to the strength distribution of similar material. Then, a location parameter is assigned to make the estimated material strength variation at a realistic level, by judging the rationality of the location parameter through the strength probability distribution thus estimated. At last, big data technique is applied to further verify the rationality of the estimated material strength distribution by testing the relation between location parameter and the minimum observation in a sample of particular size for a special three-parameter Weibull distribution.

Keywords: Material strength · Weibull distribution · Big data · Monte Carlo simulation

1 Introduction

To estimate material strength by means of experiment data, a large size sample of strength data are usually required [1–4]. In the situation of small size of sample, especially very small size of sample containing only a few observations, both distribution type identification and distribution parameter estimation are difficult. Generally, three-parameter Weibull distribution is suitable to describe material strength since its flexibility, especially the robustness of its shape parameter to the same type of products. Lots of experiences have demonstrated that the shape parameter of the Weibull material strength of a particular type of material will not be much different.

Weibull distribution has been applied to a wide range of problems [5]. It has also been employed for smaller sample sizes than any other statistical distribution [6]. However, the mostly applied is the two-parameter Weibull distribution [7–11].

E. Gdoutos and M. Konsta-Gdoutos (Eds.): ICTAEM 2020, STIN 16, pp. 44–51, 2020.
https://doi.org/10.1007/978-3-030-47883-4_8

Contrasting to two-parameter Weibull distribution, three-parameter Weibull distribution has evident advantages. However, a relatively large size of sample is required for three-parameter Weibull distribution fitting. As a typical example of small size of sample, Nagatsuka proposed a least squares procedure for parameter estimation with sample size from 7 to 20 [12]. Another study to describe material fatigue strength is based on 14 observations [13].

To estimate material strength distribution in the situation of small size sample, Yang et al. proposed a new Bayesian method based on the characteristics of the product, using a modified Weibull distribution to describe the failure rate over entire lifetime [14]. Abbasi et al. presented an approach of Artificial Neural Networks (ANN) [15]. This paper focuses on new way to estimate the parameters of three-parameter Weibull distributed material strength based on a few experiment data.

2 On the Shape Parameter of Material Strength Distribution

For a Weibull distributed material strength, the shape parameter can be determined empirically, i.e. it can be taken from similar material. For the Weibull distribution of metallic materials, a great amount of test data indicate a roughly range between 2.0 and 4.0. In the following ball bearing material strength estimation, the shape parameter is taken as 3.0.

3 On the Location Parameter

A popular expression for the three-parameter Weibull probability density function is

$$f(t) = \begin{cases} \frac{\beta(t-\gamma)^{\beta-1}}{\eta^{\beta}} \exp[-(\frac{t-\gamma}{\eta})^{\beta}] & t \geq \gamma \\ 0 & t < \gamma \end{cases} \quad (1)$$

where, β, η and γ stand for the shape parameter, scale parameter and location parameter, respectively.

Three-parameter Weibull probability density function has excellent flexibility to fit the material strengths. As to the classical statistics method, a difficulty is to estimate the location parameter in the situation of small sample of observations. Alternatively, big data technique and information fusion concept can help to find an appropriate probability density function including all its parameters to describe a random variable for engineering application.

For engineering material, the possible minimum strength must be much greater than zero, though the strength random variable may scatter on a quite large scale. If enough experiment data are available about the strength of some kind of material, its probability density function, including all the parameters, can be estimated by classical statistics method. When only a few observations, such as five, is available, relevant knowledge and information have to be applied. Included are the probability density function of similar material, especially the shape parameter of Weibull distribution, the degree of strength dispersion of similar material, etc. According to these knowledge and information, some distribution parameter can be referred, or its reasonable range can be figure out, at least.

For a Weibull distributed strength random variable, if $s_{(1,n)}$, $s_{(2,n)}$, …, $s_{(n,n)}$ are its n observations arranged in ascending order, the location parameter will be close to $s_{(1,n)}$ if n is large, especially in condition that the shape parameter is not very big.

By means of the median rank estimator of cumulative distribution function $\hat{F}(s_i)$, a sample size n_p can be figured out with which the minimal observation can be taken as the estimator of the $p \times$ 100th percentile for a specific Weibull distribution. This estimator is usually approximated as

$$\hat{F}(s_i) \approx \frac{i - 0.3}{n + 0.4} \quad (i = 1, 2, \ldots, n) \tag{2}$$

where, i is the ordinal number of the individual observations in ascending order; n is the size of the sample.

To estimate the 0.00135 percentile directly from the observations (equivalent to the lower bound of an interval of 6σ for a Gauss distribution), the necessary sample size $n_{0.00135}$ can be resolved from Eq. 2 as $n_{0.00135} = 518$. It means that if there are 518 observations, then the $s_{(1,n)}$ can be approximately taken as the location parameter.

For the situation that the sample size is much less than 518, the observation $s_{(1,n)}$ should be modified as the location parameter. To find a reasonable function to modify the minimal observation, Monte Carlo simulation is applied. The principle is that, for a given Weibull distribution, the possible differences between the location parameter γ and the minimal observation $s_{(1,n)}$ in an arbitrary sample of size n can be observed by Monte Carlo sampling. Based on the Monte Carlo simulation, the upper envelope curve of the simulated date points can be used to describe the relationship between the possible maximal difference between $s_{(1,n)}$ and the location parameter. When this difference is known, the minimum observation $s_{(1,n)}$ can be modified to be taken as the location parameter.

To perform the simulation to a given Weibull distribution W(γ, η, β), the location parameter and scale parameter have to be determined first. However, it can only be roughly chosen at this moment if the observations available are quite limited (e.g. there are only five observations). For a Weibull distribution with location parameter γ and scale parameter η, the value of the cumulative probability associated with s = γ + η equals to $1\text{-}e^{-1} = 0.632$, i.e. F(γ + η) = 0.632. For the situation of five strength observations for a material, it is easy to know by Eq. 2 that $F(s_{(4,5)}) = 0.685$, i.e. $s_{(4,5)}$ will be greater than γ + η (corresponding to a percentile of 0.632), while $s_{(3,5)}$ will be less than γ + η since $F(s_{(3,5)}) = 0.5$. Meanwhile, $F(s_{(5,5)} - F(s_{(1,5)}) = 0.741$, which is greater than the 2σ interval of 0.683 for a Gauss distribution. Therefore, $s_{(1,5)} - (s_{(5,5)} - s_{(1,5)})$ can be approximately taken as the location parameter γ, and $s_{(4,5)} - \gamma$ can be used as an approximate value of the scale parameter η. The relationship between these variates and parameters are shown in Fig. 1.

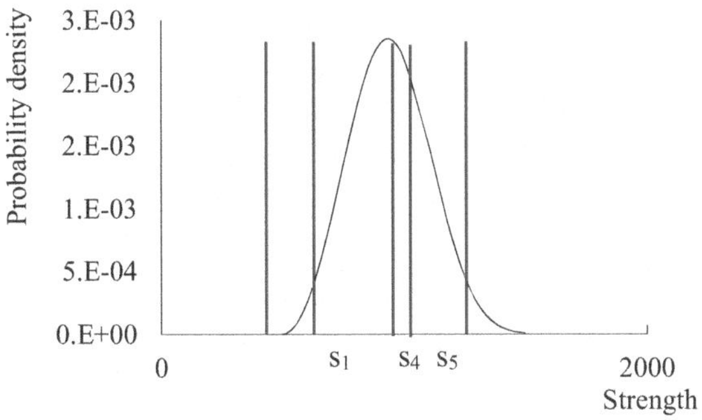

Fig. 1. The characteristic variates/parameters and their relations

Providing the five observations for a material strength are 1202, 1014, 773.3, 958.4 and 649, respectively, $s_{(1,5)} - (s_{(5,5)} - s_{(1,5)}) = 649 - (1202 - 649) = 96$ can be taken as the location parameter, and $s_{(4)} - \gamma = 1014 - 96$, i.e. 918 can be taken as the scale parameter, i.e. a Weibull distribution W(96,918,3) is estimated. From the procedure mentioned above, it is known that this is a conservative estimation.

On the other hand, the Weibull parameters can be solved in cumulation probability equations. For a Weibull distribution,

$$P(s) = 1 - e^{-(\frac{s-\gamma}{\mu})^{\beta}} \tag{3}$$

Therefore,

$$\frac{\ln(1 - P_1)}{\ln(1 - P_5)} = \frac{s_1 - \gamma}{s_5 - \gamma} \tag{4}$$

According to Eq. 2, $P_1 = 0.130$, $P_5 = 0.870$ for the situation of a sample of 5 observation. For the five observations listed above, a Weibull distribution W(267.9, 735.6, 3) is estimated. This result is not so conservative as W(96,918,3).

4 Weibull Scale Parameter Estimation by ML Method

The scale parameter of a Weibull distribution can be estimated by means of maximum likelihood method. When the shape parameter and location parameter are known, the scale parameter can be easily estimated as

$$\eta = \left(\sum_{i=1}^{n} (s_i - \gamma)^{\beta} / n\right)^{1/\beta} \tag{5}$$

For the situation of the five observations 1202, 1014, 773.3, 958.4 and 649, if the location parameter equals to 96.0 and the shape parameter equals to 3.0, then the scale parameter is estimated as 860.2; if the location parameter equals to 267.9, then the scale parameter is estimated as 699.3. If two-parameter Weibull distribution is assumed, the scale parameter is estimated as 951.4. Actually, the five observations are a sample of the Weibull distribution W(500, 500, 3) (Fig. 2).

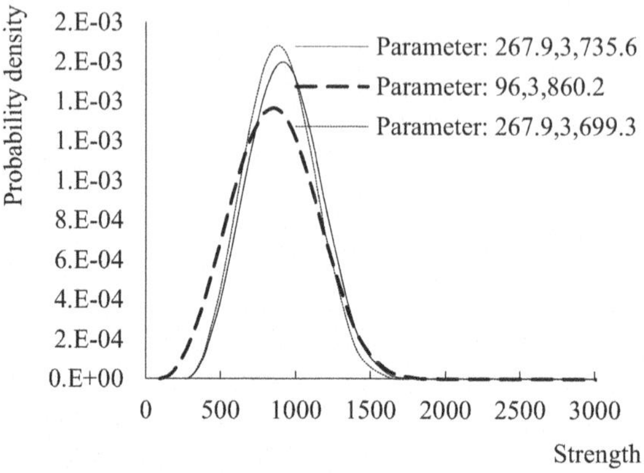

Fig. 2. The true Weibull distribution and fitting results

The rationality of the estimation results can be verified by the dispersion of the estimated distribution. Dispersion can be characterized by deviation, standard deviation or variation coefficient. The mean, deviation and variation coefficient are, respectively

$$\mu = \gamma + \eta\,\Gamma\left(1 + \frac{1}{\beta}\right) \tag{6}$$

$$\sigma^2 = \eta^2\left[\Gamma\left(1 + \frac{2}{\beta}\right) - \Gamma^2\left(1 + \frac{1}{\beta}\right)\right] \tag{7}$$

$$v = \sigma/\mu \tag{8}$$

Experiments show that for majority of metallic materials, the variation coefficient will not be greater 0.1. For instance, if a Gaussian distributed random variable has a 6σ range of $t_{min} \sim t_{max}$, and $t_{max} = 2t_{min}$, then

$$v = \frac{(t_{max} - t_{min})/6}{(t_{min} + t_{max})/2} = \frac{(2t_{min} - t_{min})/6}{(t_{min} + 2t_{min})/2} \approx 0.111$$

Meanwhile, according to Weibull cumulative distribution function, the strength associated with failure probability is

$$s_P = \eta(-\ln(1 - P))^{1/\beta} + \gamma \tag{9}$$

The lower and upper bounds of the 6σ range (probability of 99.73%) are respectively

$$t_{0.00135} = \eta(-\ln(0.99865))^{1/\beta} + \gamma$$

$$t_{0.99865} = \eta(-\ln(0.00135))^{1/\beta} + \gamma$$

The ratio of the upper bound to the lower bound, i.e. $t_{0.00135}/t_{0.99865}$ can also be applied to judge the rationality of the estimation:

$$s_{0.99865}/s_{0.00135} = \frac{\eta(-\ln(0.00135))^{1/\beta} + \gamma}{\eta(-\ln(0.99865))^{1/\beta} + \gamma} \tag{10}$$

For the above estimation results, the true situation is μ = 946.49, σ = 162.28, ν = 0.171, and the ratio = 2.59. The three-parameter Weibull estimation results are μ = 864.14, σ = 279.18, ν = 0.323, and the ratio = 8.93. The two-parameter Weibull estimation results are μ = 849.58, σ = 308.78, ν = 0.363, and the ratio = 16.93. To get a reasonable estimation, let γ = 400, and η is estimated as 578.93, then μ = 916.98, σ = 187.98, ν = 0.205, and the ratio = 3.20 (Fig. 3).

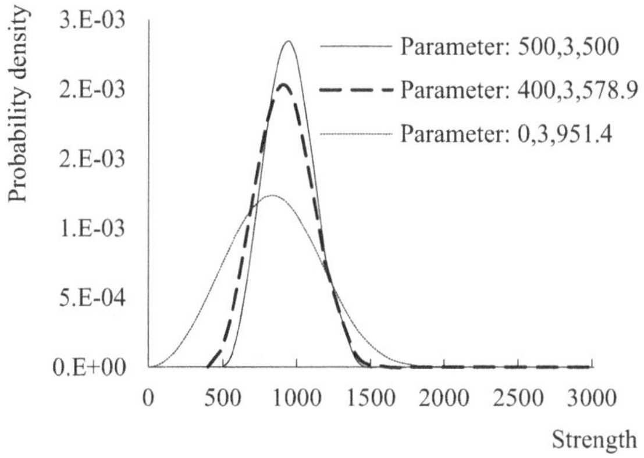

Fig. 3. The true Weibull distribution and modified fitting results

5 Monte Carlo Verification

In the following, the Weibull distributed life random variable with γ = 400, η = 578.9, and β = 3.0 is observed by Monte Carlo simulation.

According to the simulation result, for the Weibull probability density function with the only information available is n observations with the minimum $s_{(1,n)}$, the location parameter should be estimated as

$$\gamma = s_{(1,n)} + 0.67\eta \lg\left(\frac{n}{518}\right)^{1/2} \tag{11}$$

Thus, for the situation of five observation 1202, 1014, 773.3, 958.4 and 649, the location parameter can be estimated as 388.1 with the scale parameter of 578.9. It is not far from the multi-attribution estimation result of 400.

6 Conclusions

Information fusion technique is applied to material strength probability distribution estimation in the situation of small size of sample. Three-parameter Weibull distribution is used to describe material strength. To estimate a reasonable location parameter and thus a reasonable scale parameter, the knowledge and information from similar materials are utilized, included are an empirical shape parameter, a possibly large dispersion degree, as well as the experiment data. At last, the location parameter assigned is to make the estimated material strength variation at a realistic level. Besides, big data technique is applied to further verify the rationality of the estimated material strength distribution by testing the relation between location parameter and the minimum observation in a sample of particular size.

Acknowledgement. This research is subsidized by NSFC (Grant No. U1708255).

References

1. Shen, Y., Xie, M.: Nonparametric estimation of decreasing mean residual life with type II censored data. IEEE Trans. Reliab. **59**(1), 111–116 (2010)
2. McLain, A.C., Ghosh, S.K.: Nonparametric estimation of the conditional mean residual life function with censored data. Lifetime Data Anal. **17**, 514–532 (2011)
3. Soliman, A.A.: Estimation of the parameters of life for Gompertz distribution using progressive first-failure censored data. Comput. Stat. Data Anal. **56**, 2471–2485 (2012)
4. Zhao, M., Jiang, H., Liu, X.: A note on estimation of the mean residual life function with left-truncated and right-censored data. Stat. Probab. Lett. **83**, 2332–2336 (2013)
5. Elmahdy, E.E.: A new approach for Weibull modeling for reliability life data analysis. Appl. Math. Comput. **250**, 708–720 (2015)
6. Abernethy, R.B.: The New Weibull Handbook: Reliability & Statistical Analysis for Predicting Life, Safety, Survivability, Risk, Cost, and Warranty Claims, 5th edn. Gulf Publishing Company, Florida (2010)
7. Jia, X., Wang, D., Jiang, P., et al.: Inference on the reliability of Weibull distribution with multiply type-I censored data. Reliab. Eng. Syst. Saf. **150**, 171–181 (2016)
8. Mazhar, M.I.: Remaining life estimation of used components in consumer products. J. Oper. Manag. **25**, 1184–1193 (2007)
9. Ducros, F., Pamphile, P.: Bayesian estimation of Weibull mixture in heavily censored data setting. Reliab. Eng. Syst. Saf. **180**, 453–462 (2018)
10. Ahmed, M.: Bayesian. Estimator for Weibull distribution with censored data using extension of Jeffrey prior information. Procedia Soc. Behav. Sci. **8**, 663–669 (2010)

11. Sobhi, A.A.: Soliman. Estimation for the exponentiated Weibull model with adaptive type-II progressive censored schemes. Appl. Math. Model. **40**, 1180–1192 (2016)
12. Nagatsuka, H.: A study of estimation for three-parameter Weitull distribution based on doubly type-II censored data using a least square method. In: 2nd International Conference on Secure System Integration and Reliability Improvement, Yokohama, Japan, pp. 58–165. IEEE Conference Publications (2008)
13. Zhao, Y.X., Liu, H.B.: Weibull modeling of the probabilistic S-N curves for rolling contact, fatigue. Int. J. Fatigue **66**, 47–54 (2014)
14. Yang, J.W., Wang, J.H., Huang, Q.: Reliability assessment for the solenoid valve of a high-speed train braking system under small sample size. Chin. J. Mech. Eng. **31**, 47 (2018)
15. Abbasi, B.: Estimating parameters of the three-parameter Weibull distribution using a neural network. Eur. J. Ind. Eng. **2**(4), 428–445 (2008)

Performance of a Carburizing Bearing Steel After Surface Nanosized

Yanhui Wang[1,2], Fucheng Zhang[1(✉)], Zhinan Yang[1], Hezong Li[2], and Qian Yang[2]

[1] Yanshan University, Qinhuangdao 066004, China
wangyanhui@hebeu.edu.cn
[2] Hebei University of Engineering, Handan 056038, China

Abstract. Nanobainite and dispersed carbides were gained on surface layer of a modified carburizing 23Cr2Ni2Si1Mo bainitic bearing steel based on carburization, tempering at high temperature, and succedent low-temperature isothermal quenching. As compared with the conventional carburizing 20Cr2Ni4 martensitic bearing steel, the novel carburizing 23Cr2Ni2Si1Mo nanobainitic bearing steel showed a much higher wear-resisting performance and rolling contact fatigue performance. The excellent properties suggest that the novel carburizing 23Cr2Ni2Si1Mo nanostructured bainitic steel is very promising to manufacture heavy-duty bearing, especially for bearings that bear impact load and need high wear-resisting performance and high rolling contact fatigue performance during their service life.

Keywords: Performance · Carburization · Nanobainite · Bearing steel

1 Introduction

The high carbon martensitic microstructure of the traditional carburizing 20Cr2Ni4 bearing steel possesses high hardness, outstanding wear-resisting performance (WRP) and rolling contact fatigue (RCF) resistibility. However, it has poor toughness, high hydrogen embrittlement sensitivity and temper brittleness. The service life of high carbon martensite bearing steel will be greatly reduced when the service condition is bad or there is a big sudden impact load.

The nanobainite has much better toughness than the high carbon martensite [1, 2]. Meanwhile, the hardness and strength of nanobainite can match the high carbon martensite [3, 4]. The WRP of a high-Si-Al nanobainite [5] was increased by 21–51% over the tempered martensite, though the nanobainite had a lower hardness value than the tempered martensite. The RCF life of a C-1.26wt% steel sample [6] with 21 vol.% nanobainite was about 3.3 times longer than martensite sample. The prolonged RCF life should owe to the nanobainite and the carbon-rich retained austenite.

In the present work, we have studied the microstructure and property of new designed 23Cr2Ni2Si1Mo nanobainitic carburizing bearing steel, which are compared with those of the conventional 20Cr2Ni4 martensitic carburizing bearing steel.

E. Gdoutos and M. Konsta-Gdoutos (Eds.): ICTAEM 2020, STIN 16, pp. 52–57, 2020.
https://doi.org/10.1007/978-3-030-47883-4_9

2 Experimental Procedure

The experimental steels were named as 23Cr2Ni2Si1Mo and 20Cr2Ni4 in the paper.

The plate-like samples with dimensions of Φ43 mm × 6 mm and Φ52 mm × 8 mm were first machined, and then carburized. The carburizing process is the same as that in [7]. After high-temperature tempering, the 23Cr2Ni2Si1Mo steel was held at 860 °C for 1 h and austempered at 200 °C for 8 h. And the 20Cr2Ni4 steel was held at 810 °C for 1 h and then quenched in oil at once. In the end, the two steels were moved into a 200 °C furnace to temper.

The microstructure pictures were taken by a Hitachi-SU5000 SEM and a JEM-2010 TEM. The surfaces of samples were tested by a D/max-2500/PC XRD. The WRP of the experimental steels after carburizing was tested by a MMU-5G wear test machine. The test condition is dry sliding friction without lubrication at ambient temperature. The load was 1000 N, while the rotation rate was 200 rpm. The surfaces of all samples were ground and polished before the RCF test. The experiment was performed by using a TLP-1 point contact testing machine. The Hertz stress of 4500 MPa and the spin rate of 2040 rpm were chosen.

3 Results and Discussions

The carbon contents of the surfaces are ~0.83 wt% for 23Cr2Ni2Si1Mo steel and ~0.81 wt% for 20Cr2Ni4 steel. And the same carburizing depth is about ~2.8 mm.

Figure 1 shows the hardness distributions in carburized layer of the experimental steels after heat treatments. Evidently, the hardness declines continuously from surface to center for both steels, because of the carbon content keeps falling from surface to center. The surface of the 23Cr2Ni2Si1Mo steel with nanobainite achieves a hardness of 690 ± 8 HV, and its center with a low-carbon martensitic microstructure is 492 ± 6 HV. In contrast, the 20Cr2Ni4 steel has a hardness of 680 ± 10 HV in the surface with a high-carbon martensitic microstructure and 431 ± 7 HV in the center with a low-carbon martensitic microstructure. The hardness in surface of the two carburized steels have little difference in spite of different microstructures are obtained in the surfaces. The high hardness value obtained in the surface layer is helpful to improve the WRP and RCF resistance.

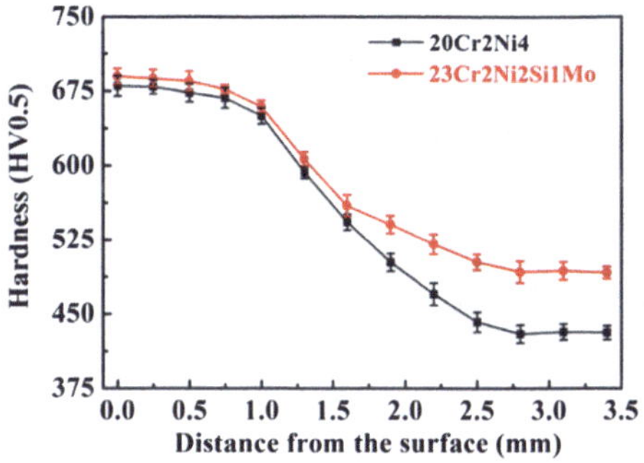

Fig. 1. Hardness distribution in carburized layer after heat treatments.

3.1 Microstructures of the Surfaces of the Experimental Steels After Carburizing

The SEM pictures of the surface are shown in Fig. 2. One can see that the tiny carbides were uniform distribution in matrix of the two steels. Figure 2(a) show that a dominant needle-like bainite was acquired on surface of the 23Cr2Ni2Si1Mo steel. The obvious network carbides, which located at the red arrows, were observed on the surface of the 20Cr2Ni4 steel from Fig. 2(b). The network carbides will increase the brittleness and drop fatigue life. After a lot of observation and measurement statistics, the average area fractions of the carbides in surfaces of the 23Cr2Ni2Si1Mo and the 20Cr2Ni4 were evaluated to be 6.3% and 6.7% for each.

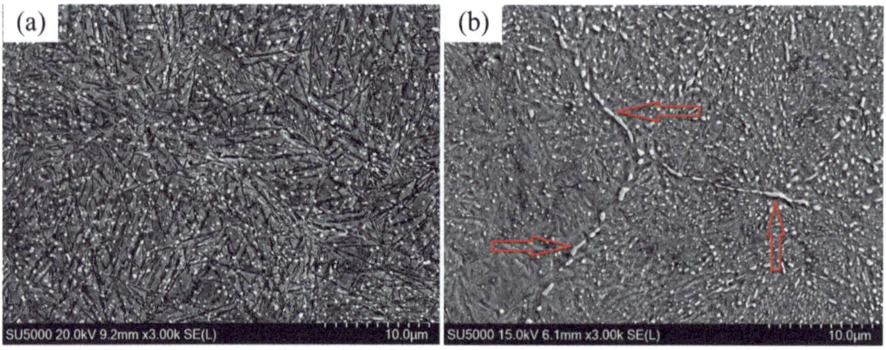

Fig. 2. SEM images in surface of carburized 23Cr2Ni2Si1Mo (a) and 20Cr2Ni4 steel (b).

The mean diameters of carbides for 23Cr2Ni2Si1Mo and 20Cr2Ni4 were evaluated to be 0.19 and 0.22 μm for each. The fine carbides embedded in matrix not only are helpful to increase the WRP [8], but also could restrain the austenite grain coarsening.

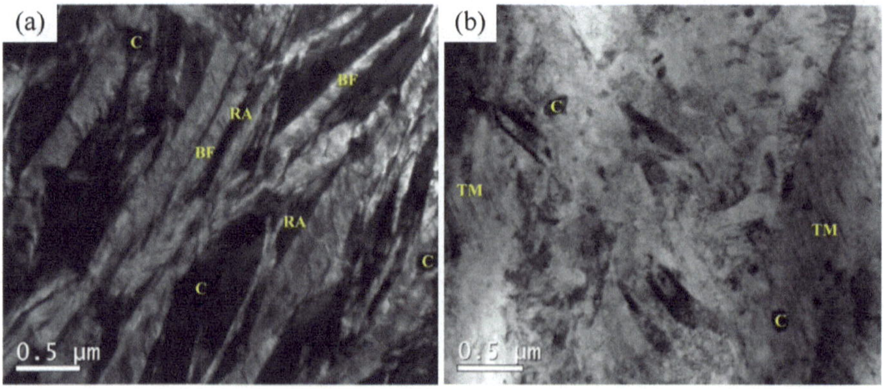

Fig. 3. TEM micrographs of the carburized surface: (a) 23Cr2Ni2Si1Mo (b) 20Cr2Ni4. Notes: RA-retained austenite, C-carbides, BF-bainitic ferrite, and TM-twinning martensite.

The typical TEM pictures are displayed in Fig. 3. The multiphase microstructures composed of bainite plate, carbides, and retained austenite were acquired in the carburized 23Cr2Ni2Si1Mo steel. Twinning martensitie is distinct in surface of the carburized 20Cr2Ni4 steel, whose presence can not only aggravate the stress concentration effect, but also decline the toughness and plasticity.

According to a large number of TEM pictures, the measured average thickness of bainite plate (t_{BF}) in the surface of carburized 23Cr2Ni2Si1Mo steel was about 71 ± 12 nm. This is a typical nanobainite.

From the XRD patterns, the volume fractions of retained austenite (V_γ) in surface of the 23Cr2Ni2Si1Mo and the 20Cr2Ni4 were confirmed to be ~18.9% and ~17.1%, respectively. Retained austenite in surface is crucial for fatigue life of carburized steels. On one hand, a part of the retained austenite of surface will occur martensitic transformation by strain during RCF testing, which causes the surface hardness to be increased further [9, 10]. On the other hand, volume expansion is inescapable when retained austenite occurs martensitic transformation, thus, a compressive stress is formed in surface [11]. Both the increased hardness and the compressive stress are beneficial to improve the RCF resistance.

3.2 WRP and RCF Performance of the Surfaces After Carburizing

Figure 4 shows weight loss with wearing time on surface of the experimental steels. Obviously, the weight loss of 20Cr2Ni4 steel after 120 min is greater than that of 23Cr2Ni2Si1Mo steel. The weight losses of the 20Cr2Ni4 steel show a linearly increasing trend with the increasing wearing time. The difference values in accumulative weight loss between the 20Cr2Ni4 steel and the 23Cr2Ni2Si1Mo steel become bigger and bigger with the increasing wearing time, which reveals that the wear rate of the 23Cr2Ni2Si1Mo steel is less marked than that of the 20Cr2Ni4 steel. We can consider the relative WRP of the 20Cr2Ni4 steel as one unit, thus the relative WRP of the 23Cr2Ni2Si1Mo steel could be denoted by using the ratio of weight loss for 20Cr2Ni4 steel to 23Cr2Ni2Si1Mo steel. From Fig. 4 the relative WRP of the 23Cr2Ni2Si1Mo steel is calculated to be 1.61, implying that the WRP of 23Cr2Ni2Si1Mo is greater than 20Cr2Ni4 by 61%.

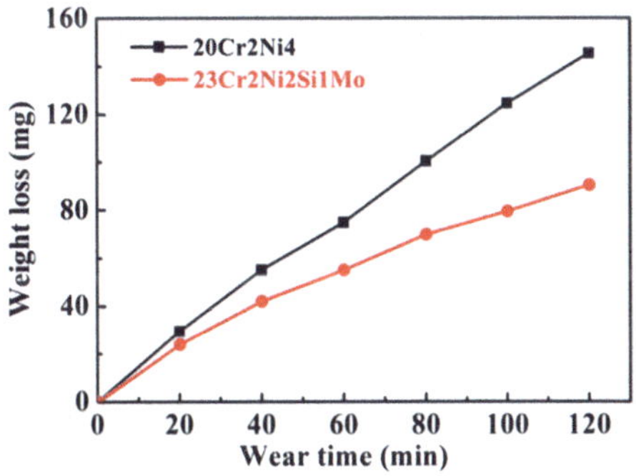

Fig. 4. Weight loss with wearing time on surface of the experimental steels.

The wearing results show that although the initial hardness values of different microstructures are almost the same, they show different WRP. The initial hardness of the microstructure is only one factor affecting the WRP. Microstructure also has an important effect on the WRP. The average thickness of bainite plate (t_{BF}) in carburized 23Cr2Ni2Si1Mo surface was about 71 ± 12 nm, which was much thinner than the martensite plate in carburized surface of 20Cr2Ni4. The surface microstructure refinement of 23Cr2Ni2Si1Mo could directly enhance the plasticity, hardness and strength and indirectly increase the WRP. In addition, the nanocrystals were easily to produce on the nanobainite surface in the process of severe friction and wear [5, 12], which can enhance the WRP further.

Figure 5 shows Weibull distribution curves of RCF life of the 23Cr2Ni2Si1Mo steel and the 20Cr2Ni4 steel. Table 1 generalizes the Weibull fatigue lives. Apparently, the RCF life of 23Cr2Ni2Si1Mo is much longer than that of 20Cr2Ni4 steel in spite of the V_γ and the hardness value of surface are nearly parallel for the two steels. The L_{10}, L_{50}, and characteristic life of 23Cr2Ni2Si1Mo are 3.10, 3.47, and 3.54 times longer than those of 20Cr2Ni4, respectively. The remarkable RCF performance of 23Cr2Ni2Si1Mo owes to the development of nanobainite in surface. Nanobainite could prolong the RCF life by restraining the nucleation and propagation of micro crack [6]. Stress concentration effect on hard phase in the process of RCF could be relieved by the nanobainite, which is relatively soft. Therefore, the micro crack nucleation becomes difficult. Furthermore, the stress on micro crack tip could be relieved by nanobainite, as a consequence it will demand more energy for the propagation of micro crack.

Table 1. RCF lives of 23Cr2Ni2Si1Mo and 20Cr2Ni4 steel.

Steel	L_{10}/(× 10^7)	L_{50}/(× 10^7)	V_s/(× 10^7)	B
23Cr2Ni2Si1Mo	1.3751	4.5339	5.7143	1.58
20Cr2Ni4	0.4439	1.3076	1.6136	1.74

L_{10} : rated life, L_{50} : median life, V_s : characteristic life, β : Weibull slope.

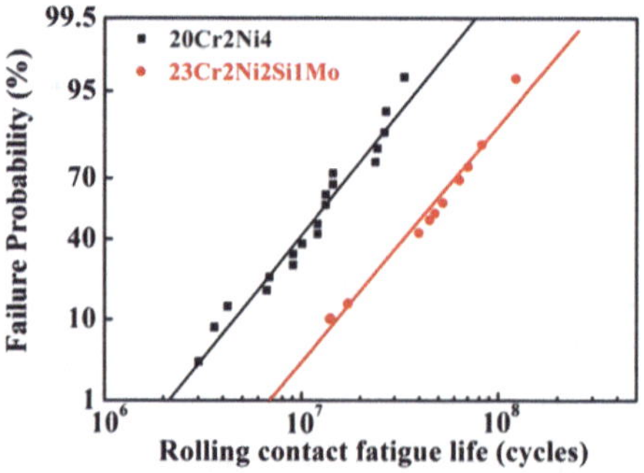

Fig. 5. Weibull distribution curves of RCF life of 23Cr2Ni2Si1Mo and 20Cr2Ni4 steel.

4 Conclusions

1. The microstructure, consisting of nanobainite with 71 ± 12 nm thickness bainitic ferrite plates and dispersed carbides, with a hardness of 690 ± 8 HV was obtained on surface layer of the modified carburizing 23Cr2Ni2Si1Mo bainitic bearing steel based on carburization, tempering at high temperature, and succedent low-temperature isothermal quenching.
2. The WRP and the characteristic life of RCF of the modified carburized 23Cr2Ni2Si1Mo nanobainitic bearing steel are about 61% and 254% greater than those of the conventional carburized 20Cr2Ni4 martensitic bearing steel. The excellent WRP and RCF performance of the nanobainitic steel owe to the ultrafine microstructure.

Acknowledgement. The work in this paper thanks the support of the Natural Science Foundation of Hebei Province (E2019402433), and the University Science and Technology Research Project of Hebei Province (QN2018078), P.R. China.

References

1. Bhadeshia, H.K.D.H.: Nanostructured bainite. Proc. R. Soc. A **466**, 3–18 (2010)
2. Yoozbashi, M.N., Yazdani, S., Wang, T.S.: Design of a new nanostructured, high-Si bainitic steel with lower cost production. Mater. Des. **32**, 3248–3253 (2011)
3. Hasan, H.S., Peet, M.J., Avettand-Fènoël, M.-N., et al.: Effect of tempering upon the tensile properties of a nanostructured bainitic steel. Mater. Sci. Eng. A **615**, 340–347 (2014)
4. Amel-Farzad, H., Faridi, H.R., Rajabpour, F., et al.: Developing very hard nanostructured bainitic steel. Mater. Sci. Eng. A **559**, 68–73 (2013)
5. Yang, J., Wang, T.S., Zhang, B., et al.: Sliding wear resistance and worn surface microstructure of nanostructured bainitic steel. Wear **282–283**, 81–84 (2012)
6. Liu, H.J., Sun, J.J., Jiang, T., et al.: Improved rolling contact fatigue life for an ultrahigh-carbon steel with nanobainitic microstructure. Scr. Mater. **90–91**, 17–20 (2014)
7. Wang, Y.H., Yang, Z.N., Zhang, F.C., et al.: Microstructures and mechanical properties of surface and center of carburizing 23Cr2Ni2Si1Mo steel subjected to low-temperature austempering. Mater. Sci. Eng. A **670**, 166–177 (2016)
8. Chakraborty, J., Bhattacharjee, D., Manna, I.: Development of ultrafine bainite + martensite duplex microstructure in SAE 52100 bearing steel by prior cold deformation. Scr. Mater. **61**, 604–607 (2009)
9. Zheng, C.L., Dan, R., Zhang, F.C., et al.: Effects of retained austenite and hydrogen on the rolling contact fatigue behaviours of carbide-free bainitic steel. Mater. Sci. Eng. A **594**, 364–371 (2014)
10. Wang, Y.H., Zhang, F.C., Yang, Z.N., et al.: Rolling contact fatigue performances of carburized and high-C nanostructured bainitic steels. Materials **9**, 960 (2016)
11. Bhadeshia, H.K.D.H.: Steels for bearings. Prog. Mater. Sci. **57**, 268–435 (2012)
12. Wang, T.S., Yang, J., Shang, C.J., et al.: Sliding friction surface microstructure and wear resistance of 9SiCr steel with low-temperature austempering treatment. Surf. Coat. Technol. **202**, 4036–4040 (2008)

Residual Strength and Toughness of Nano and Micro Scale Fiber Reinforced Cementitious Composites

Panagiotis A. Danoglidis[1](✉), Maria S. Konsta-Gdoutos[1], and Emmanuel E. Gdoutos[2]

[1] Center for Advanced Construction Materials, The University of Texas at Arlington, Arlington, TX 76019, USA
panagiotis.danoglidis@uta.edu
[2] Academy of Athens, 106 79 Athens, GR, Greece

Abstract. This study reports a significant synergistic effect of using combined networks of nanoscale fibers (CNTs) and micro scale fibers (PPs) on the key mechanical properties of OPC mortars that define a uniquely tough and energy-absorbing material. The flexural strength at the first crack, δ, were firstly investigated by conducting three point close loop bending tests on prismatic specimens of mortars reinforced with 0.1 wt% CNTs and/or 0.73 wt% PPs. To further assess the mechanical performance of the proposed nanocomposites after the formation of the first-crack, the residual strength and the energy absorption capability of the nanocomposite mortars are also investigated at three post-crack stages, 3δ, 5.5δ and 10.5δ, according to the ASTM C1018. The experimental results showed that a combination of nano and micro scale fiber reinforcement yields a composite with a significantly enhanced load-carrying capacity at the elastic stage, as defined by the approximately 100% higher first crack strength. The multi scale fiber reinforcement has also an exceptional impact on improving the tensile strain-capacity of resultant composite after the formation of the first crack. The observed 90% increased residual strength and post-crack toughness of CNT-PP reinforced mortars indicate an improved ductility over the mortars reinforced with microscale PP fibers alone.

Keywords: Mortar · Carbon nanotubes · Polypropylene · Residual strength · Post-crack toughness

1 Introduction

Fiber reinforced concrete presents enhanced ductility overcoming the limited strain capacity of the quasi brittle cementitious materials. The introduction of fibers at the micro scale, such as polypropylene fibers, PPs, polyvinyl alcohol fibers, PVAs, and polyethylene fibers, PEs, into the cementitious matrix results in post-elastic mechanical property changes [1, 2]. Although, the micro scale fiber reinforcement's ability to bridge the microcracks and delay the formation of a through-specimen macrocrack, the initiation of cracks at the nanoscale, which affects the fundamental mechanical response of concrete cannot be prevented. The use of fibers at the nanoscale, such as carbon

E. Gdoutos and M. Konsta-Gdoutos (Eds.): ICTAEM 2020, STIN 16, pp. 58–63, 2020.
https://doi.org/10.1007/978-3-030-47883-4_10

nanotubes CNTs, in cementitious matrices offers unprecedented improvements in the mechanical properties of cement based materials and concrete [3, 4]. The effective incorporation of individual CNTs and the formation of a complete nanoscale fiber network into the cementitious matrix significantly improves the nanocomposite's flexural load carrying capacity; hence the proportionality of stress to strain before the formation of the "first crack" [5]. As a result, nanoscale fiber reinforced composites were successfully developed, exhibiting 100% increases in their "first crack" strength and toughness, modulus of elasticity, energy absorption capability and fracture characteristics [6].

While the effect of micro or nano scale fiber addition on the mechanical performance of cementitious materials and concrete is thoroughly reported, there is a lack of information about the synergistic effect of the simultaneous addition of nano and micro scale on the load carrying capacity and energy absorption capability of hybrid cement based composites at the "first crack" and the post-crack stages. The purpose of the present study is to investigate the mechanical performance including (i) the "first crack" strength and toughness; (ii) the residual strength; and (iii) the post-crack energy absorption capability of mortars reinforced with an optimum network of CNTs and PPs. Prismatic mortar specimens, reinforced with 0.1 wt% CNTs and/or 0.73 wt% PPs, were subjected to a three point close loop bending test. The combination of CNTs and PPs provide significantly enhanced mechanical performance at both the first-cracking and post-cracking over the mortars reinforced with polypropylene fibers alone. The hybrid CNT and PP reinforcement increases the first crack strength, +98%, and improves the residual strength, +88%, and toughness, +55%, of micro scale fiber mortar composites.

2 Experimental Work

2.1 Materials and Specimens

Mortar composites were produced by using OPC Type I (w/c = 0.485), standard sand (s/c = 2.75), multiwalled carbon nanotubes (CNTs) and polypropylene staple fibers (PPs). The nano and micro scale fibers present aspect ratios of 307 and 480, respectively. The uniform dispersion of CNTs in aqueous suspensions was achieved by applying ultrasonication energy and using a polycarboxylate based surfactant [7]. Mortar mixes containing (a) 0.73 wt% PPs; and (b) 0.1 wt% CNTs and 0.73 wt% PPs were prepared following the ASTM C305, and was cast in 4 × 4 × 16 cm oiled molds for three point bending experiments. Following demolding, the samples were cured in lime-saturated water for 28 days.

2.2 Experimental Determination of First Crack Strength, Residual Strength and Post-crack Toughness

Three-point bending tests were conducted on 40 × 40 × 160 mm prismatic specimens. The tests were performed using a 25 kN MTS servo-hydraulic, closed-loop testing machine under displacement control. The rate of displacement was kept as

0.1 mm/min. A typical load-deflection response of a 28d mortar reinforced with 0.1 wt% CNTs and 0.73 wt% PPs is presented in Fig. 1. The first-crack strength and toughness at the deflection δ; and the residual strength and toughness at the post-crack deflections 3δ, 5.5δ and 10.5δ were evaluated by following the analysis described by the ASTM C1018 [8] and the ACI report on Fiber Reinforced Concrete [9].

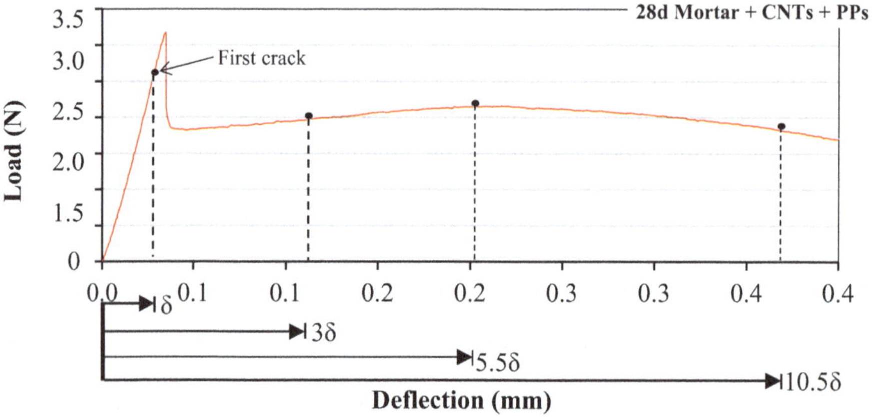

Fig. 1. Typical load-deflection curve of 28d mortar reinforced with 0.1 wt% CNTs and 0.73 wt% PPs.

3 Results and Discussion

Flexural strength of the 28d mortars reinforced with 0.1 wt% CNTs and/or 0.73 wt% PPs at the "first crack" deflection (δ) and at deflections of 3.0, 5.5 and 10.5 times the first-crack deflection are presented in Fig. 2. The nano and micro scale fiber reinforced mortars exhibit 98% higher first crack strength over the PP reinforced mortars. From the previous studies, it is already observed that mortars reinforced with 0.1 wt% CNTs exhibit an increase first-crack strength, compared to the unreinforced matrix, due to an due to an enhanced crack-bridging mechanism, resulting in a higher load transfer efficiency between the nanotubes and the matrix. Interestingly, and despite the fact that the geometrical characteristics of CNTs are in the scale of nanometers, their addition has a great impact on the residual strength of nanocomposites compared to the singly micro scale fiber reinforced mortars. The mortars reinforced with the nano and micro scale fibers are able to sustain much higher loads after the formation of first crack, exhibiting higher residual strengths. The simultaneous incorporation of CNTs and PPs leads to higher flexural strength, up to 88%, at the three post-first crack stages, 3δ, 5.5.δ and 10.5δ.

To further examine the synergistic effect of nano and micro scale fibers on the mechanical performance of mortar composites at the first-crack and post-crack stages, the energy absorption capability, represented by the flexural toughness, at the deflections δ, 3δ, 5.5δ and 10.5δ are presented in Fig. 3. The hybrid mortars exhibit an

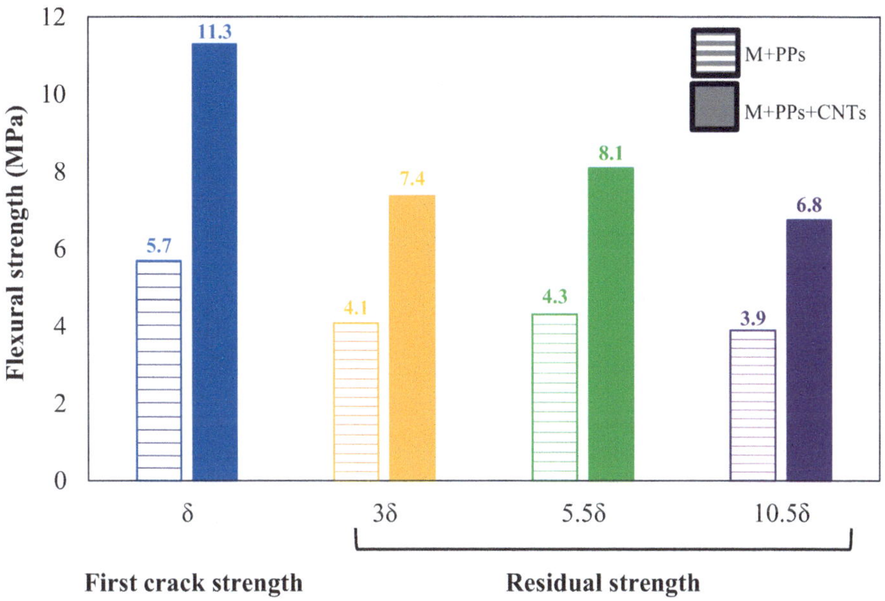

Fig. 2. First crack and residual strengths of 28d mortars reinforced with CNTs and/or PPs

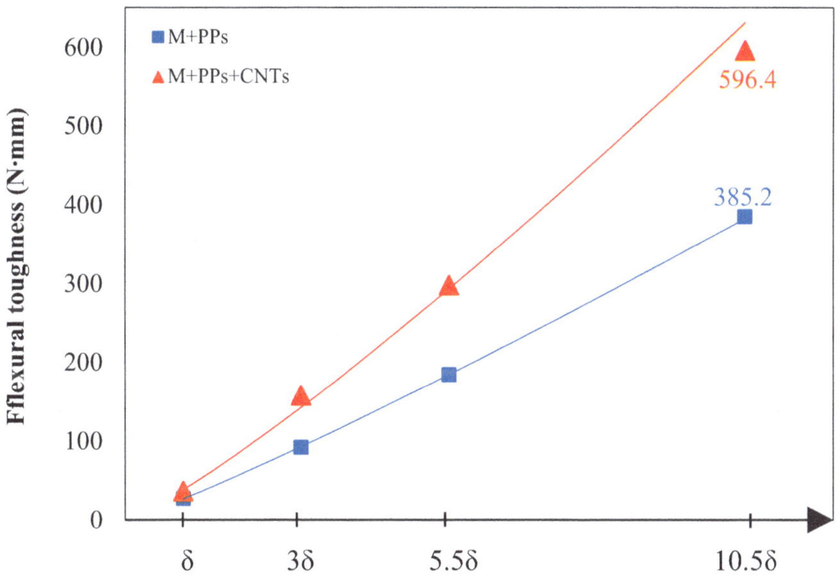

Fig. 3. First crack and residual strengths of 28d mortars reinforced with CNTs and/or PPs

improved energy absorption capability at all the stages after the formation of the first crack. The flexural toughness of nanomodified mortars at the post-crack stages up to a deflection of 3.0, 5.5 and 10.5 times the first-crack deflection is 88%, 90% and 93% higher than the post-crack toughness of singly PP reinforced mortar. Recently it was found that the incorporation of CNTs in the PP reinforced polymers leads to a higher interfacial shear strength due to the better adhesion between the micro scale fiber and the matrix [10]. The strengthen interfacial bonding between the fiber and the matrix increases the demand of energy for crack propagation, improving the energy absorption capacity of hybrid mortars.

4 Conclusions

In this study, it is demonstrated that a significantly higher load carrying capacity and energy absorption capability, at the first and the post cracking area, can be attained by incorporating combined reinforcement of nano and micro scale fibers into the cementitious matrix. From the analysis of the three point bending load-deflection curves, it is presented that the CNT and PP reinforced mortars exhibit extremely higher first crack strength, +98%, compared to the singly PP reinforced mortars. The synergistic CNT-PP reinforcing mechanism in mortars also leads to an even greater strength and toughening behavior at the post-crack area. The enhanced bridging mechanism, after the formation of the "first crack", by the simultaneous use of nano and micro scale fibers exceptionally improves the composites' load carrying capacity and post-crack energy absorption capability as reflected by the 88% and 55% higher residual strength and post-crack flexural toughness, respectively.

References

1. Lawler, J.S., Zampini, D., Shah, S.P.: Permeability of cracked hybrid fiber reinforced mortar under load. ACI Mater. J. **99**(4), 379–385 (2002)
2. Yu, K.-Q., Yu, J.-T., Dai, J.-G., Lu, Z.-D., Shah, S.P.: Development of ultra-high performance engineered cementitious composites using polyethylene (PE) fibers. Constr. Build. Mater. **158**, 217–227 (2018)
3. Danoglidis, P.A., Konsta-Gdoutos, M.S., Shah, S.P.: Relationship between the carbon nanotube dispersion state, electrochemical impedance and capacitance and mechanical properties of percolative nanoreinforced OPC mortars. Carbon **145**, 218–228 (2019)
4. Konsta-Gdoutos, M.S., Danoglidis, P.A., Shah, S.P.: High modulus concrete: effects of low carbon nanotube and nanofiber additions. Theor. Appl. Fract. Mech. **103**, 102295 (2019). https://doi.org/10.1016/j.tafmec.2019.102295
5. Danoglidis, P.A., Konsta-Gdoutos, M.S., Gdoutos, E.E., Shah, S.P.: Strength, energy absorption capability and self-sensing properties of multifunctional carbon nanotube reinforced mortars. Constr. Build. Mater. **120**, 265–274 (2016)
6. Gdoutos, E.E., Konsta-Gdoutos, M.S., Danoglidis, P.A.: Portland cement mortar nanocomposites at low carbon nanotube and carbon nanofiber content: a fracture mechanics experimental study. Cem. Concr. Compos. **70**, 110–118 (2016)

7. Shah, S.P., Konsta-Gdoutos, M.S., Metaxa, Z.S.: Highly dispersed carbon nanotube reinforced cement based materials. United States Patent US9,365,456 (B2), 14 June 2016
8. ASTM C1018-97: Standard Test Method for Flexural Toughness and First-Crack Strength of Fiber-Reinforced Concrete (Using Beam With Third-Point Loading) (1997)
9. American Concrete Institute: Report on Fiber Reinforced Concrete, ACI 544.1R-96 (1996)
10. Yumitoril, S., Araoz, Y., Tanakaz, T., Naito, K., Tanaka, K., Katayama, T.: Increasing the interfacial strength in carbon fiber/polypropylene composites by growing CNTs on the fibers. In: Computational Methods and Experimental Measurements XVI, vol. 55, p. 275 (2013)

Experimental and Numerical Strategy for the Determination of Mechanical Properties Related to Human Cortical Bone Fracture

J.-L. Tailhan[1(✉)], Y. Godio-Raboutet[2], and C. Boulay[3]

[1] MAST-EMGCU, Université Gustave Eiffel, IFSTTAR, 77447 Marne-la-Vallée, France
jean-louis.tailhan@ifsttar.fr
[2] Aix-Marseille Université, Université Gustave Eiffel, IFSTTAR, LBA, 13016 Marseille, France
[3] CME, Faverelles, France

Abstract. The paper presents an experimental test of macrocrack propagation in bones associated with a numerical strategy to determine from this test some mechanical properties by inverse identification. A 3 point bending test is performed on a notched segment of long human bone, and the load vs notch opening displacement is measured. The compliance method is applied in the context of a realistic FE simulation of the test. A first application of this methodology is described and shows its feasibility and capabilities. The ultimate objective of this research project is to develop a probabilistic modelling of the macrocracking processes in human long bone tissues based on an experimental database of mechanical parameters fed up by this methodology.

Keywords: Bone cracking · Experiment · Numerical model · Compliance method

1 Introduction

Fracture modes of bone structures prediction is essential for both the comprehension and the prevention of injuries and their surgical treatment. While many works on bone properties and model developments are proposed in the literature [1], the ability to predict more accurately their cracking behaviour remains a major scientific challenge. The ultimate objective of this research project is to develop a probabilistic modelling of the cracking processes in human long bone tissues, taking into account the effect of the material structural heterogeneity on these processes. In that sense, the present work is deeply inspired by other works originally developed for concrete structures [2–4].

Many natural or human-made materials exhibit hierarchical inner-structure on one or more length scales [5]. This structure plays an essential role in their physical or mechanical properties. It also has an essential part to play in their cracking processes, especially for quasi-brittle materials. It is precisely the case for concrete or bone tissue [6], which have particularly complex and heterogeneous inner-structures.

Stress concentrations drive cracking processes. If one excepts potential geometrical singularities on the shape of a considered object of study, subjected to uniform loading

E. Gdoutos and M. Konsta-Gdoutos (Eds.): ICTAEM 2020, STIN 16, pp. 64–69, 2020.
https://doi.org/10.1007/978-3-030-47883-4_11

conditions, the phenomenon of stress concentration can appear in its constitutive material for two different reasons: because of material heterogeneity (in the sense of differences in elastic properties – moduli – of the inner components of the material itself), or because of internal discontinuities (flaws or initial micro-cracks) [3, 7]. Cracking processes can be conceptually separated into two distinct phenomena: crack creation and crack propagation. Flaws and heterogeneity have a real impact on both phenomena because they directly affect the strength and the toughness of the material.

Establishing a (statistical) link between heterogeneity/defects and material mechanical properties to be used for the numerical modelling of cracking processes in human bone requires the elaboration of a substantial database of experimental results involving both mechanisms of crack initiation and propagation. The global strategy proposed for establishing this link consists briefly in (1) the design of an experimental test involving cracking processes in bones, (2) realistic numerical simulations of the test for the inverse determination of model parameters, (3) the characterisation of the level of heterogeneity of the material in the specimen.

Given the high complexity of this strategy, and as a first step of this research program, the paper focuses mainly on the two first points. It presents, in a first section, an experimental test of crack propagation in a notched piece of human femur subjected to a three-point bending. In this test, only the cortical tissue is considered. The second section presents the main steps of the inverse determination of both elastic modulus and fracture energy (the most often used properties in modelling) from realistic numerical simulations of the test. The final section will then be devoted to results, comments and conclusions.

2 Experimental Test

Long human bones are considered here as "structural elements". This point sets the scale of study. Considering full bones leads to consider both cortical and trabecular tissues. For simplicity, this study focuses only on the cortical bone. Cracking behaviour of cancellous bone will be studied in future works.

In the sake of producing enough samples at this scale, mainly to feed statistical post-treatment results, several samples (generally three) of about 6 cm length are extracted from the diaphysis of femurs from human cadavers. Note that, if present on the sample, the remaining cancellous bone is systematically removed.

At first, the work will mainly focus on pure mode I cracking, favouring thus local tensile failures. Direct tension is particularly difficult to be applied experimentally (even if the geometry of the specimen is perfectly regular), mostly because of the control of the boundary conditions. Nevertheless, indirect tension can be obtained with a bending test, more comfortable to perform, although it remains delicate because of the geometrical irregularity of the specimens. As the ratio height (about 6 cm) over length (about 3 cm) of the specimens is too small, and to overcome the critical effect of the shear force, it is necessary to extend the sample by adding two prismatic extensions, made of steel, at each end. The bone segment is then sandwiched between the two extensions, and the bond is ensured by glueing. A particular procedure to prepare and assemble the sandwiched components has been elaborated to guaranty their quite perfect alignment and minimize any geometrical defects. This solution has the advantage of better control of the boundary conditions on the machine.

The purpose of the test shown in Fig. 1(a) is to follow the evolution under loading of a localized "structural" macro-crack. A notch (of 500 µm width) is made in the central cross-section of the sample, right under the loading point. Its purpose is to localize the crack initiation in this section of the specimen. The presence of the notch that induces a stress concentration at its tip also leads to a stress level reduction at the bone-extension bonds. Note, that this last point can facilitate the glueing solution. A pre-dimensioning numerical simulation of the test (considering the bone as a perfect cylinder of 3 cm of external diameter and 3 mm of thickness) has been carried out to estimate the stress level at the bone-extension bonds according to the notch length: as a result, the longer the notch is, the lower the tensile stress level is. An estimation of the maximum tensile stress level of about 20–25 MPa was obtained for a notch length equal to half the diameter of the bone. The glue has been chosen taking into account this level of stress. Two types of glue have been eventually selected: a methacrylate and an epoxy resin. Their respective characteristics ensure to bear the estimated stress level. Glueing components is usually a complicated matter, and a particular attention has also been paid to the defatting pre-treatment of the bone itself to avoid any incompatibility with the glue.

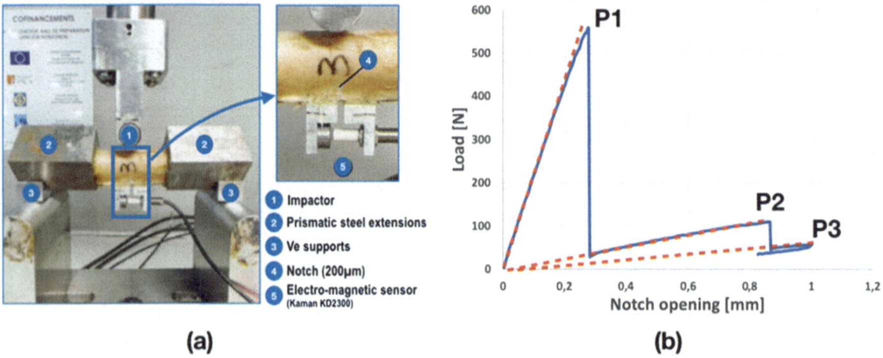

Fig. 1. Experimental setup (a) and an example of test results (b). On the curve, *P*1, *P*2 are applied loads corresponding to the macrocrack propagations. *P*3 is an unloading point.

The bending test, Fig. 1(a), is performed on a Landmark 370.10 hydraulic press (MTS Systems corporation, MN, USA). The notch opening is measured with a non-contacting linear proximity measuring system (Kaman KD2300). The sensor and its target are maintained by two aluminium supports to avoid any electro-magnetic interferences on the measurements. The supports are directly glued on the bone surface, on each side of the notch allowing the sensor to bridge it. The load applied by the jack is monitored by the notch-opening displacement measured. This solution has the advantage to better control the test monitoring and to assess at each time the opening of the macro-crack. The test result is given in terms of load-notch opening displacement curves. Figure 1(b) shows an example of result obtained on a piece of femur taken from a cadaver of an 83 years old woman.

3 Numerical Simulations of the Experiment

The longitudinal modulus of elasticity and the fracture energy can be determined from this test by an indirect determination procedure, and a "realistic" numerical simulation of the test is performed for this purpose. The methodology is described as follow: the 3D-geometry of the tested bone is reconstructed from scans (obtained with a medical scanner before the glueing operation) using specialized pieces of software (slicer-3D version 4.8.1, Rhinoceros® Version 5.5.3, McNeel & Assoc. Edt.). The spatial position of the bone relative to its metallic extensions is reproduced identically in Rhino3D by using different tracking techniques based on photographic shots of the sample. Thus oriented in space, the geometry of the bone can then be imported into a finite element mesh generator. By adding the geometries representing the extensions, the sensor supports and other additional parts used in the test as well as the notch itself, the complete mesh of the test specimen can be generated (Fig. 2(a)). Boundary and loading conditions can also be added.

3.1 Determination of the Longitudinal Modulus of Elasticity

The principle of the methodology is elementary. A simulation of the test is performed in elasticity considering the value of the modulus as the unknown and searching it so that the numerical response fits the experimental one.

3.2 Determination of the Fracture Energy

The methodology is here more complex, but its bases remain relatively classical [8, 9]. The compliance method is based on the Irwin–Kies relation:

$$G_I = \frac{P^2}{2B}\frac{dC_\delta}{da} \tag{1}$$

Where G_I represents the fracture energy in mode I, P is the applied load, B is the width of the cracked section, $C_\delta = \delta/P$ is the compliance (δ being the displacement of the loading point), and a is the crack length.

Equation (1) requires the determination of the width B and the compliance C_δ as functions of the crack length a. For the considered tested bone, whose geometry is purely irregular, the relationship $B(a)$ is determined with a special feature of Rhinoceros®. The relationship $C_\delta(a)$ is obtained numerically with a series of elastic simulations of the bending test taking into account the presence of an idealized crack of varying length in the central section of the bone as depicted on Fig. 2(b). Note that the displacement of the loading point δ is not measured experimentally, and as a consequence, the experimental compliance C_δ^{exp} cannot be determined. This difficulty is overcome by considering besides the quantity $C_\omega = \omega/P$ (ω being the notch opening), which can be assessed experimentally. Numerically, this quantity can also be calculated and expressed as a function of a, i.e. $C_\omega(a)$.

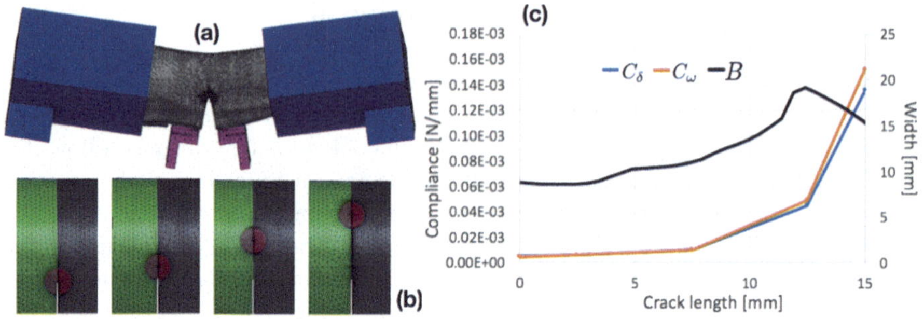

Fig. 2. Meshes for the application of the compliance method, and compliance curves

Finally, knowing the experimental value of the quantity C_ω, for a given load level corresponding to the propagation of the crack in the material, the point corresponding to C_ω^{exp} on the numerical curve $C_\omega(a)$ gives the length a of the crack. The value of the derivative $\frac{d}{da}C_\delta(a)$ and the one of B are then given by the numerical curves $C_\delta(a)$ and $B(a)$ respectively (Fig. 2(c)). The corresponding value of G_I is, therefore, calculated.

4 Results, Discussions and Conclusion

An experimental result is given in Fig. 1(a). The test led to the creation of a macro crack starting from the notch and propagating quickly and vertically towards the loading point. The curve gives the evolution of the applied load versus the notch opening. This result clearly shows a relatively linear pre-peak behaviour with a slight non-linearity of the curve in the peak vicinity. This non-linearity appears between approximately 85% and 90% of the maximum load, corresponding to a micro-cracking process established at the notch tip. However, remaining relatively small at the scale of the overall curve this phenomenon must be indicative of a limited process. The post-peak part of the curve has a saw-tooth shape indicating successive propagations and blockages mechanisms of the macrocrack. The different reloading branches consist of a linear part and a non-linear part near the local peak, significant of a limited evolution of the microcracking at the macrocrack tip before propagation. Besides the tangents to the linear parts of the curve pass almost through the origin of the reference system. This latter point shows that there appears to be little or no permanent deformation. The post-peak brittle failure of the specimen can be explained by two possible hypotheses. The first is that of age because due to the evolution of the internal structure of the material (osteoporosis, bone remodelling, …) toughness decreases with age [10, 11]. The second hypothesis is that of the conservation conditions imposed to the specimen before the test, which here probably led to an excessive drying of the specimen.

The elastic modulus of the bone is assessed following the described methodology and considering the material isotropic. This seemingly rough hypothesis has a minimal consequence on the value obtained since the longitudinal behaviour is mainly mobilized in the test. The value obtained is 11.2 GPa, and remains reasonable compared to values found in the literature.

Due to the high brittleness of the test, the fracture energy can be determined by the compliance method for only two points (marked as P1 and P2 on Fig. 1). Values obtained are 0.25 N/mm^2 and 1.06 N/mm^2. These values, again consistent with the literature on the subject, correspond to equivalent crack lengths of 0.86 mm and 13.6 mm, respectively.

Even if the quantity of post-peak information obtained remains modest, this first test shows a possible methodology for determining mechanical properties relating to the propagation of a macrocrack in a human cortical bone. Possible leads for improvement are linked to (1) mastering bonding and (2) maintaining an acceptable level of the bone humidity. Repeating these testing and post-treating procedures will produce statistical data essential for the development of a probabilistic cracking model.

References

1. Schwarcz, H.P., Abudeia, D., Jasiuk, I.: The ultrastructure of bone and its relevance to mechanical properties. Front. Phys. **5**, 39 (2017). https://doi.org/10.3389/fphy.2017.00039
2. Rossi, P., Richer, S.: Numerical modelling of concrete cracking based on a stochastic approach. Mater. Struct. **20**, 334–337 (1987)
3. Rossi, P., Wu, X., Le Maou, F., Belloc, A.: Scale effects on concrete in tension. Mater. Struct. **27**, 437–444 (1994)
4. Tailhan, J.-L., Dal Pont, S., Rossi, P.: From local to global probabilistic modelling of concrete cracking. Ann. Solid Struct. Mech. **1**(2), 103–115 (2010)
5. Lakes, R.: Materials with structural hierarchy. Nature **361**, 511–515 (1993)
6. Rho, J.-Y., Kuhn-Spearing, L., Zioupos, P.: Mechanical properties and the hierarchical structure of bone. Med. Eng. Phys. **20**, 92–102 (1997)
7. Currey, J.D.: Stress concentration in bones. J. Cell Sci. **3**(61), 111–133 (1962)
8. Rossi, P.: Fissuration du béton: du matériau à la structure. Application de la mécanique linéaire de la rupture. PhD thesis, ENPC, December 1986
9. Morais, J.J.L., de Moura, M.F.S.F., Pereira, F.A.M., Xavier, J., Dourado, N., Dias, M.I.R., Azevedo, J.M.T.: The double cantilever beam test applied to mode I fracture characterization of cortical bone tissue. J. Mech. Behav. Biomed. Mater. **3**, 446–453 (2010)
10. Koester, K.J., Barth, H.D., Richie, R.O.: Effect of aging on the transverse toughness of human cortical bone: evaluation by r-curves. J. Mech. Behav. Biomed. Mater. **4**, 1504–1513 (2011)
11. Phelps, J.B., Hubbard, G.B., Wang, X., Agrawal, C.M.: Microstructural heterogeneity and the fracture toughness of bone. J. Biomed. Mater. Res. **51**(4), 735–741 (2000)

Effect of External Field Treatment on the Microstructure and Deformation Behavior of Nickel-Base Superalloy

Yang Liu(✉), Lei Wang, Xiu Song, and Yao Wang

Key Laboratory for Anisotropy and Texture of Materials, Northeastern University, Shenyang 110819, People's Republic of China
liuyang@mail.neu.edu.cn

Abstract. The electrostatic field and electropulsing were applied to the nickel-base superalloy and the effect and mechanism of the external field on the microstructure were investigated in the present study. The results showed that the electrostatic field has significant effect on the vacancy movement and atomic diffusion. As a consequence, a lot of twins can be obtained in the alloy by electrostatic field treatment and both the precipitation and growth of the strengthening phase changed. The plastic deformation resistance of the nickel-based superalloy can be effectively reduced by the adding of pulse current as well as the improvement of the plastic deformation ability. At the same time, the activation energy of the recrystallization can also be reduced by the adding of the pulse current, and the promotion of the nucleation and growth of the recrystallization of the alloy.

Keywords: External field treatment · Superalloy · Microstructure · Deformation behavior

1 Introduction

Recently, the combination of materials science and electromagnetism have been developed into a new field, named Electromagnetic Processing of Materials (EPM) [1–3]. The Electric field treatment (EFT) and electropulsing treatment (EPT), as a powerful and controllable external field, have been paid highly attention in recent years [4–7]. Electric field is another important energy field except the temperature field and stress field, and it shows many special coupling effects when it is combined with the other fields. Properties of a material are said to be related with the microstructure and defects (including vacancy, dislocation, grain boundary and so on) of the material. Therefore, the improving properties of the material can be achieved by controlling the microstructure and defects. In the present study, the microstructure evolution of the nickel-base wrought superalloys after EFT and EPT were investigated. The mechanisms of the effects were also discussed. The aim of present study is to provide the bases for further application of electromagnetic technology on superalloys.

E. Gdoutos and M. Konsta-Gdoutos (Eds.): ICTAEM 2020, STIN 16, pp. 70–75, 2020.
https://doi.org/10.1007/978-3-030-47883-4_12

2 Experimental Procedures

The chemical compositions of GH4169 and GH4199 alloys used in the present study are listed in Table 1.

Table 1. Chemical compositions of GH4169 and GH4199 alloy (wt%)

GH4169	Fe	Cr	Nb	Mo	Ti	Al	Si	C	S	Ni
	19.25	18.78	4.96	3.09	1.00	0.65	0.082	0.04	0.003	Bal.
GH4199	Fe	Cr	W	Mo	Ti	Al	Si	C	S	Ni
	3.95	25.60	14.50	1.45	0.65	0.40	0.65	0.10	0.01	Bal.

The plate type specimen for EFT with a dimension of 120 mm × 20 mm × 1 mm was machined along the rolling direction of the bar by spark cutting. The specimens were solution treated at 1200 °C for 8 min, followed by air cooling (AC), and then, the aged at 500 °C to 900 °C in the specific electric field, respectively. The specimen was connected to the anode of a direct current voltage power supply, and a stainless steel plate was connected to the cathode. The normal aging treatment was also carried out for comparing to EFT. The EPT of the alloy was carried out on MTS 810 material testing system at 800 °C. The electrode of HPC-5 pulsed power was connected with the specimens by the self-made insulation clamping. The temperature was controlled by the high temperature testing chamber system to keep the specimen at 800 °C.

The microstructures were examined by an optical microscope (OLYMPUS GX71), a laser scanning confocal microscope (OLYMPUS LEXT 3100) and a field emission scanning electron microscope (JEOL 7001). The morphology of precipitation was observed by a JEOL JEM-2100F transmission electron microscope. Positron annihilation lifetime (PAL) and vacancy intensity percentage of specimens were measured by positron annihilation spectroscopy. Radioactive 22NaCl source was adopted, and the measurement was conducted for 4 h at room temperature.

3 Results and Discussions

Figure 1 shows microstructure evolution of GH4199 alloy with and without EFT. With EFT, the annealing twins occurred, and with the increasing of treating time, the number of annealing twins increased [8]. While the variation of the grain size and morphology of precipitation with and without EFT was not obvious.

Figure 2 shows the grain boundary character distributions of the alloy with EFT for different time [9]. It is clear that there are lots of annealing twins inside the grains in the alloy with EFT. With the increasing of treatment time, the number of annealing twin increases and the interface spacing of annealing twins decreases obviously.

It has been reported that the formation of twins during the growth of the grain boundary can decrease the grain boundary energy and increase the fraction of coherent interfaces [9]. While the interfaces with low energy such as twin boundary can inhibit

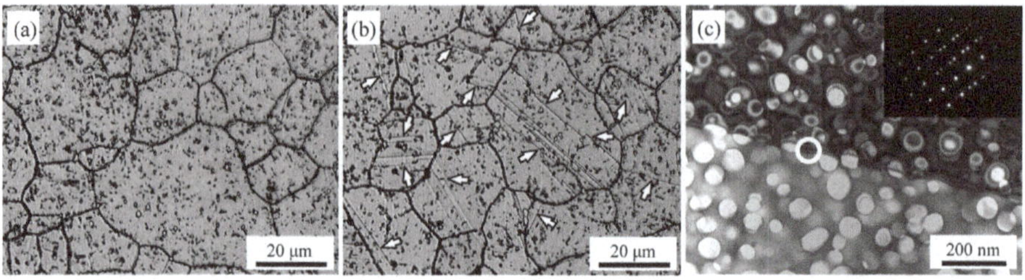

Fig. 1. Microstructure of specimens aged at 820 °C for 5 h: (a) without EFT, (b) with EFT (4 kV/cm), (c) with EFT (4 kV/cm): TEM image and the corresponding SAED

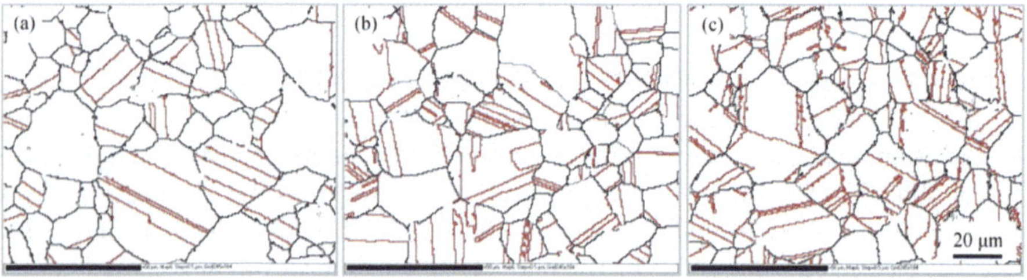

Fig. 2. OIM-reconstructed grain boundaries of the alloy with EFT for: (a) 0 h, (b) 10 h, (c) 20 h

the depletion of alloying elements at high angle grain boundary and improve the corrosion resistance of the alloys effectively [8]. During the formation and growth of the annealing twins in the alloy with EFT, the redistribution of elements at the crossing of the original high angle grain boundaries and annealing twin occurred. The exhaustion of alloying elements can be locally improved and the number of such kind of interface increases with the treatment time of EFT. The original high angle grain boundaries will be replaced and some of continuously distributed original grain boundaries will be partitioned [10]. And also, it has been found [11] that the direction of dislocation slip has been changed by the boundaries of annealing twins during deformation. The direction of microcrack propagation has also been changed on crossing the boundaries of the annealing twins. It causes the increasing plastic deformation work and delays fracture, these are considered as the reasons of the increasing ductility of the alloy by EFT.

Figure 3 shows the stress-strain curves of GH4169 alloy tested at 800 °C with different EPT current density. It can be noticed that there is a remarkable change in the flow stress with the increasing of current density. The deformation resistance in the elastic deformation range has no obvious change under the pulse current comparing with that without the pulse current. However, both the elastic limit and the deformation resistance in the plastic deformation range decreases significantly, thus the plastic deformation ability increases. The strength decreases more than 20% and the plasticity increases more than 100% under the pulse current with the current density of 1 kA/mm^2 comparing with that without the pulse current. And the strength decreases more than 75%, and the plasticity increases 700%, with the current density of 4 kA/mm^2 [11–13].

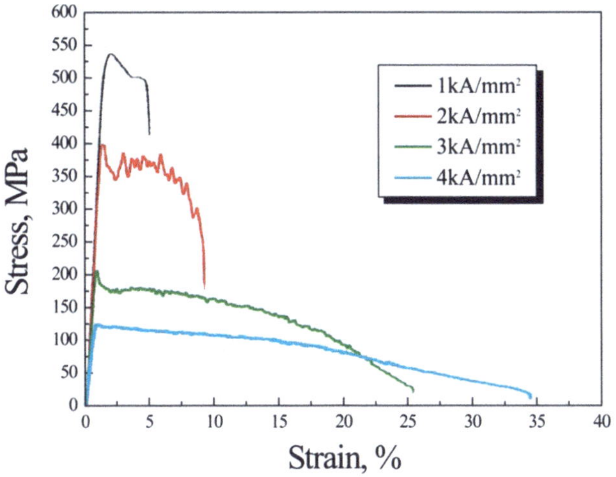

Fig. 3. Stress-strain curves of GH4169 alloy tensile test at 800 °C with different EPT current densities

In order to investigate the effect of the non-thermal effects on the recrystallization behavior of GH4169 alloy, the temperature compensation and nitrogen cooling were employed in the present study to ensure the same temperature of 650 °C of the alloy under different EPT conditions.

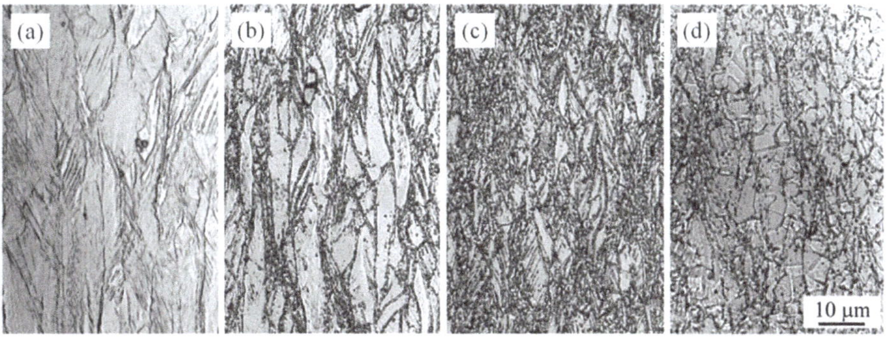

Fig. 4. Microstructures of GH4169 alloy: (a) at 650 °C for 0.5 h, (b) EPT (4 kA/mm^2, 20 Hz, 30 μs), (c) EPT(4 kA/mm^2, 30 Hz, 30 μs), (d) EPT(4 kA/mm^2, 40 Hz, 30 μs)

Figure 4 shows the microstructure of the GH4169 alloy with different treat processes. It can be seen that the microstructure of the alloy treated at 650 °C for 0.5 h (a) is still the cold rolled microstructure. While under the EPT of 4 kA/mm^2, 20 Hz and 30 μs at 650 °C for 0.5 h (b), the recrystallization characteristics and the precipitation of δ phase at the original high angle grain boundary can be found, the twin grain boundary and the deformation band although the microstructure is still cold rolling deformed microstructure. And as the EPT frequency increases, this trend gradually increases. It can be seen that most of the deformed grains in the matrix had been swallowed by a great deal of fine and uniform recrystallization grains when the pulse frequency increases to 40 Hz.

Because the distribution of solute atoms in the alloy aged at different temperatures is roughly similar, the distribution of solute atoms in the specimen with EFT is shown as an example in Fig. 5. It is clear that the distribution of Nb, Ti, Al, and Mo atoms is homogeneous in the matrix, but the obvious inhomogeneous distribution of Fe and Cr atoms can be found, and the concentration peak, as shown in Fig. 5(b).

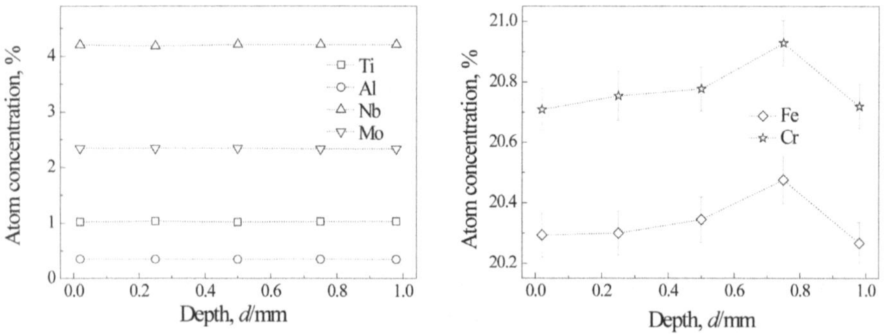

Fig. 5. The distribution of main atoms at various layers in GH4169 alloy with EFT at 800°C.

The vacancy formation energy, diffusion activation energy, or radius of Mo atom is much higher than those of Fe and Cr atoms, so that it is difficult to diffuse even though EFT is applied. The vacancy formation energy of Al atom is much lower than that of Fe or Cr atom, and the diffusion activation energy of Al or Nb atom is within the data range of Fe and Cr atoms.

4 Conclusions

(1) Under the EPT, the deformation resistance decreased and the elongation increased significantly during tensile test at 800 °C. And this effect increased with the increasing of EPT energy. Since the dislocation motion can be promoted by the electroplastic effect of the EPT. The initial temperature of recrystallization can be reduced by the electric effect of the pulse current and the dynamic recrystalllization nucleation is promoted. Therefore the dynamic recrystallization would occur at lower temperature. This is the main reason for the decreasing of deformation resistance and the increasing of plasticity of the alloy under the EPT.

(2) EFT enhances the atomic vibration in the superalloy, and it leads to the increasing of both quantity and average size of vacancies. The vacancy flow consisting of monovacancies generates induced, and monovacancies transform to vacancy clusters under the EFT during aging, which causes the coarsening of precipitins in

superallpy. Meanwhile, the short-distance diffusion of Al and Nb atoms is accelerated with EFT, resulting in the increasing of the diffusion coefficients of these atoms, and which are about 1.6–5.0 times larger than those without EFT.

References

1. Hsieh, C.H., Lo, S.L., Hu, C.Y., et al.: Thermal detoxification of hazardous metal sludge by applied electromagnetic energy. Chemosphere **71**, 1693–1700 (2008)
2. Tournier, R.F., Beaugnon, E., Noudem, J., et al.: Materials processing in a magnetic force opposed to the gravity. J. Magn. Magn. Mater. **226–230**, 2094–2100 (2001)
3. Karaca, H.E., Karaman, I., Basaran, B., et al.: On the stress-assisted magnetic-field -induced phase transformation in Ni2MnGa ferromagnetic shape memory alloys. Acta Mater. **55**, 4253–4269 (2007)
4. Ho, P.S., Kwok, T.: Electromigration in metals. Report Progress Phys. **52**, 301–348 (1989)
5. Conrad, H., Yang, D.: Effect of an electric field on the plastic deformation kinetics of electrodeposited Cu at low and intermediate temperatures. Acta Mater. **50**, 2851–2866 (2002)
6. Baranov, Y.V.: Effect of electrostatic fields on mechanical characteristics and structure of metals and alloys. Mater. Sci. Eng., A **287**, 288–300 (2000)
7. Conrad, H.: Influence of an electric or magnetic field on the liquid-solid transformation in materials and on the microstructure of the solid. Mater. Sci. Eng., A **287**, 205–212 (2000)
8. Wang, L., Liu, Y., Wang, S., et al.: Effects of electric field treatment on corrosion behavior of a Ni-Cr-W-Mo superalloy. Mater. Trans. **50**, 1644–1648 (2009)
9. Krupp, U., Kane, W.M., Liu, X.Y., et al.: The effect of grain-boundary- engineering-type processing on oxygen-induced cracking of IN718. Mater. Sci. Eng., A **349**, 213–217 (2003)
10. Schuh, C.A., Kumar, M., King, W.E.: Analysis of grain boundary networks and their evolution during grain boundary engineering. Acta Mater. **51**, 687–700 (2003)
11. Liu, Y., Wang, L., Liu, H.H., et al.: Effects of pulse current on dynamic recrystallization behavior of GH4169 superalloy. Mater. Trans. **53**, 1400–1404 (2012)
12. Zhou, Y.Z., Qin, R.S., Xiao, S.H., et al.: Reversing effect of electropulsing on damage of 1045 steel. J. Mater. Res. **15**, 1056–1061 (2000)
13. Tang, D.W., Zhou, B.L., Cao, H., et al.: Thermal stress relaxation behavior in thin films under transient laser-pulse heating. J. Appl. Phys. **73**, 3749–3752 (1993)

Dynamic Speckle Contrast Imaging for Surface Defect Detection

Rishikesh Kulkarni[1](✉), Shanta Patil[1], and Pramod Rastogi[2]

[1] Department of Electronics and Electrical Engineering, Indian Institute of Technology Guwahati, Guwahati, India
rishi.k@iitg.ernet.in

[2] Applied Computing and Mechanics Laboratory, Ecole Polytechnique Federale de Lausanne, Lausanne, Switzerland

Abstract. We propose to explore the dynamic speckle contrast imaging (DSCI) modality for non-contact surface defect detection. In this approach, the object under investigation is subjected to a continuous and monotonic loading. A number of speckle images are recorded during the continuous deformation of object surface. Speckle contrast analysis algorithm is applied to the stack of images which utilizes the fact that the variations on the object surface is reflected in significant changes in the speckle intensity variations. A simulation study was performed with the speckle images and with specklegrams recorded in the speckle interferometry and Shearographic setup. The results motivate the applicability of the DSCI technique in surface defect detection.

Keywords: Non-contact defect detection · Dynamic speckle contrast imaging · Correlation measure · Singular value decomposition

1 Introduction

Optical interferometric techniques such as electronic speckle pattern interferometry (ESPI), holographic interferometry and shearography are commonly used for submicron level deformation analysis applications [1, 2]. In these techniques, intensity of an interference pattern, *specklegram*, is recorded which is generated due to the superposition of a light beam scattered from the object surface and an undisturbed reference light beam. Both the light beams are derived from the same coherent laser source. A sinusoidally varying intensity pattern, usually termed as *fringe image,* is obtained from the subtraction of two specklegrams corresponding to the un-deformed and deformed state of the object. The deformation of the object surface mainly changes the phase of the scattered light beam. Consequently, fringe analysis for phase estimation plays crucial role in the optical measurement techniques. Apart from quantitative deformation analysis, shearography has found a wide range of applications in industry due to its robustness against environment disturbances for defect detection. Although these techniques have found to be useful in quantitative assessment of deformation process, the major bottleneck of these techniques is the experimental stability requirement which is difficult to meet outside laboratory environment.

E. Gdoutos and M. Konsta-Gdoutos (Eds.): ICTAEM 2020, STIN 16, pp. 76–81, 2020.
https://doi.org/10.1007/978-3-030-47883-4_13

Dynamic speckle contrast imaging (DSCI) has been widely used in bio-imaging applications mainly due to the simplicity involved in the experimental setup which allows its use outside laboratory conditions [3]. In this technique, the object under investigation is illuminated with a coherent laser light source. The speckle pattern generated by the optically rough object surface is recorded using a CCD or CMOS camera. In bio-imaging applications, the bio-activity within the object is the main source of speckle intensity variations. Analysis of these dynamic speckle patterns allows to obtain speckle contrast images. A number of numerical and graphic speckle analysis techniques have been reported in the literature [4]. Speckle correlation technique has also been applied in material characterization [5, 6]. In this work, we propose to explore speckle correlation for non-contact surface defect detection.

2 Dynamic Speckle Contrast Imaging

In the context of deformation analysis, when the object surface is subjected to external loading, defective surface points undergo larger deformation compared to the nondefective points. In ESPI and shearography, this fact results into the variation in the fringe density which becomes a signature for the presence of defect. Note that only two specklegrams are recorded one for the un-deformed and one for the final deformed state of the object. In the paper, we propose a scheme wherein a number speckle images are recorded while the object under investigation is subjected to continuously and monotonically changing load. In this case, variations in the speckle intensities are caused due to the object surface deformation. We propose to utilize these variations in the speckle intensity for locating the surface defects.

Consider that K number of speckle images of size $M \times N$ are recorded during the continuous loading of the object surface. Thus, we have an image stack I of size $M \times N \times K$. In the present study, since we are interested in identification of defect location, three graphical speckle contrast analysis techniques are discussed. Fujii [7] and generalized difference (GD) [8] are two commonly used measures for segmentation of object surfaces based on some local activity variations. These are computed as

$$Fujii(m,n) = \sum_{k=1}^{K} \frac{|I_k(m,n) - I_{k+1}(m,n)|}{|I_k(m,n) + I_{k+1}(m,n)|}$$

$$GD(m,n) = \sum_{k} \sum_{l} |I_k(m,n) - I_{k+l}(m,n)|$$

We propose a new algorithm for speckle contrast analysis using the singular value decomposition (SVD). Each speckle image in the stack is divided into a number of image patches each of size $M_p \times N_p$. The patch is converted into a vector of size $M_pN_p \times 1$. All the vectors associated with same patches in different speckle images are appended together to obtain a matrix I_p of size $M_pN_p \times K$. The SVD of I_p is computed as

$$I_p = U_p \Lambda_p V_p^T,$$

where, U_p and V_p are left and right singular matrices of size $M_pN_p \times M_pN_p$ and $K \times K$, respectively. Λ_p is the diagonal matrix of singular values arranged in an ascending order. Depending on the surface activity in a given image patch, the speckle intensity variations change and so does the distribution of singular values in the SVD of I_p. We define this change in singular values distribution over patches as spatio-temporal speckle correlation (STC) measure as follows

$$STC = 1 - \frac{\Lambda_1}{\sum_i \Lambda_i}$$

where, Λ_1 is the first (and highest) singular value and $\sum_i \Lambda_i$ represents sum of all singular values. Speckles are more de-correlated at the location of defect as the deformation will be higher compared to non-defective location. Accordingly, the first singular value Λ_1 corresponding to the image patch at defect will have lower magnitude compared to the same at non-defective location. The STC is computed at each patch and the two dimensional correlation map of the size same as that of the speckle image is computed using interpolation.

3 Results

A simulation study was performed to demonstrate the applicability of the DSCI technique for defect detection. The speckle simulation was performed based on the technique proposed in [9] with adjustable speckle size. We considered $K = 50$ speckle images generated with the phases corresponding to continuous object surface deformation. One such speckle image is shown in Fig. 1(a). Apart from these speckle images, the DSCI technique is also applied to simulated *specklegrams* corresponding to ESPI and Shearography setup. For the purpose of illustration, the ESPI and Shearographic fringe images corresponding to the final surface deformation are shown in Figs. 1(b) and (c), respectively. The speckle contrast images obtained using Fujii, GD and STC methods are shown in second, third and fourth columns, respectively, in Fig. (1). For the purpose of comparison, the speckle contrast images have been normalized. It can be observed that the STC method provides high contrast images compared to Fujii and GD methods with which the defect locations can be clearly identified.

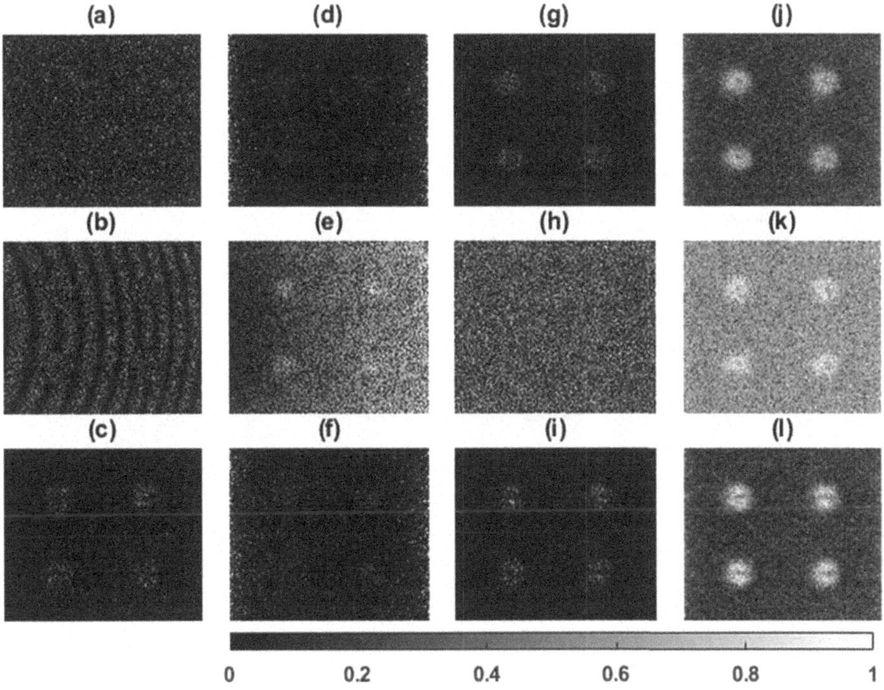

Fig. 1. (a) Speckle image in DSCI setup. Fringe images corresponding to surface deformation in (b) ESPI and (c) Shearography setup. The speckle contrast images obtained using Fujii, GD and STC methods are shown in second, third and fourth columns, respectively.

A study on the effect of number of speckle images and speckle sizes on the speckle contrast images is performed and the results are shown in Fig. 2. The first row shows the speckle contrast images obtained using the proposed STC for increasing number of speckle images (K) from left to right. It can be noted that the speckle contrast does not vary significantly with K. The second row shows the speckle contrast images obtained using the proposed STC with increasing speckle sizes from left to right. Expectedly, the speckle contrast decreases with the increase in speckle size.

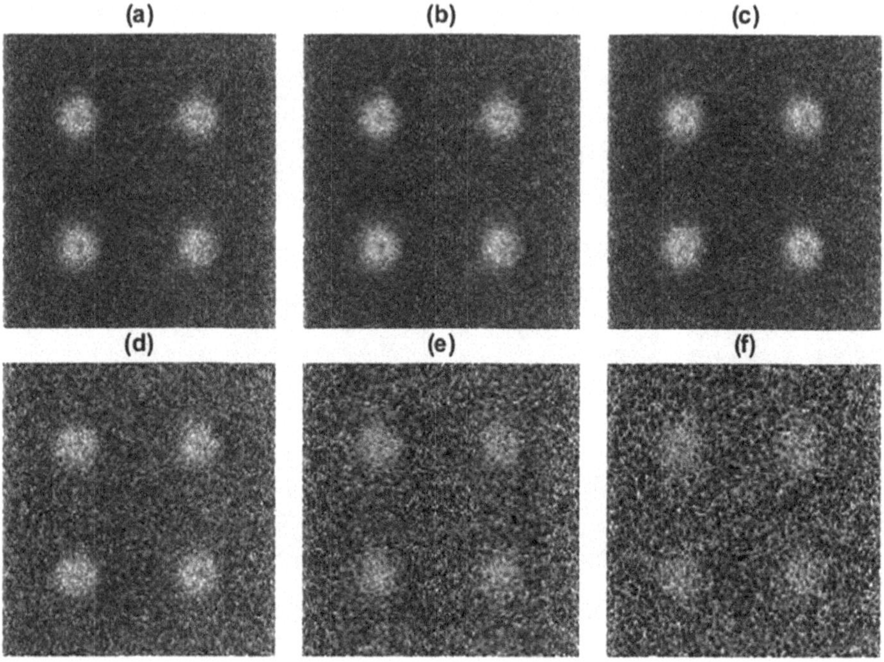

Fig. 2. The speckle contrast images are obtained using the STC method. First row: (a) K = 10 (b) K = 25 and (c) K = 75. Second row corresponds to speckle images with increasing speckle sizes from left to right.

4 Conclusion

The simulation results clearly indicate the applicability of DSCI technique for the surface defect detection. The experimental setup in this technique is much easier simpler compared to the interferometric techniques. The proposed STC algorithm based on the singular value decomposition is found to provide higher contrast images compared to the Fujii and GD methods. In fact, other speckle contrast algorithms have also been compared with STC method. The STC seem to provide better contrast images compared to most of them.

References

1. Jacquot, P.: Speckle interferometry: a review of the principal methods in use for experimental mechanics applications. Strain **44**, 57–69 (2008)
2. Kulkarni, R.D., Rastogi, P.: Single and Multicomponent Digital Optical Signal Analysis: Estimation of Phase and Its Derivatives. Institute of Physics publishing (2017)
3. Rabal, H.J., Braga, R.A. (eds.): Dynamic Laser Speckle and Applications. CRC Press, Boca Raton (2019)

4. Braga, R.A., Rivera, F.P., Moreira, J.: A Practical Guide to Biospeckle Laser Analysis, Theory and Software (2016)
5. Thomas, F.G., Gerd, G., Klaus, D.H., Karen, W.: Corrosion monitoring with speckle correlation. Appl. Opt. **38**, 5948–5955 (1998)
6. Yamaguchi, I., Yokota, M., Ida, T., Sunaga, M., Kobayashi, K.: Monitoring of paint drying process by digital speckle correlation. Opt. Rev. **14**, 362–364 (2007)
7. Fujii, H., Nohira, K., Yamamoto, Y., Ikawa, H., Ohura, T.: Evaluation of blood flow by laser speckle image sensing: part 1. Appl. Opt. **26**, 5321–5325 (1987)
8. Arizaga, R., Cap, N.L., Rabal, H.J., Trivi, M.: Display of local activity using dynamical speckle patterns. Opt. Eng. **41**, 287–294 (2002)
9. Guillermo, H.F., Gustavo, E.G.: Speckle noise reduction in television holography fringes using wavelet thresholding. Opt. Eng. **35**(1), 9–14 (1996)

Experimental Investigation of the Elastic Modulus of Timber by Using Density and Fiber Orientation

Wei-Chung Wang(✉) and Tzu-Yu Kuo

National Tsing Hua University, No. 101, Sec. 2, Kuang Fu Rd, Hsinchu 30013, Taiwan, Republic of China
wcwang@pme.nthu.edu.tw, asmit.com@msa.hinet.net

Abstract. Wood is a material found in nature with distinct applications in construction engineering. For the sake of greenhouse gas emissions, wood has been widely used. Two essential factors, i.e. fiber orientation and density, of the timber were adopted in the prediction model proposed in this paper to obtain the modulus of elasticity (MOE) of the timber. Besides, a bending test was used to verify the accuracy of the prediction model. Good agreement was found between the results obtained from the prediction model and static bending test along the longitudinal direction. Therefore, the proposed model presented in this paper has an excellent feasibility of predicting the MOE of the timber to ensure the reliability and safety of wood construction.

Keywords: Timber · Modulus of Elasticity · Tracheid effect · Fiber orientation

1 Introduction

Wood is one of the essential natural materials as a composite material found in nature with diverse applications in civil, architectural, and constructional engineering [1]. However, modulus of elasticity (MOE) of the timber is dependent on the density, fiber orientations, and position of knots, etc. Hence the values of MOE vary at each location of the timber [2, 3]. Wood research in the Photomechanics Laboratory of National Tsing Hua University (NTHU), Taiwan, Republic of China, started with an international collaboration with Prof. Ying-Hei Chui, University of New Brunswick, Canada, in 2010. In that collaboration, three dimensional (3D) digital image correlation (DIC) technique was first used to measure the shrinkage of the timber in three directions on Jack pine [4]. From 2012 to 2019, a series of foundational research work from micro- to macro-scale were performed to investigate the surface characteristics and mechanical behavior of timber [5–13]. Based on the previous research outcomes, this paper is focused on the establishment of a model to predict MOE of the timber by using the density and fiber orientation.

E. Gdoutos and M. Konsta-Gdoutos (Eds.): ICTAEM 2020, STIN 16, pp. 82–85, 2020.
https://doi.org/10.1007/978-3-030-47883-4_14

2 Material Species and the Specimen

The specimen of Japanese cedar (*Cryptomeria Japonica*) was obtained in Hsinchu County of northern Taiwan. For the dimensional stability and moisture content adjustment, the specimen was placed in the lab with temperature controlled at 25°C and relative humidity controlled at 55% for more than one year before the experiment. The dimension of the specimen with a knot is $558 \times 46 \times 17$ mm^3.

3 Prediction Model for the Modulus of Elasticity

A least-squares fit linear relationship with the coefficient of determination, i.e. $R^2 = 0.9$, between MOE and density of Japanese cedar was reported [12]. By following ASTM D2395 [14], the density of the specimen was measured as 435 kg/m^3. As listed in Table 1, material properties, i.e. Young's modulus (E), shear modulus (G) and Poisson's ratio (v), of Japanese cedar were measured at density of 435 kg/m^3.

Table 1. Material properties of Japanese cedar at density of 435 kg/m^3.

Density	E_l	E_r	E_t	G_{lr}	G_{rt}	G_{tl}	v_{lr}	v_{rt}	v_{tl}
435 kg/m^3	10.2	0.31	0.39	0.78	0.23	0.73	0.41	0.84	0.61

1. Unit: GPa.
2. Subscripts: l: longitudinal, r: radial, t: transverse.

By using the tracheid effect, a self-developed optical fiber orientation measurement system was used to determine fiber orientations from the elliptic shape formed by the laser scattering pattern projected to the surface of the timber [15]. The in-plane fiber direction was measured by the angle between the major axis of the elliptic shape and the global longitudinal axis of the specimen. The diving angle was determined by the ratio of the minor to major axes of the laser scattering pattern. Measurement of fiber orientations on four surfaces of the timber are needed to investigate the variation of MOE at each location of the timber.

The proposed prediction model of MOE for the timber was based on the orthotropic elasticity relating strains to stresses. The compliance matrix S_{ij} ($i, j = 1, \ldots, 6$) was obtained by the fiber orientations transformed to the global longitudinal direction. The effective local MOE in the longitudinal direction was obtained from the parameter S_{11} of the compliance matrix. By using the rule of mixtures, the global MOE was obtained from all fiber orientations in four surfaces of each position along the longitudinal direction.

By considering both density and fiber orientations, a prediction model was proposed to obtain the MOE. A static bending test was used to verify the correctness of the MOE calculated from the prediction model.

4 Results and Discussions

It should be pointed out here that the linear relationship between MOE and density of Japanese cedar reported in [12] was based on measurement results of the clear wood with fiber orientations parallel to the longitudinal direction. However, the specimen used in this paper has a knot in the middle, in addition, the fiber orientation is not perfectly parallel to the longitudinal direction. Furthermore, fiber orientations in the vicinity of the knot are random. Hence, the MOE value obtained from the relationship between MOE and density reported in [12] is higher than that obtained from the static bending test. By using the optical scanning system developed in this paper, real fiber orientation can be obtained so that the position and the size of the knot can be determined. The density value obtained from the linear relationship between MOE and density of Japanese cedar reported in [12] can then be modified. Therefore, all parameters of the compliance matrix will be re-calculated by using the modified density. The modified S_{11} will be used to calculate the modified MOE in the longitudinal direction.

5 Conclusions

A relationship between MOE and density was first adopted to calculate material properties in the compliance matrix. With the consideration of the effect of knot on the density and fiber orientation, an optical fiber scanning system was developed and used in this paper to measure fiber orientations on surfaces of the timber. Parameters of the compliance matrix were then modified to obtain the modified MOE. Results obtained from the prediction model proposed in this paper and static bending test along longitudinal direction are well matched. Based on the results obtained from this paper, it is clear that the global MOE is strongly related to the density, fiber orientation, and position as well as size of the knot. By using the proposed prediction model and the optical scanning system, the value of MOE can be obtained nondestructively. The reliability and safety of wooden structures can be ensured and greenhouse gas emissions can be reduced.

References

1. Matthews, F.L., Rawlings, R.D.: Composite Materials: Engineering and Science. CRC Press, Boca Rathon (1999)
2. Briggert, A., Olsson, A., Oscarsson, J.: Three-dimensional modelling of knots and pith location in norway spruce boards using tracheid-effect scanning. Eur. J. Wood Wood Prod. **74**(5), 725–739 (2016)
3. Kandler G., Lukacevic M., Füssl J.: From the knot morphology of individual timber boards to the mechanical properties of glued laminated timber. In: World Conference on Timber Engineering, WCTE 2016, 22–25 August 2016, Vienna, Austria (2016)

4. Peng, M., Ho, Y.C., Wang, W.C., Chui, Y.H., Gong, M.: Measurement of wood shrinkage in jack pine using three dimensional digital image correlation (DIC). Holzforschung **66**(5), 639–643 (2012)
5. Kuo T.Y., Wang W.C., Chu C.I., Chen J.H., Hung T.H., and Chang J. Y.: Buckling measurement of cylindrical shells by digital image correlation method. In: ASME 2013 International Mechanical Engineering Congress and Exposition, San Diego, California, 15–21 November 2013
6. Kuo, T.Y., Wang, W.C.: Experimental and numerical investigation of effects of fiber orientation of wood stiffness. In: Emerging Challenges for Experimental Mechanics in Energy and Environmental Applications, Proceedings of the 5th International Symposium on Experimental Mechanics and 9th Symposium on Optics in Industry, ISEM-SOI, 17–21 August 2015, Guanajuato, Mexico, pp. 249–254 (2015)
7. Kuo, T.Y., Cheng, I.F., Wang, W.C.: Experimental Investigation on Macroscopic Fracture Behavior of Wood Plates under Tensile Load Using Digital Image Correlation Method. In: International Digital Imaging Correlation Society - Proceedings of the First Annual Conference, Philadelphia, PA, 2017, pp. 247–250, 7–10 November 2016
8. Kuo, T.Y., Wang, W.C.: Determination of effective modulus of elasticity of wood from proportion of latewood. In: World Conference on Timber Engineering, WCTE 2016, 22–25 August 2016, Vienna, Austria (2016)
9. Kuo, T.Y., Wang, W.C.: Experimental and numerical investigation of grain orientation of timber with knot by using digital image analysis. In: ECCOMAS Thematic Conference on Computational Methods in Wood Mechanics - from Material Properties to Timber Structures, CompWood 2019, Vienna, Austria, 17–19 June 2017
10. Kuo, T.Y., Wang, W.C., Lin, C.H., Yang, T.H.: Experimental investigation of cross-laminated timber shear wall under shear force by using digital image correlation method. In: Proceedings of the First International Conference on Theoretical, Applied and Experimental Mechanics, Paphos, Cyprus, pp. 95–97, 17–20 June 2018
11. Cheng Y.J., Kuo T.Y., Wang W.C.: Determination of global modulus of elasticity of timber by using fiber orientation and proportion of latewood. In: International Conference on Computational Methods in Wood Mechanics - from Material Properties to Timber Structures. ECCOMAS Thematic Conference, CompWood 2019, Växjö, Sweden, 17–19 June 2019
12. Kuo, T.Y., Wang, W.C.: Determination of elastic properties of latewood and earlywood by digital image analysis technique. Wood Sci. Technol. **53**(3), 559–577 (2019)
13. Kuo, T.Y., Wang, W.C., Huang, H.S., Wang, J.S.: Determination of moduli of elasticity of latewood and transition latewood of Japanese cedar by using digital image analysis. In: International Conference on Computational Methods in Wood Mechanics - from Material Properties to Timber Structures. ECCOMAS Thematic Conference, CompWood 2019, Växjö, Sweden, 17–19 June 2019
14. ASTM D2395-17: Standard Test methods for density and specific gravity (relative density) of wood and wood-based materials. In: ASTM International, West Conshohocken, PA (2017)
15. Nyström, J.: Automatic measurement of fiber orientation in softwoods by using the tracheid effect. Comput. Electron. Agri. **41**(1), 91–99 (2003)

Effect of pH on Microstructure and Properties of Ultrasonic-Assisted Electroless Ni-P Coatings

Jingjing Cao(✉), Xiaoyu Wang, Yifan Zhao, and Guofang Kong

Hebei University of Engineering, Handan 056038, China
Caojingjing1105@163.com

Abstract. Ni-P coatings with various pH values (4.5 ~ 5.4) were fabricated on Ti6Al4V titanium alloy by ultrasonic-assisted electroless deposition. The effect of pH on the deposition rate and phosphorus content was studied, as well as pH on the surface morphology, phase structure and microhardness of the deposited Ni-P coatings. The results showed that the increasing pH enhanced the deposition reaction, and ultrasonic cavitation and mass transfer effect improved the catalytic activity of the substrate surface and the reaction system. So the deposition rate was increased. But the higher pH decreased the stability of the bath and precipitated $NiHPO_3$, resulting in the decrease of deposition rate. With the increase of pH, the phosphorus content decreased at first and then kept stable (17 ± 0.45wt%),while the microhardness enhanced at first and then kept stable (605 ± 2HV). The ultrasonic-assisted electroless Ni-P coatings had the typical cauliflower-like morphology and microcrystalline structure characteristics.

Keywords: Ni-P coating · Ultrasonic-assisted · pH value

1 Introduction

The electroless coatings possess good corrosion resistance, high hardness, good electrical conductivity and high tribological properties [1]. These excellent properties make it popular in many engineering community, such as electronic components, aerospace, petrochemical and others [2]. However, the autocatalytic electroless plating is a relatively slower process. The stability of the plating solution will decreases and the solution generate evaporation in the long time plating. So, it is vital to seek some method to promoting the autocatalytic action or the deposition rate and decreasing the deposition temperature. Many researchers studied the effect of ultrasonic assistance on the microstructure and properties of electroless plating [3, 4]. They found the ultrasonic assistance was beneficial to the increase of deposition rate and the improvement of properties. But the ultrasonic can decrease the plating bath stability and even cause the self-decomposition of plating solution [5]. During the electroless, various factors can affect the autocatalytic reaction, including the main salt, reducing agents, complex agents, temperature and pH value [6]. But the most important is pH that controls the rate and direction of the deposition reaction. Thus, we proposed to fabricate Ni-P coatings by ultrasonic-assisted electroless on titanium alloy substrate, and performed

E. Gdoutos and M. Konsta-Gdoutos (Eds.): ICTAEM 2020, STIN 16, pp. 86–87, 2020.
https://doi.org/10.1007/978-3-030-47883-4_15

zinc double immersion and alkaline pre-plating Ni-P film to avoid the oxidation of titanium substrate causing the decrease of adhesion strength between coatings and substrate. And this work aims to study the effect of pH on the microstructure and properties of electroless Ni-P coatings in case of the ultrasonic field.

2 Experimental Details

2.1 Specimen Preparation

The titanium alloy (Ti6Al4V) was used as the substrate, in the shape of sheet with 20 mm × 20 mm × 2 mm size. A ϕ1.5 mm hole was cut at the center of the edge with laser beam machining machine to make it easy to be hanged. The specimen was grinded using SiC paper followed by ultrasonic cleaning in alcohol. The specimen was first cleaned in alkaline solution to degrease. Then the specimen was etched to remove oxide layers. Next, the zincating was conducted to obtain Zn films. The zincated titanium alloy plate was dipped into alkaline Ni-P solution for pre-plating. And it was worth noting that sufficient water washing was required after each step.

After pre-plating process, ultrasonic-assisted electroless plating was performed and the pH was adjusted in the range of $4.5 \sim 5.4 \pm 0.02$. The schematic diagram of ultrasonic-assisted electroless plating was shown in Fig. 1).

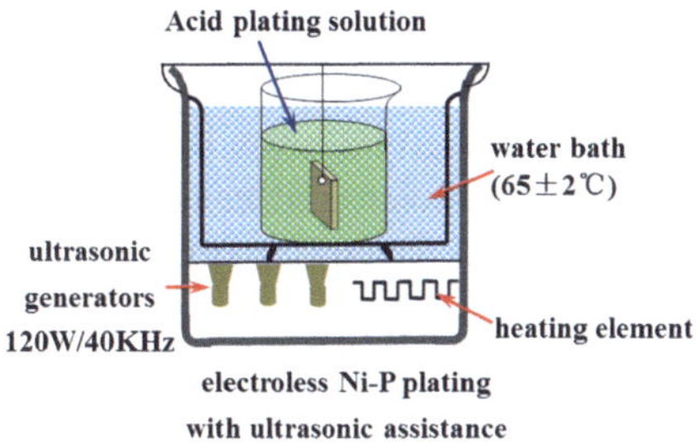

Fig. 1. Schematic diagram of ultrasonic-assisted electroless Ni-P coatings.

References

1. Hari Krishnan, K., John, S., Srinivasan, K.N.: An overall aspect of electroless Ni-P depositions - a review article. Metall. Mater. Trans. A **37A**, 1917–1926 (2006)
2. Loto, C.A.: Electroless nickel plating - a review. Silicon **8**, 177–186 (2016)
3. Jiang, J.B., Chen, H.T., Wang, Y.H.: Effect of ultrasonication and Na_2MoO_4 content on properties of electroless Ni-Mo-P coatings. Surf. Eng. **35**(10), 873–882 (2019)
4. Vitry, V., Bonin, L.: Effect of temperature on ultrasound-assisted electroless nickel-boron plating. Ultrason. Sonochem. **56**, 327–336 (2019)
5. Kobayashi, K., Chiba, A.A., Minami, N.: Effects of ultrasound on both electrolytic and electroless nickel depositions. Ultrasonics **38**(1/8), 676–681 (2000)
6. Guo, L., Zhao, X.J., Xiao, L.R.: Kinetic study of electroless cobalt deposition on WC particles. J. Alloy. Compd. **750**, 774–780 (2018)

Tailoring the Toughness and Electrochemical Capacitance of CNT Reinforced Mortars

Maria S. Konsta-Gdoutos, Panagiotis A. Danoglidis(✉), and Myrsini E. Maglogianni

Center for Advanced Construction Materials, The University of Texas at Arlington, Arlington, TX 76019, USA
panagiotis.danoglidis@uta.edu

Abstract. To unlock the contribution of carbon nanotubes in potential applications it is necessary to take into consideration the degree of their dispersion. In this work, a simple technique has been used for characterizing the degree of CNT dispersion in mortar based on values of the capacitive reactance measured using AC impedance spectroscopy. The relationship between capacitance values and fracture toughness provides valuable information on the actual CNT distribution in the matrix.

Keywords: Fracture toughness · Capacitance · Carbon nanotubes · Mortars

1 Introduction

CNT distribution in a cementitious matrix is a key parameter affecting the mechanical and electrical properties of the nanoreinforced cementitious materials and concrete [1, 2]. Besides the resistive phase, a nanotube-mortar composite contains capacitive characteristics, mainly due to the localized nanotube/mortar/nanotube, similar to that of an electrical capacitor [3]. This study characterizes CNT dispersion using AC impedance spectroscopy and fracture mechanics tests [3, 4]. It was observed that the capacitive reactance of the nanoreinforced mortars can accurately evaluate the degree of CNTs dispersion. The relationship among the capacitance measurements and fracture toughness provides information on the actual CNT distribution in the matrix.

2 Results

CNTs were added in mortar mixtures at amounts of 0.08%, 0.1%, 0.2%, 0.3% and 0.5 wt% of cement. CNT dispersion was achieved by applying ultrasonic energy with the use of a surfactant [4]. Figure 1 illustrates the capacitance and fracture toughness of mortars as a function of the CNT amount. When the CNTs are added in the mixture at amounts of 0.08% and 0.1 wt% an effective nanotube network is established in the matrix and the capacitance is decreased. This indicates a continuous nanotube-nanotube structure, in which the electrical charge cannot be stored. At the same time, 0.08% and 0.1 wt% CNT/mortars exhibit the highest improvements in fracture toughness, 70% and 89%, respectively. The addition of CNTs at amounts higher than 0.1 wt% results in

E. Gdoutos and M. Konsta-Gdoutos (Eds.): ICTAEM 2020, STIN 16, pp. 88–89, 2020.
https://doi.org/10.1007/978-3-030-47883-4_16

lower increases of the toughness and at the same time higher capacitance. This is indicative of poor CNT dispersion, which consequently leads to a lower toughening mechanism and a higher amount of energy stored in the nanoreinforced structure.

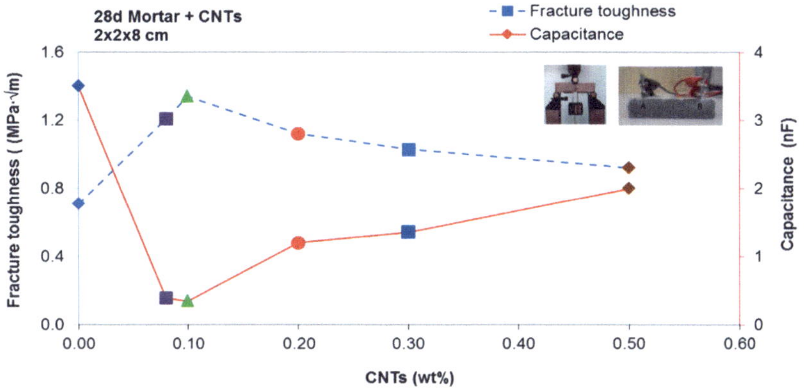

Fig. 1. Fracture toughness and Capacitance of 28d mortars reinforced with CNTs at amounts of 0.08%, 0.1%, 0.2%, 0.3% and 0.5 wt%

3 Conclusions

The degree of CNT dispersion was successfully evaluated through electrochemical spectroscopy and fracture mechanics tests. The relation between the capacitive reactance and fracture toughness of the nanoreinforced mortars provides accurate information about the degree of nanotube dispersion. The addition of highly dispersed CNTs leads in an effectively connected carbon nanotube network in the matrix. Mortars reinforced with up to 0.1 wt% exhibit high fracture toughness values and decreasing resistance and capacitance values. Addition of CNTs at higher amounts result in lower toughness values and a relative increase in the ability of the nanotube network to store electrical energy.

References

1. Konsta-Gdoutos, M.S., Danoglidis, P.A., Shah, S.P.: High modulus concrete: effects of low carbon nanotube and nanofiber additions. Theoret. Appl. Fract. Mech. **103**, 102295 (2019). https://doi.org/10.1016/j.tafmec.2019.102295
2. Konsta-Gdoutos, M.S., Danoglidis, P.A., Shah, S.P.: Tailoring the piezoresistive sensing of CNT reinforced mortar sensors. ACI Spec. Publ. **335**, 12–26 (2019)
3. Danoglidis, P.A., Konsta-Gdoutos, M.S., Shah, S.P.: Relationship between the carbon nanotube dispersion state, electrochemical impedance and capacitance and mechanical properties of percolative nanoreinforced OPC mortars. Carbon **145**, 218–228 (2019)
4. Danoglidis, P.A., Konsta-Gdoutos, M.S., Gdoutos, E.E., Shah, S.P.: Strength, energy absorption capability and self-sensing properties of multifunctional carbon nanotube reinforced mortars. Constr. Build. Mater. **120**, 265–274 (2016)

Fracture

The Creep Resistance of Short-Fibre Reinforced Metal Matrix Composites

Vaclav Sklenicka[1,2](✉), Kveta Kucharova[1], Milan Svoboda[1,2], Petr Kral[1,2], Marie Kvapilova[1,2], and Jiri Dvorak[1,2]

[1] Institute of Physics of Materials, Academy of Sciences of the Czech Republic, Zizkova 22, 616 62 Brno, Czech Republic
sklen@ipm.cz

[2] CEITEC – IPM, Institute of Physics of Materials AS CR, Zizkova 22, 616 62 Brno, Czech Republic

Abstract. The creep resistance of two discontinuously reinforced with 20 vol.% alumina short-fibres magnesium composites were examined by performing a comparison between the creep properties of these composites and their magnesium matrix alloys AZ 91 and QE 22. It was found that the magnesium AZ 91 composite exhibits an improved creep resistance arising mainly from an efficient load transfer effect and the existence of a threshold stress, respectively. By contrast, the beneficial effect of reinforcement and the improved creep resistance of the QE 22 composite is significantly influenced by the creep loading conditions.

Keywords: Magnesium composites · Creep resistance · Reinforcement

1 Introduction

The creep resistance of magnesium alloys is rather limited at elevated temperatures. However, a marked improvement in the creep properties of magnesium monolithic alloys can potentially be achieved through the production of composite materials, where the matrices consist of conventional magnesium alloys which are strengthened through the introduction of non-metallic fibres or particulates to form metal matrix composites (MMCs) [1]. The most fundamental issue of the creep behaviour of these composites is the determination of the mechanism(s) by which the creep life of the composite is increased by reinforcing the creep matrix with less-creeping non-metallic short-fibres or particulates. Once the mechanism is clarified, one can design new MMC with higher creep resistance by tailoring constituent parameters of the matrix alloy and short fibre phases.

This work reports the experimental results obtained in an investigation of the high temperature creep behaviour of short-fibre reinforced AZ 91 and QE 22 matrix-based composites and their monolithic alloys. The objective of the present research is a further attempt to clarify the creep strengthening and creep resistance in short-fibre reinforced magnesium alloys.

E. Gdoutos and M. Konsta-Gdoutos (Eds.): ICTAEM 2020, STIN 16, pp. 93–99, 2020.
https://doi.org/10.1007/978-3-030-47883-4_17

2 Experimental

All the experimental materials used in this work were fabricated at the Technical University of Clausthal, Germany. Short-fibre reinforced and unreinforced blocks of the common AZ 91 alloy (Mg-9 wt%Al-1 wt%Zn-0.3 wt%Mn) and the high strength silver-containing QE 22 alloy (Mg-2.5 wt%Ag-2.0 wt%Nd rich rare earths-0.6 wt%Zr) were processed by squeeze casting [2]. The fibre preform consisted of planar randomly distributed δ-alumina short fibres (Saffil Al_2O_3 fibres ~5 μm in diameter with varying lengths up to an estimated maximum of ~150 μm). The final fibre fraction after squeeze casting in both composites was about 20vol.%. For convenience, the composites are henceforth designated AZ 91 + Saffil and QE 22 + Saffil. All materials were subjected to a T6 heat treatment [2].

Flat creep specimens with a 25 mm gauge length and 3×3.2 mm^2 cross-section were machined from the blocks so that the longitudinal specimen axes parallel to the plane in which the long axes of the fibres were situated. Constant stress tensile creep tests were carried out at temperatures from 423 K to 523 K and at applied stresses from 20 to 200 MPa. The creep elongations were measured using a linear variable differential transducer (W2K from Hottinger-Baldwin Co., Germany) and continuously recorded digitally and computer-processed.

Metallographic and fractographic investigations were conducted after creep testing using scanning electron microscope Lyra 3 TESCAN.

3 Creep Strengthening of the Composite

At present it is generally accepted that creep deformation in metal matrix composites is controlled by flow in the matrix materials [2]. Creep strengthening of the magnesium composites may occur by either direct or indirect mechanisms [1] as it is illustrated in Fig. 1. Direct strengthening is due to a load transfer from the matrix to the reinforcements. Thus, load transfer is accompanied by a redistribution of stresses in the matrix and this reduces the effective stress for creep. Indirect strengthening occurs when the presence of the reinforcements or the process used to manufacture the composite influence the matrix microstructure which could modify the creep resistance. Potential microstructural effects include changes in the dislocation arrangements, accelerated ageing, microstructural decomposition, a matrix chemical compositional variation, and/or a reinforcement transformation. As a result, the microstructural changes in the matrix of the composite can retard and/or inhibit dislocation motion and lead to a threshold for creep that increases the creep resistance.

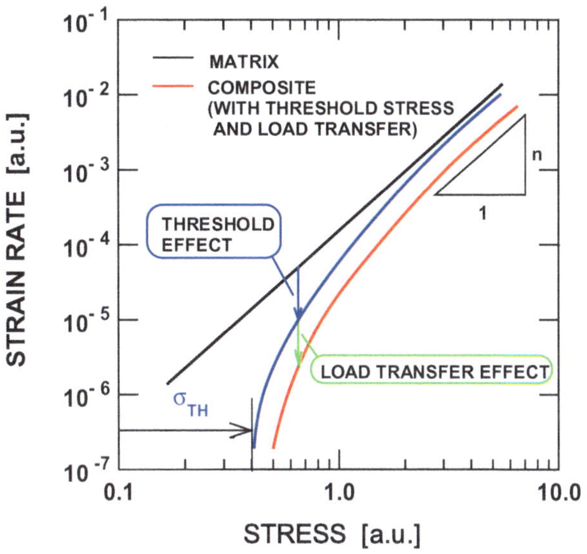

Fig. 1. A schematic illustration of the power law creep in conjunction with the effects of the threshold stress and load transfer.

4 Results and Discussion

4.1 Creep in Monolithic and Fibre-Reinforced AZ 91 and QE 22 Alloys

The most important creep data for the AZ 91 and QE 22 alloys and their composites at a testing temperature of 473 K are shown in Fig. 2, where the minimum creep $\dot{\varepsilon}_m$ rate $\dot{\varepsilon}_m$ and the time to fracture t_f are plotted against the applied tensile stress σ on a double logarithmic scale. Inspection of the creep data in Fig. 2(a) leads to two observations. First, the AZ 91 + Saffil composite exhibits an improved creep resistance (that means a substantial decrease of the minimum creep rate $\dot{\varepsilon}_m$), typically by >2 orders of magnitude in comparison with the monolithic alloy and, therefore, the trends of these plots are different because the unreinforced alloy exhibits a decreasing value of the true stress exponent of the creep rate $n = (\partial \ln\dot{\varepsilon}_m/\partial \ln\sigma)_T$ at the lower stresses, whereas the composite exhibits a higher value of n with decreasing stress. Second, the creep resistance of the QE 22 + Saffil composites is also somewhat better than that the unreinforced alloy but at the low stresses ($\sigma \leq$ 100 MPa) only.

Figure 2(b) shows the variation of the time to fracture with the stress for the same specimens tested in Fig. 2(a). The results for AZ 91 alloy and its composite demonstrate the creep lifetimes of the composite are up to one order of magnitude longer than that for the monolithic alloy at low stresses, although this difference decreases with increasing applied stress so that ultimately there is very little difference at stresses $\sigma >$ 100 MPa. By contrast, the creep life of the QE 22 + Saffil composite is markedly shorter than that of the unreinforced alloy at stresses > 100 MPa.

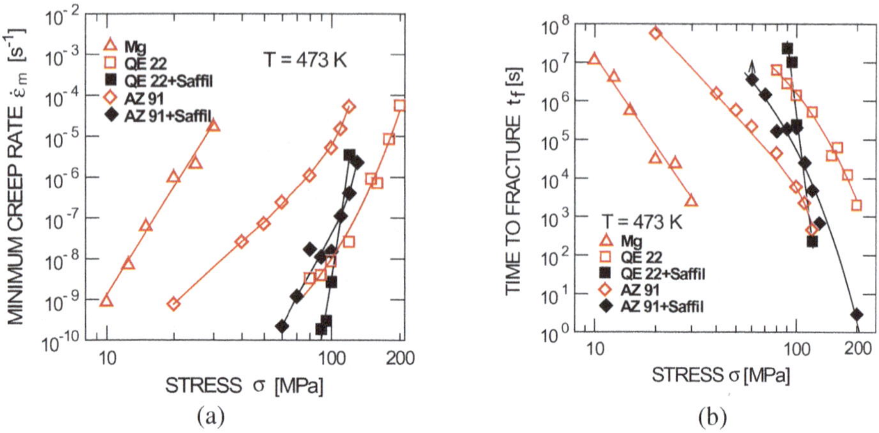

Fig. 2. Stress dependences of (a) minimum creep rate and (b) time to fracture for the monolithic magnesium alloys and their short-fibre composites.

In conclusion, it was found that the presence of reinforcement leads to a substantial decrease in the overall ductility of the monolithic alloys. Thus, the values of the strain to fracture ε_f in both composites were only ≈1–2%. By contrast, the strains to fracture in the monolithic alloys were significantly higher, typically up to 15% in the AZ 91 alloy and up to 30% in the QE 22 alloy.

Despite the general similarity of the creep behaviour of both composites, the presence of the same composition and amount of a short-fibre reinforcement in the unreinforced AZ 91 and QE 22 alloys results in a significantly different improvement of the creep resistance of composite by comparison with the matrix alloy. Thus, it will only be appropriate here to analyse differences that can be inferred from a comparison between the creep strengthening mechanisms of both composites and the unreinforced alloys (Fig. 1).

4.2 Direct Strengthening Due to a Load Transfer

As already shown in Fig. 2(a), the presence of short fibre reinforcement could lead to a reduced creep rate in both composites. Such effect can arise when significant load transfer partitions the external load between the matrix and the reinforcement. In the presence of load transfer, the creep data may be successfully reconciled by putting the ratios of the minimum creep rates of the composite and the rates of the matrix alloy at the same loading conditions, equal to a factor given by $(1-\alpha)^n$, where α is a load transfer coefficient having values lying within the range from 0 (no load transfer) to 1 (full-load transfer). Thus, the values of α inferred from the data for the unreinforced AZ 91 and its composite in Fig. 2(a) using $n = 3$ are within the range of 0.8 to 0.9. The predicted values α are in reasonable agreement with the theoretical prediction according to an analytical treatment based on a modified shear-lag model. Unfortunately, values of the load transfer coefficient α, for the QE 22 composite cannot be rigorously obtained due to insufficient creep data inferred from this study. The indicative experimentally estimated values of α, for which the rate of the composite $\dot{\varepsilon}_c < \dot{\varepsilon}_m$ were found to be 0.11 and 0.26 for 423 and 523 K, respectively.

4.3 Indirect Strengthening Due to a Threshold Stress σ_0

Having only a small number of experimental points it was appropriate to make use of the procedure in which it is possible to estimate the magnitude of σ_0 by plotting the data on linear axes as $(\dot{\varepsilon})^{1/n}$ against σ and linearly extrapolating the data to give the threshold stresses at a zero strain rate. A plot of this type requires, a priori, a judicious selection of the appropriate value of the true stress exponent, n. Figure 3 shows the determination of the threshold stress for the unreinforced AZ 91 alloy (Fig. 3(a)) and its composite (Fig. 3(b)) in double linear coordinates. Detailed inspection of the individual plots shows that a true stress exponent of 3 yields the best linear fit – Fig. 3(b). The finding of $n = 3$ provides support for adopting the suggestion that the AZ 91 composite exhibits creep behaviour which is consistent with the behaviour anticipated for magnesium solid solution alloys. Similar results for QE 22 alloy and its composite show that, while no threshold stress was found in the case of monolithic QE 22 alloy, the values of the threshold stress of the composite at 473 K and 523 K were estimated as 22 MPa and 18 MPa, respectively.

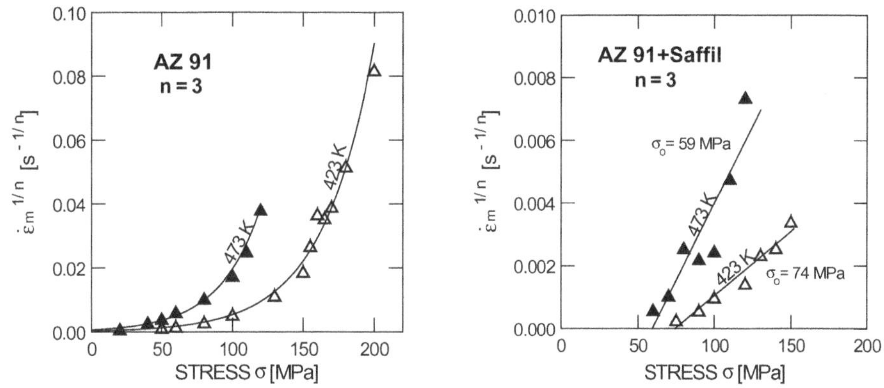

(a) (b)

Fig. 3. Procedure for determining the threshold stress σ_0 with its estimated values: (a) AZ 91 alloy (no threshold stress), (b) AZ 91 composite.

It was mentioned earlier, that indirect strengthening may be caused by the microstructural changes in the material. In our previous studies [2] we found that the threshold stress values are closely connected with the development of a continuous precipitate morphology of $Mg_{17}(Al, Zn)_{12}$ particles. Further, it was observed that the fibres in the QE 22 composite can act as nucleation centre in the precipitation process, promoting precipitation of Al_2Nd, $Mg(Al)_{12}Nd$ and Mg_3Ag phases.

4.4 Matrix/Reinforcement Bonding

The creep behaviour of the composites may be substantially influenced by the matrix/fibre interfaces. Weak matrix/reinforcement bonding may lead to inferior creep properties of the composite. Thorough fractographic investigations of the AZ 91 composite did not reveal either substantial creep fibre cracking and breaking or any substantial debonding at the interfaces between the fibres and the matrix due to creep. By contrast, and intensive debonding was revealed at the fracture surfaces at the QE 22 composite. Thus, in the limit of zero interface strength, where no stress can be transmitted to the fibres, the composite may be weaker than the matrix alone. Detailed fractographic investigation revealed enhanced precipitation of secondary phases at the interfaces in the AZ 91 composite, which can favourably affect the strength of the matrix/fibre interface and thus the creep behaviour of the composite (Fig. 4).

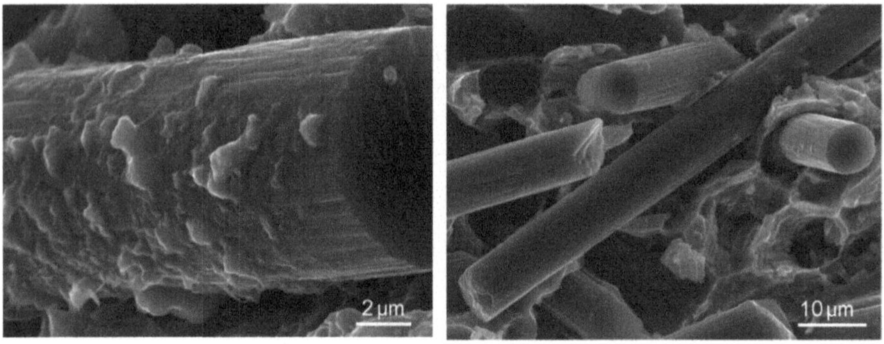

(a) (b)

Fig. 4. Creep fracture surface: (a) good interface bonding due to an occurrence of precipitates at the fibre surface (AZ 91 composite), (b) weak cohesion of the interface due to clean fibre surface (QE 22 composite).

5 Conclusions

Creep strengthening in the composite arises mainly from the load transfer and the existence of a threshold stress. The load transfer occurs by plastic deformation of the matrix being limited by the intact bond of the fibre and enhanced precipitation of secondary phases.

Acknowledgement. This research was carried out under the project CEITEC 2020 (LQ1601) with financial support from the Ministry of Education, Youth and Sports of the Czech Republic under the National Sustainability Programme II.

References

1. Mileiko, S.T.: Metal and Ceramic Based Composites. Elsevier, Amsterdam (1997)
2. Sklenicka, V., Pahutova, M., Kucharova, M., Svoboda, M., Langdon, T.G.: Creep processes in magnesium alloys and their composites. Metall. Mater. Trans. **33**, 883–889 (2002)

Adhesive Joint Fracture Under Combined Pulsed and Vibrational Loading

Yuri Petrov[1,2], Andrey Logachev[1(✉)], Nickolay Granichin[1], and Grigory Volkov[1]

[1] Saint Petersburg State University, Universitetskaya nab. 7-9, Saint Petersburg 199034, Russia
a.n.logachev@mail.ru

[2] Institute of Problems of Mechanical Engineering, Russian Academy of Sciences, Bolshoj pr. 61, V. O, Saint Petersburg 199178, Russia

Abstract. The strength of the adhesion joint under the combined vibrational-pulsed loading is considered. Beam and membrane models on elastic foundation is applied to simulate behavior of adhesion zone. The incubation time criterion is used to determine the threshold parameters of a load required to adhesive zone fracture. The solution of the differential equation describing the behavior of the adhesive zone is obtained by the Fourier method. Different types of combination of pulsed loading with high-frequency external fields are considered. It demonstrates that background vibration fields of even low intensity at certain frequencies can significantly reduce these critical values. The dependence of the threshold value of the external pulse on the model parameters is shown. The possible way to control to adhesive joint strength by background vibrational fields is suggested.

Keywords: Adhesion zone · Incubation time criterion · Elastic foundation · Vibrational field

1 Introduction

High-frequency harmonic fields has a significant impact on the strength properties of materials. Microelectromechanical Systems (MEMS) is a promising area for applying this effect. Previously, this effect has already been investigated on the model of the adhesion zone based on the string approximation [1]. Winkler base and its modifications are the most widely used approach for describing the adhesion joint [2–4]. The simplest one-dimensional model of beam and 2D-model of membrane on the elastic foundation are considered. The influence of external harmonic fields on the strength of these models is subject of the study.

E. Gdoutos and M. Konsta-Gdoutos (Eds.): ICTAEM 2020, STIN 16, pp. 100–105, 2020.
https://doi.org/10.1007/978-3-030-47883-4_18

2 Incubation Time Criterion

The fracture criterion based on the concept of incubation time is used. This criterion is originally formulated in terms of stress. On the contrary, equations describing the behavior of the adhesive zone and boundary conditions are written using displacements. The assumption about elastic behavior is therefore accepted. It makes possible to rewrite the criterion in terms of displacements:

$$\max_{t\in[0,t_c]} \frac{1}{\tau} \int\limits_{t-\tau}^{t} u(\eta)d\eta \leq u_c, \tag{1}$$

where $u(t)$ is displacement, u_c is a limit elongation of the link under quasistatic loading, τ is fracture incubation time, t_c is upper border of considered time period.

3 Adhesive Zone Models

3.1 Membrane Approximation

The adhesive zone of a circular membrane on an elastic base is considered. Only resistance to tensile forces is taken into account. The model is described by the following equation:

$$\frac{\partial^2 \mathrm{u}}{\partial r^2} + \frac{1}{\mathrm{r}}\frac{\partial u}{\partial r} - \frac{1}{c^2}\frac{\partial^2 \mathrm{u}}{\partial t^2} - \frac{\omega^2}{c^2}u = -f(r,t), \tag{2}$$

where r is distance from the center of the membrane, c is the wave velocity, ω is the characteristic of the elastic foundation of rigidity, f is the external force. Boundary conditions correspond to fixed edge of the membrane and zero initial conditions is considered:

$$u(r,0) = \frac{\partial u}{\partial t}(r,0) = 0, \qquad u(R,t) = 0. \tag{3}$$

The solution of the problem (2, 3) is obtained by Fourier method [5]: $u(r,t) = \sum\limits_{k=1}^{n} f_k(r)g_k(t)$:

$$u(r,t) = \frac{2c^2}{R^2} \cdot \sum_{k=1}^{n} \frac{J_0\left(\frac{\alpha_{0k}\cdot r}{R}\right)}{\Omega(k)\cdot\left(J_0'(\alpha_{0m})\right)^2} \int\limits_0^t \left(\int\limits_0^R f(\varepsilon,\eta)\cdot\varepsilon\cdot J_0\left(\frac{\alpha_{om}\cdot\varepsilon}{R}\right)d\varepsilon \right) \cdot \sin(\Omega(k)(t-\eta))d\eta, \tag{4}$$

where J_0 – is the zero-order Bessel function, α_{0m} – are the zeroes of the zero Bessel function, $\Omega(k) = \sqrt{c^2 \cdot \lambda_k^2 + \omega^2}$, $\lambda_k = R \cdot \alpha_{ok}$.

Constant and Pulse Loading

Concentrated constant force P and a background vibration field of frequency v is considered:

$$f(r,t) = P\left(H(t)\frac{\delta(r)}{R^2 \cdot r} + \gamma \cdot sin(vt)\right),$$

where δ(r) is the Dirac delta function, H(t) is the Heaviside function, γ is a relative intensity of the background vibration field.

Fracture is considered at center point. Threshold force amplitude at which delamination of the adhesion zone occurs founded by Eq. (4) and fracture criterion (1). The dependence of the threshold amplitude P versus the background field frequency v at various relative intensities γ for this type of loading is shown in Fig. 1. Based on these results a significant decrease of the threshold amplitude is observed at certain frequencies.

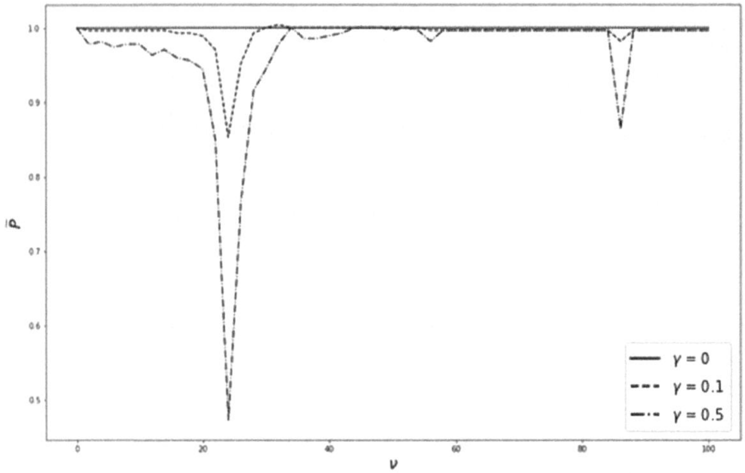

Fig. 1. Dependence of the load threshold amplitude on the background field frequency, where load function is the combination of constant force and infinite vibrational field

Another considered type of loading is combination of impulse of finite duration T and background vibration field:

$$f(r,t) = P\left((H(t) - H(t-T))\frac{\delta(r)}{R^2 \cdot r} + \gamma \cdot sin(vt)\right).$$

The critical force change with increasing pulse duration is also investigated at a certain frequency of the background field $v = \frac{\Omega(1)+\Omega(2)}{2}$. The results are presented on Fig. 2. The plot demonstrates a decrease of the threshold amplitude with the increasing of the pulse duration until $T = 0.1$. Almost constant dependence is observed after this value.

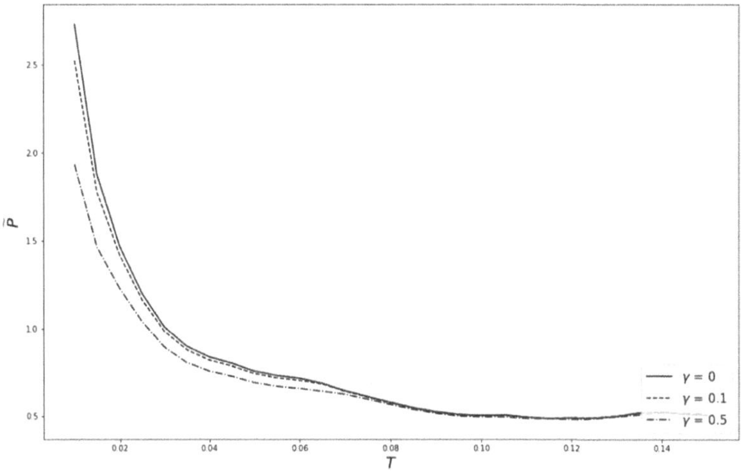

Fig. 2. Dependence of the pulse load critical amplitude on its duration in the continuous background vibrational field

3.2 Beam Approximation

Similar results are obtained for the model based on beam approximation. Bernoulli-Euler beam on elastic foundation is considered. Tensile forces are neglected. In this case, the adhesion zone will be described by following equation [6]:

$$\frac{\partial^4 w}{\partial x^4} + \frac{1}{c^2}\frac{\partial^2 w}{\partial t^2} + \frac{\omega^2}{c^2} w = f(x,t), \tag{5}$$

where x is the coordinate along axis of the beam, w is vertical deflection.

Zero initial conditions:

$$w(x,0) = \frac{\partial w}{\partial t}(x,0) = 0. \tag{6}$$

Boundary conditions of 3 types are considered

- Hinged ends of the beam

$$w(0,t) = w(l,t) = 0, \quad \frac{\partial^2 w}{\partial x^2}(0,t) = \frac{\partial^2 w}{\partial x^2}(l,t) = 0, \tag{7}$$

- Clamped ends of the beam

$$w(0,t) = w(l,t) = 0, \quad \frac{\partial w}{\partial x}(0,t) = \frac{\partial w}{\partial x}(l,t) = 0, \tag{8}$$

- Beam with free ends

$$\frac{\partial^2 w}{\partial x^2}(0,t) = \frac{\partial^2 w}{\partial x^2}(l,t) = 0, \quad \frac{\partial^3 w}{\partial x^3}(0,t) = \frac{\partial^3 w}{\partial x^3}(l,t) = 0. \tag{9}$$

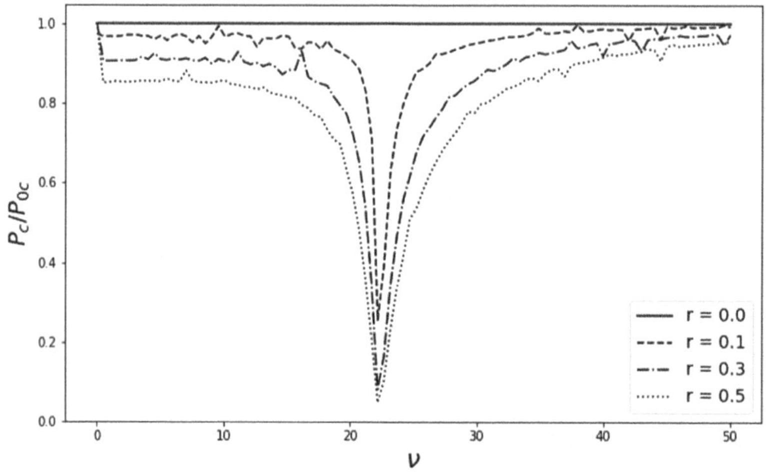

Fig. 3. Dependence of the loading threshold amplitude on the background field frequency for the clamped beam

The solution of the problem (5–9) is obtained as follows:

$$w(x,t) = c^2 \sum_{k=1}^{n} \frac{U_k(x)}{\Omega_k \|U_k\|^2} \int_0^t \left(\int_0^l f(\xi,\eta) U_k(\xi) d\xi) \right) \sin(\Omega_k(t-\eta))d\eta, \tag{10}$$

where $\Omega_k = \sqrt{\omega^2 + c^2\lambda_k^4}$, $\|U_k\|^2 = \int_0^l (U_k(x))^2 dx$, U_k and λ_k is solution of spectral problem $U^{IV} - \lambda^4 U = 0$ and expressions of these auxiliary parameters are given in the Table 1.

Table 1. Auxiliary parameters for different types of boundary conditions.

	λ_k	U_k
Hinged beam	$\pi k/l$	$\sin \lambda_k x$
Clamped beam	Roots of $\mathrm{ch}(\lambda l)\cos(\lambda l) = 1$	$K_4(\lambda_k x) - \frac{K_4(\lambda_k l)}{K_3(\lambda_k l)} K_3(\lambda_k x)$
Free ends	Roots of $\mathrm{ch}(\lambda l)\cos(\lambda l) = 1$	$K_2(\lambda_k x) - \frac{K_4(\lambda_k l)}{K_3(\lambda_k l)} K_1(\lambda_k x)$

where K_i are the functions of Krylov:

$$K_1(z) = (\mathrm{ch}\, z + \cos z)/2 \quad K_2(z) = (\mathrm{sh}\, z + \sin z)/2,$$
$$K_3(z) = (\mathrm{ch}\, z - \cos z)/2 \quad K_4(z) = (\mathrm{sh}\, z - \sin z)/2.$$

Constant Load

External load of the following form is considered:

$$f(x,t) = P\left(H(t)\delta\left(x - \frac{l}{2}\right) + rsin(vt)\right). \tag{11}$$

The loading is similar to the previous one. Solution is obtained in the same way as in the case of membrane. Figure 3 shows the dependence of vibrational frequency and critical force amplitude at point $x = l/2$. The graph demonstrates the same effect of the threshold amplitude decrease as for membrane approximation.

4 Conclusion

The observed effect of reducing the threshold load at certain frequencies appears even for harmonic fields of relatively small amplitude. In addition, the effect appeared for both models considered (membrane and beam approximation). This phenomena could be used in MEMS application with ultrasound as a control factor to shift volatile regimes of nano electro-mechanical switches.

References

1. Gorbushin, N., Petrov, Y., Zhao, Y.-P., Zhang, Y.: Threshold characteristics of short-pulsed loads combined with the ultrasound field causing dynamic delamination of adhesive joints. Theor. Appl. Mech. Lett. **8**(1), 28–31 (2018)
2. Dillard, D.A., Mukherjee, B., Karnal, P., Batra, R.C., Frechette, J.: A review of Winkler's foundation and its profound influence on adhesion and soft matter applications. Soft Matter **14**(19), 3669–3683 (2018)
3. Godara, R.K., Joglekar, M.M.: Suppression of contact bounce in beam-type microelectromechanical switches using a feedforward control scheme. JVC/J. Vib. Control **24**(23), 5502–5513 (2018)
4. Farokhi, H., Ghayesh, M.H., Gholipour, A., Tavallaeinejad, M.: Stability and nonlinear dynamical analysis of functionally graded microplates. Microsyst. Technol. **24**(5), 2109–2121 (2018). https://doi.org/10.1007/s00542-018-3849-7
5. Polyanin, A.D., Zaitsev, V.F.: Handbook of Nonlinear Partial Differential Equations. Chapman and Hall/CRC (2016)
6. Zhang, Y., Zhao, Y.P.: Determining both adhesion energy and residual stress by measuring the stiction shape of a microbeam. Microsyst. Technol. **21**(4), 919–929 (2015)

Study on the Fracture Mechanism of the P91 Steel During Small Punch Tensile Testing

Hezong Li[1,4(✉)], Hao Chen[2], H. K. Al-Abedy[3,4], and Wei Sun[4]

[1] Hebei University of Engineering, Handan 056038, China
Lhzong@hebeu.edu.cn
[2] University of Nottingham Ningbo China, Ningbo 315100, China
[3] Wasit University, Wasit, Iraq
[4] University of Nottingham, Nottingham NG7 2RD, UK

Abstract. This paper investigates the fracture generation processing of the P91 steel during a small punch tensile test. P91 disc specimens, 8 mm in diameter and 0.5 mm in thickness, were tested in a small punch test rig at 600 °C using a constant displacement rate of 2 µm/s. Interrupted tests were performed to investigate the crack generation and evolution of P91 in small punch tensile tests. The disc-shaped specimens were deformed into domes of different depths during the tests. Cross-sectional microstructure analysis revealed that thinning and circumferential necking started to occur at the edge of contact before the maximum load was reached, and then significant point necking occurred in the circumferential necking area after the maximum load. The crack formed during the tests aligns with the plastic flow of the specimen. Alignment of crack propagation occurred in the plastic deformation direction, before achieving the maximum small punch load.

Keywords: Small punch tensile test · P91 steel · Deformation regimes · Crack generation · Fracture mechanism

1 Introduction

In the last decade, testing techniques using miniature samples have received ever-increasing attention from the electricity generation industry for situations where the amount of material used in the test is limited [1, 2]. The Small Punch Tensile Test (SPTT) is widely used as a miniature specimen test method to characterise the tensile mechanical properties for ductile metallic materials, with the specimen being a disc of ~8 mm diameter and ~0.5 mm thickness [3]. The punch was applied to the disc at a given constant displacement rate. The very small specimen size denotes that the SPTT can be considered as a limited volume and non-destructive methodology to provide a real evaluation of mechanical behaviours when compared with the conventional tensile test [4–7]. Moreover, the SPTT specimen can be taken directly from in-service components without significantly affecting their structural integrity [8], allowing the mechanical behaviour to be examined using SPTT [9, 10]. In particular, the SPTT exhibits unique advantages in characterising power plant steels, of which the high temperature mechanical behaviour is crucial to the durability of pressure vessels.

E. Gdoutos and M. Konsta-Gdoutos (Eds.): ICTAEM 2020, STIN 16, pp. 106–111, 2020.
https://doi.org/10.1007/978-3-030-47883-4_19

One of the power plant steels which is of great interest is the 9Cr-1Mo steel (commercially known as P91). It is widely recognised that fracture in P91 steels occurs through progressive nucleation, growth and eventual coalescence of voids. Meanwhile, most studies are based on uniaxial tests, using standard specimens where necking appears in a location that cannot be controlled. The understanding and prediction of strain and damage interactions in P91 steels remain a challenging topic. At present, the bi-axial loading state of SPTT appears to have the potential to examine the above physical phenomena to reveal the deformation and fracture mechanisms. However, work concerned on this aspect, especially in addressing the microscopic evidence with the deformation and fracture mechanisms of SPTT, does not appear to have been detailed until now.

Therefore, considering the significance of microstructure characteristics on the degradation of P91 during SPTT, the aim of the work reported in this paper is to investigate specifically the microstructural evolution of P91 during the SPTT, in an attempt to reveal the deformation mechanisms in different regimes of the SPTT. The results of a physical characterisation of the high-temperature deformation and fracture process of the P91 steel via SPTT at 600 °C are presented and the deformation and fracture mechanisms in different SPTT regimes are discussed.

2 Material and Experimental Procedure

The as-received P91 steel was machined into a cylinder shape with 8 mm in diameter and was then sliced into disc SPTT specimens with a thickness slightly larger than 0.5 mm using wire electro-discharge machining. The specimens were then ground down to a final thickness of 0.5 mm using 1200-grade silicon carbide papers.

A Tinius Olsen H5KS single column material testing machine was used to perform the SPTTs. A 1.0 kN load cell, which has a loading accuracy of ±0.5% of the applied load, is installed to measure the load transmitted to the specimen through a hemispherical punch head. The displacement of the punch was recorded as the average reading of two linear variable differential transformers, which have an accuracy of ±0.01 mm, a resolution of 1 μm and a speed resolution of 1 μm/min. A constant displacement rate of 2 μm/s was applied to the punch when the specimen has failed or when the displacement of the punch has reached the setting values (for interrupted tests) and a backoff speed of 4 mm/s was applied.

3 Mechanical Behaviour

3.1 Load-Displacement Curves

A schematic diagram of the applied force, F, and displacement, u, of SPTT for a ductile material is shown in Fig. 1a. According to Abendroth and Dymacek [11, 12], the F-u curve obtained from the present tests can be divided into six stages, representing different deformation regimes during SPTT. Thus, to investigate the physical nature of the above deformation process and the fracture behaviour of the P91 steel at 600 °C, a series of interrupted SPTTs have been carried out. Firstly, two specimens are used to perform the repeated SPTTs for the whole stroke until fracture. The load-displacement curves obtained are shown in Fig. 1b.

To investigate the deformation mechanisms of P91 in SPTT, the key interrupted displacements were determined based on Fig. 1a. The main objective of interrupted SPTTs is to provide "In Situ" investigations into the evolution of deformation at each stage of the SPTT until fracture. Seven interrupted SPTTs were conducted on the P91 steel at 600 °C in Fig. 1b. The interrupted displacements are set as 0.26, 0.60, 1.00, 1.25, 1.50, 1.65 and 1.74 mm respectively to represent the different deformation regimes in Fig. 1a.

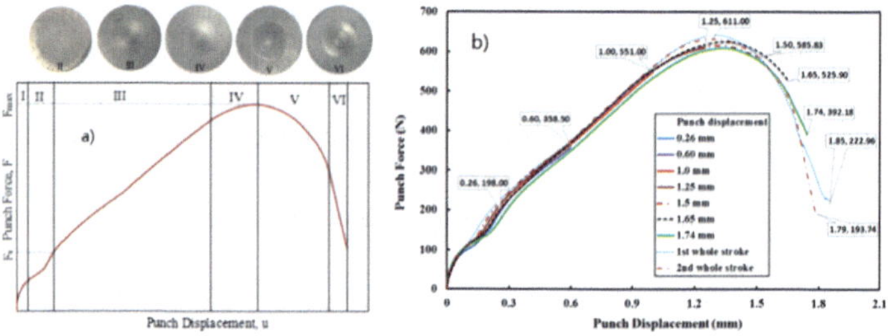

Fig. 1. The schematic diagram of the SPTT load-displacement curve (a), the load-displacement curves at interrupted displacements and whole stroke (b).

3.2 Physical Characterization of the Deformation Process

The cross-sectional microstructure of the above deformed P91 specimen is shown in Fig. 2. The small deformation after a displacement of 0.26 mm in Fig. 2a exhibits the dome shape is about to form due to the punch movement. It can be seen that the thickness of the specimen remains the same without obvious material thinning in Fig. 2a. The primary phase that can be seen is the prior austenitic phase in Fig. 2b. No large voids have formed yet for a displacement of 0.26 mm. The dome shape becomes more evident when the deformation of specimen increases. The alignment of the grains is not found in Fig. 2b and 2d, but such evidence is seen at large punch displacements in Fig. 2f and 2h. This indicates that the grain alignment will tend to occur when the plastic deformation becomes dominant. The beginning of necking occurs, as shown in Fig. 2e and g, due to the bending and stretching stress at the edge of contact between the punch and specimen.

Figure 3 shows the cross-sectional microstructure of the P91 after achieving the maximum load. The dome shape has developed further due to the increased punch displacements. It is noticed that the necking and thinning at the edge of contact progress in these specimens, where the thickness of the specimen has been further reduced, as shown in Fig. 3c due to the ongoing bending and stretching stresses under the bi-axial loading state. Meanwhile, void nucleation and elongation are found to exist in Fig. 3b. Due to the large plastic deformation in these punch displacements, the voids have been deformed into elongated shapes of which the long axis is pointing in the direction of the plastic flow. The phases also align with the plastic flow in Fig. 3b and 3d. Moreover, it is found from Fig. 3d that the austentic and martensitic phases have been elongated along the plastic flow direction due to the extensive plastic deformation at the edge of contact. A number of voids, approximately 3-5 μm across, are found in Fig. 3d. The morphology of the voids has been further stretched and elongated. Figure 3e and 3f depict the microstructure of the fracture point at the edge of contact. It can be seen that the phases have also been elongated due to the large plastic

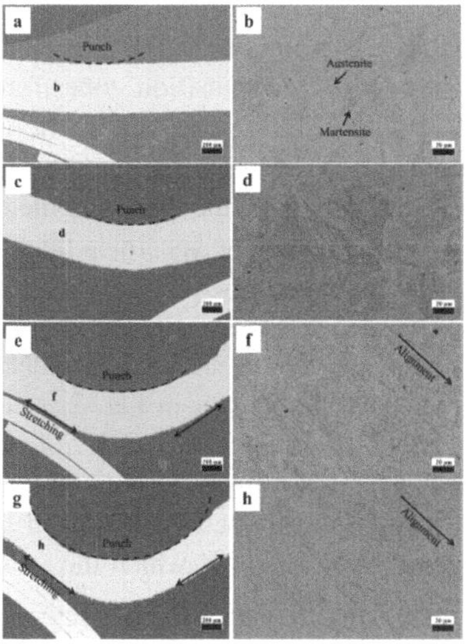

Fig. 2. The cross-sectional microstructure of the P91 specimen at interrupted displacements of 0.26 mm (a & b), 0.6 mm (c & d), 1.0 mm (e & f) and 1.25 mm (g & h).

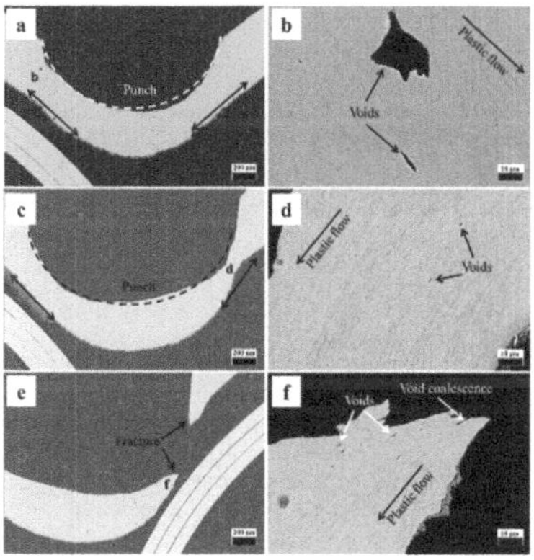

Fig. 3. The cross-sectional microstructure of the P91 specimen at interrupted displacements of 1.5 mm (a & b), 1.65 mm (c & d) and 1.74 mm (e & f).

deformation in Fig. 3f. Some discrete and ellipsoidal voids are found in this region, especially close to the fracture site. Noting that significant material necking and thinning have occurred during this stage, it is possible that the thinning of the specimen at the edge of contact allows the coalescence of voids to occur, resulting in the fracture of the specimen. Eventually, the tearing of the material occurs, producing the rough fracture surface.

4 Deformation and Facture Mechanisms

Figure 4 gives the schematic diagram of deformation of P91 in SPTT at 600 °C at different stages. The first stage, I, is related to the elastic deformation, which is not shown in Fig. 4. Stage II represents the transition zone from elastic to plastic deformation, in which at the beginning, the plastic deformation occurs in the local area outside of the contact zone of punch and specimens. And then it will expand to the materials in the gap between the punch and the die along the circle. Stage III represents the homogeneous plastic deformation and the specimen thickness will reduce relatively uniformly in the deformation zone (as shown Fig. 2a, c, e from punch displacement approximately 0.26 mm to 1.0 mm). In stage III, the energy consumed is more than 40% of the total energy. The thickness of specimen will reduce at local circle band near the punch contact area in stage IV (as shown in Fig. 2g), in which a circumferential necking will occur and the energy consumed rate is 20.2%, in which a circumferential necking will occur and the energy consumed rate is 20.2%. In stage V, the plastic deformation will focus on a point until crack occurring in the local circle band happened in stage IV (as shown in Fig. 3a, c), in which the voids will have a nucleation, growth, coalescence and creates a macroscale crack in the point necking area and the energy consumed 29.4% or so. Finally, the crack will extend and the material will be teared ductility until failure in state VI (as shown in Fig. 3e), in which the energy consumed is only 4.1% of the total energy due to the connect area reducing down to immediate failure.

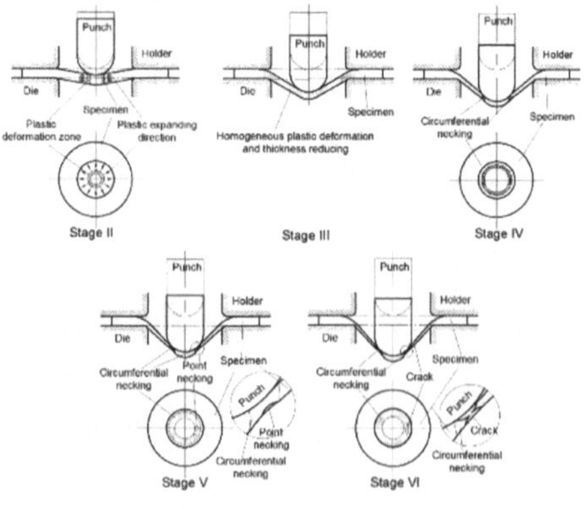

Fig. 4. The schematic diagram of deformation regimes of P91 in SPTT at different stages.

5 Conclusions

The interrupted small punch tensile tests of P91 were performed at 600 °C to investigate the deformation and fracture mechanisms of P91.

The dome shape of the SPTT specimen gradually develops when the punch displacements increase from 0.26 to 1.74 mm. It is found that the voids are likely to form after the maximum load is achieved, which indicates that the fracture of the P91 specimen is starting to occur. From the microstructure evolution, it is shown that significant necking and thinning occur after the maximum load.

References

1. Morris, A., Cacciapuoti, B., Sun, W.: The role of small specimen creep testing within a life assessment framework for high temperature power plant. Int. Mater. Rev. **63**, 102–137 (2018)
2. Hyde, T.H., Sun, W., Williams, J.A.: Requirements for and use of miniature test specimens to provide mechanical and creep properties of materials: a review. Int. Mater. Rev. **52**, 213–55 (2007)
3. Blagoeva, D., Hurst, R.: Application of the CEN (European Committee for Standardization) small punch creep testing code of practice to a representative repair welded P91 pipe. Mater. Sci. Eng., A **510**, 219–223 (2009)
4. Panayotou, N., Puigh, R., Opperman, E.: Miniature specimen tensile data for high energy neutron source experiments. J. Nucl. Mater. **104**, 1523–1526 (1981)
5. Madia, M., Foletti, S., Torsello, G., Cammi, A.: On the applicability of the small punch test to the characterization of the 1CrMoV aged steel: mechanical testing and numerical analysis. Eng. Fail. Anal. **34**, 189–203 (2013)
6. Lucas, G.: The development of small specimen mechanical test techniques. J. Nucl. Mater. **117**, 327–339 (1983)
7. Ha, J.S., Fleury, E.: Small punch tests on steels for steam power plant (I). KSME Int. J. **12**, 818 (1998)
8. Rodríguez, C., Fernández, M., Cabezas, J., García, T., Belzunce, F.: The use of the small punch test to solve practical engineering problems. Theoret. Appl. Fract. Mech. **86**, 109–116 (2016)
9. Rodríguez, C., García, J., Cárdenas, E., Belzunce, F., Betegón, C.: Mechanical properties characterization of heat-affected zone using the small punch test. Weld Jt. **88**, 188–192 (2009)
10. Fernández, M., Rodríguez, C., Belzunce, F., García, T.: Use of small punch test to estimate the mechanical properties of powder metallurgy products employed in the automotive industry. Powder Metall. **58**, 171–177 (2015)
11. Dymáček, P.: Recent developments in small punch testing: applications at elevated temperatures. Theoret. Appl. Fract. Mech. **86**, 25–33 (2016)
12. Abendroth, M.: FEM analysis of small punch tests. Key Eng. Matet. **734**, 23–36 (2017)

Damage Accumulation and Growth Models for the Creep-Fatigue Interaction

A. V. Tumanov(✉) and V. N. Shlyannikov

Institute of Power Engineering and Advanced Technologies, FRC Kazan Scientific Center, Russian Academy of Sciences, Kazan, Russia
tymanoff@gmail.com

Abstract. The residual life prediction based on damage accumulation models for the BT3-1 titanium alloy are present. Mechanical properties were obtained at temperature 370 °C. Life prediction was based on the Kachanov–Rabotnov damage model for creep and the Duyi-Zhenlin model for fatigue. The contribution of each type of loading to the generalized damage parameter was calculated according to the law proposed by Skelton and Grandy. Damage accumulation laws were integrated into ANSYS software. Cyclic loading with a trapezoidal form cycle was simulated. The integrated model of damage accumulation under creep-fatigue interaction allows obtaining a fatigue life of structural elements. The model is recommended to be used for cases when the scalar damage parameter is justified.

Keywords: Creep-fatigue interaction · Damage accumulation · Life prediction

1 Introduction

It is well known that the creep damage accumulation is different from damage caused by cyclic loading. At the moment, there are many models allow to take into account the damage accumulation, both during creep and cyclic loading independently.

The Hayhurst model [1] one of the most generalized models for the creep damage accumulation. In order to describe the damage accumulation according to this model, it is necessary to use six material constants. Moreover, the methods for determining these constants are mainly indirect. In this regard, the Rabotnov-Kachanov model one of the most popular due to the direct method of material constants.

One of the most perspective model for the fatigue damage accumulation prediction is the Lemaitre model [2]. Using the tensor damage parameter in similar phenomenological models leads to considerable difficulties for material constants determination. In the identical creep laws if the stress state is unchanged and scalar, vector, and tensor parameters give almost identical results. It is correct not only for a "pure" complex stress state, but for solving some boundary value problems In this regard, models with scalar damage parameters can be considered as a simplified alternative [3–6].

The main problem of the residual life prediction under the creep-fatigue interaction is the complex nature of their interaction [7–10]. In this direction the most common approach is the statistical approximation of the experimental results by any functions. The basic procedure for taking into account the creep-fatigue interaction isn't

E. Gdoutos and M. Konsta-Gdoutos (Eds.): ICTAEM 2020, STIN 16, pp. 112–116, 2020.
https://doi.org/10.1007/978-3-030-47883-4_20

introduced in any finite element software. Simple summation is used at most cases. In this regard, there is a need to develop damage accumulation models taking into account a complex nature of creep-fatigue interaction.

2 Damage Accumulation Model

According to [7] it is recommended to use the following expression to calculate the total level of accumulated damage under creep-fatigue interaction:

$$\Delta\omega = \frac{\Delta\omega_{cr}}{1 - \Delta\omega_f} + \frac{\Delta\omega_f}{1 - \Delta\omega_{cr}}, \tag{1}$$

where $\Delta\omega_{cr}$ - creep damage increment, $\Delta\omega_f$ - fatigue damage increment. At the beginning of each iteration, the level of accumulated damage from fatigue and creep is determined independently. General integration is carried thru taking into account the both contribution.

The creep damage is proposed to be established by the classical Kachanov-Rabotnov power law. According to this model, the strain rate, taking into account the accumulation of damage, is written as follows:

$$\dot{\varepsilon}_e = B\left(\frac{\sigma_e}{1 - \omega_{cr}}\right)^n. \tag{2}$$

where the creep damage accumulation rate is:

$$\frac{d\omega_{cr}}{dt} = D\left(\frac{\sigma_e}{1 - \omega_{cr}}\right)^m. \tag{3}$$

B, n, D, m – material properties.

The creep damage increment can be obtained by integrating the expression (3):

$$\Delta\omega_{cr} = 1 - [(-D\sigma^m\Delta t - C_3)(m+1)]^{\frac{1}{m+1}}. \tag{4}$$

where C_3 - integration constant which can be determined from the previous iteration.

$$C_3 = -\frac{(1 - \omega_0)^{m+1}}{m+1} - D\sigma_e^m t_0. \tag{5}$$

It is well known that the creep damage accumulation is different from damage caused by fatigue. Usually, the one of the main characteristic in models with a scalar damage parameter is the number of cycles before fracture N_f at the stress amplitude σ_a. In the simplest case of uniaxial tension-compression, the relationship between them is follow:

$$2N_f = \left(\frac{\sigma_a}{\sigma'_f}\right)^{\frac{1}{b}}, \tag{6}$$

where σ'_f and b are fatigue constants.

The damage parameter at constant stress amplitude is determined from [3]

$$\omega_f = -\frac{\omega_{f_{N-1}}}{\ln N_f}\ln\left[1 - \frac{N}{N_f}\right]. \tag{7}$$

where $\omega_{f_{N-1}}$ - the critical value of the damage parameter in the penultimate cycle and it is associated with static toughness U_{T0} by the following relationship:

$$\omega_{f_{N-1}} = 1 - \frac{\sigma_a^2}{2EU_{T0}}. \tag{9}$$

where E - Young's modulus.

3 Mechanical Properties and Finite Element Modeling

Mechanical properties were obtained during static, cyclic and creep tests (Table 1).

Table 1. Mechanical properties of BT3-1 for 370 °C

E, MPa	ν	B, $(MPa^n \cdot h)^{-1}$	n	σ_{02}, MPa	C, $(MPa^n \cdot h)^{-1}$	M
97553	0.3	1e−15	1.934	699	1e−10	3

In order to predict the residual life at creep-fatigue interaction, the above models were integrated into the ANSYS finite element method software as user programmable features. The finite element model of the specimen geometry recommended by ASTM E466 was used. The three-dimensional model of the specimen consists of two hundred thousand of second order finite elements. In numerical simulation, the loading was applied in the form of a trapezoidal loading cycle with different holding time (6, 60, 600 and 3600 s).

4 Results and Discussion

The stress-strain fields in cylindrical specimen were obtained by finite element method with taking into account the accumulated damage on each loading cycle. It is assumed that the failure is occur if the value of the generalized damage parameter exceeded 0.9 at any point in specimen volume.

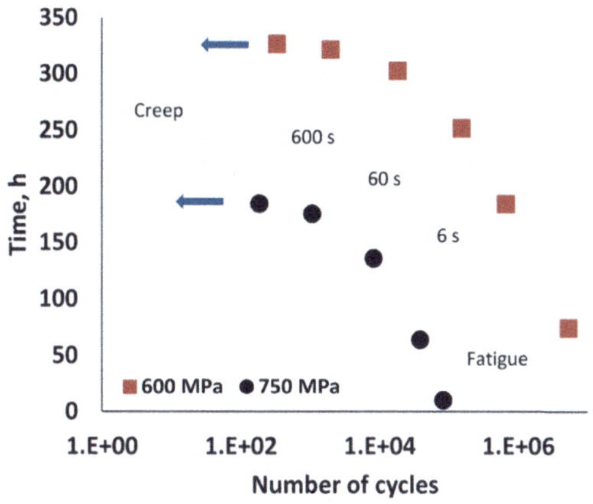

Fig. 1. Relationship between the fatigue life and the time to failure

Figure 1 shows the numerical results of the fatigue life prediction at various combinations of applied load levels and holding time within each loading cycle. It is shown that the integrated damage accumulation model allows to cover the wide spectrum of load parameters for creep-fatigue interaction.

The relative damage accumulation rate in each loading cycle can be considered as a parameter characterizing the contribution of load type to the total damage. The dependences of relative damage rate on the holding time for the penultimate loading cycle are shown in Fig. 2. It is clear that the increasing load level leads to situation when fatigue makes a greater contribution to the overall damage of the material. Accordingly, the nominal stresses decreasing leads to increasing of creep contribution to the damage.

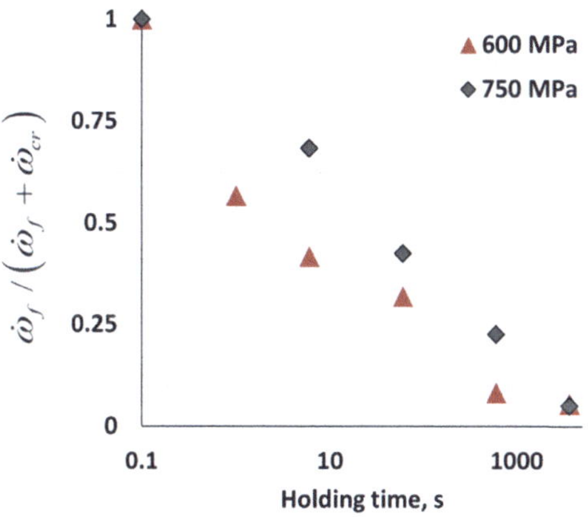

Fig. 2. Relationship between the holding time and the relative damage accumulation rate

5 Conclusions

The model allows to take into account the nonlinear contribution of damage accumulation caused by creep-fatigue interaction is integrated into the ANSYS software as user programmable features. Integrated model of damage accumulation allow, to predict the macro damage initiation during the creep-fatigue interaction.

References

1. Hayhurst, D.R.: Creep rupture under multi-axial states of stress. J. Mech. Phys. Solids **20**, 381–390 (1972). https://doi.org/10.1007/978-3-642-84455-3_43
2. Lemaitre, J.: Coupled elasto-plasticity and damage constitutive equations. Comput. Methods Appl. Mech. Eng. **51**, 31–49 (1985). https://doi.org/10.1016/0045-7825(85)90026-X
3. Ye, D.Y., Wang, Z.L.: A new approach to low-cycle fatigue damage based on exhaustion of static toughness and dissipation of cyclic plastic strain energy during fatigue. Int. J. Fatigue **23**, 679–687 (2001). https://doi.org/10.1177/1056789512456030
4. Lv, Z., Huang, H.-Z., Zhu, S.-P., Gao, H., Zuo, F.: A modified nonlinear fatigue damage accumulation model. Int. J. Damage Mech **24**, 168–181 (2015). https://doi.org/10.1177/1056789514524075
5. Zhang, J., Xi, F., Lin, J., Liu, Z., Liu, N., Bin, W.: Study on damage accumulation and life prediction with loads below fatigue limit based on a modified nonlinear model. Materials **11**, 1–16 (2018). https://doi.org/10.3390/ma11112298
6. Huang, T., Ding, R.-C., Li, Y.-F., Zhou, J., Huang, H.-Z.: A modified model for nonlinear fatigue damage accumulation of turbine disc considering the load interaction effect. Metals **9** (919), 1–10 (2019). https://doi.org/10.3390/met9090919
7. Skelton, R.P., Gandy, D.: Creep – fatigue damage accumulation and interaction diagram based on metallographic interpretation of mechanisms. Mater. High Temp. **25**, 27–54 (2008). https://doi.org/10.3184/096034007X300494
8. Esztergar, E.P.: Creep-fatigue interaction and cumulative damage evaluations for type 304 stainless steel. OAK Ridge National Laboratory, Oak Ridge, Tennessee, p. 131 (1972)
9. Barbera, D., Chen, H., Liu, Y.: On creep fatigue interaction of components at elevated temperature. J. Press. Vessel Technol. Trans. ASME **138**, 28 (2015). https://doi.org/10.1115/1.4032278
10. Wright, J.K., Carroll, L.J., Sham, T.-L., Lybeck, N.J., Wright, R.N.: Determination of the creep–fatigue interaction diagram for alloy 617. In: Proceedings of the ASME 2016 Pressure Vessels and Piping Conference. PVP2016-63704, pp. 1-10 (2016). https://doi.org/10.1115/PVP2016-63704

Crack Growth and Plastic, Fracture Process and Damage Zones Behavior Under Mixed Mode Loading

Valery Shlyannikov(✉) and Daria Fedotova

FRC Kazan Scientific Center of Russian Academy of Sciences, Kazan 420111, Russia
shlyannikov@mail.ru

Abstract. Fatigue crack paths for inclined cracks are studied through experiments and computations under different mixed-mode loading. The experimental study of fatigue crack growth in the aluminum and titanium alloys and two types of the steels is performed on compact tension shear specimens. The cyclic fracture process and damage zone concepts are introduced and used for numerical analysis of the crack tip fields. The elastic K_1, K_2 and plastic K_P stress intensity factors are calculated by finite element method to characterize the fracture resistance along the curvilinear trajectories in compact tension shear specimens. The influence of mode-mixity and elastic-plastic material properties on the behavior of three crack tip regions as a function of cyclic loading is demonstrated.

Keywords: Crack path · Fracture process and damage zones · Mixed mode

1 Introduction

To date, several of models have been presented in the literature for the purpose of analyzing the process of fatigue crack growth from the viewpoints of micromechanics, fracture and damage mechanics. Many fracture mechanics models are based on a critical distance local to the crack tip. It is considered a fundamental characteristic parameter that distinguishes damage at the microscopic and macroscopic scale levels. This characteristic distance is often identified with a fracture process zone or fracture damage zone. A general assumption regarding the distance criterion under elastic–plastic cyclic loading conditions is that a crack increment occurs when the fracture resistance parameter (stress, strain, or energy) reaches a critical value at a characteristic distance from the crack tip. It is well known that mixed-mode conditions appear when the direction of the applied loading does not coincide with the orthogonal K_I-K_{II}-K_{III} space. The main feature of the mixed-mode fracture is that the crack growth no longer takes place in a self-similar manner and does not follow a universal trajectory. Various fracture criteria and phenomenological approaches can predict the behavior of cracks in

E. Gdoutos and M. Konsta-Gdoutos (Eds.): ICTAEM 2020, STIN 16, pp. 117–123, 2020.
https://doi.org/10.1007/978-3-030-47883-4_21

brittle and ductile materials loaded in combined modes. Even though many studies have been carried out in the literature to determine fracture characteristics of ductile materials under cyclic loading, it is important to investigate the local crack tip zones behavior due to low cycle fatigue for quasi-brittle and moderate ductile materials under mixed mode conditions.

2 Material Properties and Loading Conditions

Compact tension shear specimens (CTS, Fig. 1a) are made from P2M and 34X steels as well as aluminum Al7050 and titanium Ti-6Al-4 V alloys and used for cyclic mixed-mode fracture test with the loading direction variation from 90° (pure Mode I) to 0° (pure Mode II) with respect to the initial crack plane. Main static and low-cyclic mechanical properties of the tested materials are listed in Tables 1 and 2. The mixed-mode parameters expressed through the elastic and plastic stress intensity factors (SIFs) were varied by changing the load direction. Both Mode I and Mode II elastic SIFs K_1 and K_2 as well as general plastic SIF K_P are evaluated along experimental crack path of each tested material based on the series of numerical calculations. On the CTS was realized the full range of mixed-mode cyclic fracture from tensile to shear loading. Figure 1b represents experimental fatigue crack growth trajectories with initial pure Mode II loading for tested steels, aluminum and titanium alloys.

Table 1. The static properties of materials.

Material	E (GPa)	σ_0 (MPa)	σ_b (MPa)	σ_u (MPa)	N	α
P2M steel	226.9	362.4	636.0	1190.0	4.13	4.141
34X steel	216.21	714.4	1040.0	1260.4	7.89	0.529
Al-7050	70.57	471.6	524.4	701.0	10.85	1.570
Ti-6Al-4 V	118.0	118.0	963.8	1289.6	12.59	1.225

Table 2. The cyclic properties of materials.

Material	σ'_f	B	ε'_f	c	K′	n′
P2M steel	1785.0	−0.087	0.424	−0.58	1963.5	0.15
34X steel	1890.6	−0.087	0.381	−0.58	2079.6	0.15
Al-7050	1170.7	−0.095	0.35	−0.69	1128.6	0.11
Ti-6Al-4V	2153.6	−0.095	0.35	−0.69	2076.2	0.11

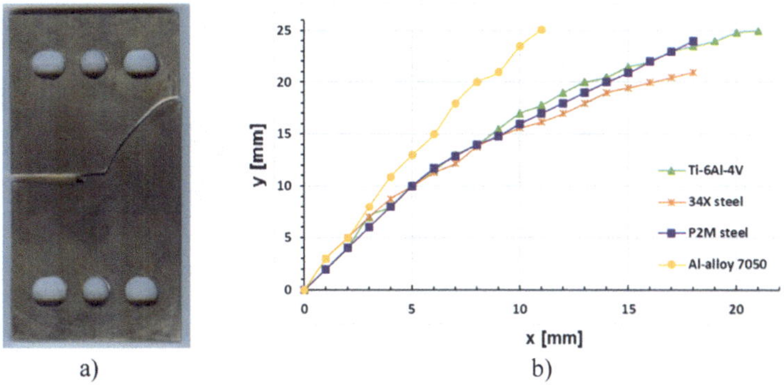

Fig. 1. Experimental crack paths in CTS specimens at pure mode II.

Crack growth in the CTS specimen with the applied load angle β = 0° has its own characteristics. Initially, the deformation mode corresponds to the pure shear with $M_E = M_P = 0$. With an increase in the number of loading cycles *N*, the contribution of mixed modes of deformation and fracture gradually increases. As follows from Fig. 1, the crack path in an aluminum alloy Al7050 differs significantly from the experimental trajectories for other materials under the same loading conditions.

3 Plastic, Cyclic Fracture Process and Damage Zones Behavior

In front of a stable growing crack, we may distinguish three regions. The first outer contour represents the reversed plastic zone where the current values of stresses and deformations exceed the yield stress or strain of the material. The second region is fatigue fracture process zone. The third small area exists immediately ahead of the crack tip which is called the damage zone. For a strain-hardening material the elastic-plastic boundary around the crack tip can be obtained using the equivalent von Mises stress as a function of principal stress ratio $\lambda = \sigma_2/\sigma_1$

$$\sigma_e = \sigma_1 \eta_i \tag{1}$$

where for plane stress $\eta_i = \sqrt{1 - \lambda + \lambda^2}$ and plane strain $\eta_i = \sqrt{1 - \lambda + \lambda^2 + \nu(\nu - 1)(1 + \lambda)^2}$.

Full-field 3D FE-analyses are carried out to determine stress-strain fields, elastic and plastic SIFs along the curvilinear crack path in the CTS subjected to different mixed mode loadings. Figure 2 shows that, for pure mode I (Fig. 2a) and pure mode II (Fig. 2b) the plastic zones have a symmetrical shape while for mixed mode (Fig. 2c) they is non-symmetrical with respect to the initial crack plane. The dimensions of the plastic zones in Fig. 2 for titanium alloy are satisfied to the conditions of small-scale yielding. Similar shapes of plasticity zones are observed for other tested materials.

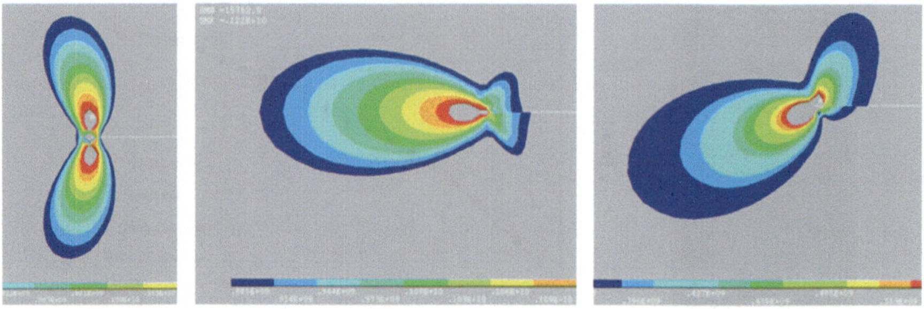

a) b) c)

Fig. 2. Plastic zone under different mode mixity in CTS specimen for Ti6Al4 V.

The variable critical distance is one of important parameter in modern fracture mechanics, denoted to as the fracture process zone (FPZ) size. A general assumption regarding the distance criterion under elastic–plastic cyclic loading is that a crack increment occurs when the fracture resistance parameter (stress, strain, or energy) reaches a critical value at a characteristic distance from the crack tip. The following expression was proposed by Shlyannikov [1] for normalized fracture process zone size based on the nonlinear crack-tip stress field and the main low cycle fatigue material properties

$$\bar{r}_f = \left(\frac{\sigma_n}{\sigma_0}\right)^2 (\bar{S}_1 + S_P) / \left[4 \frac{\sigma_f' \varepsilon_f' E}{\sigma_0^2} (2N_f)^{(b+c)}\right] \tag{2}$$

$$\bar{S}_1 = \frac{(1+v)}{2}\left(a_{11}Y_1^2 + a_{12}Y_1Y_2 + a_{22}Y_2^2\right); \ \bar{S}_P = \frac{\alpha n'}{(n'+1)} \frac{\pi \tilde{\sigma}_e^{n'+1}}{I_n} \left(\frac{K_{eqv}}{\sigma_n \sqrt{\pi a}}\right)^2 \tag{3}$$

where n' is low cycle fatigue hardening exponent, ε_f' and σ_f' are the fatigue ductility and strength coefficients, respectively, a is the crack length, σ_u is the ultimate true tensile strength, σ_0 is the yield stress, σ_n is the nominal stress, $\tilde{\sigma}_e(\theta)$ is dimensionless equivalent stress angular function, I_n is the governing parameter of the nonlinear crack tip field [1], N_f is the number of cycles to failure, Y_1 and Y_2 are geometry dependent for SIF correction factors. The equivalent elastic stress intensity factor (SIF) for mixed mode fracture can be determined by using energy release rate criterion [2]

$$K_{eqv} = \cos\frac{\theta^*}{2}\sqrt{K_1^2(1+\cos\theta^*) - 4K_1K_2\sin\theta^* + K_2^2(5 - 3\cos\theta^*)} \tag{4}$$

$$K_1 = \sigma_n\sqrt{\pi a}\cdot Y_1(a/w); \ K_2 = \sigma_n\sqrt{\pi a}\cdot Y_2(a/w) \tag{5}$$

where, θ^* is the crack deviation angle, w is the cracked body width. The most general criterion of mixed mode crack growth direction is obtained [3] on the basis of the limiting state theory under multiaxial loading.

During low-cycle fatigue, the accumulation and growth of damage occurs in the local fracture process zone that covers the crack tip. The size of the fracture process zone is typically regarded as a distance-dependent parameter that physically represents the increase in crack size during the basic cyclic fracture process. Figure 3 shows the dependence of the both size and shape of the fracture process zone, according to Eqs. (2–5), on the accumulated number of loading cycles for the pure mode (Fig. 3a), pure mode II (Fig. 3b) and mixed mode (Fig. 3c, d) loading. It can be clearly observed that the FPZ contour behavior is not constant-size and were dependent on the accumulated loading cycles, mode mixity and both the static and low-cycle fatigue material properties. Figure 3c presents the process of changing the FPZ in the 7050 aluminum alloy according to the stages of cyclic loading, which are accompanied by corresponding changes in the mixed modes of fracture. Figure 3d give a comparison of the dimensions of the mixed-mode fatigue fracture process zone in the CTS specimen for all tested materials having different crack paths at final the number of cycles of loading, $N/N_f = 1$, as shown in Fig. 1b. Initially, owing to the same pure mode II plastic deformation in the crack tip region in several first cycles, the size of the low-cycle fatigue fracture process zone due to mode mixities is changed in different degree as a function of the tested material main mechanical properties.

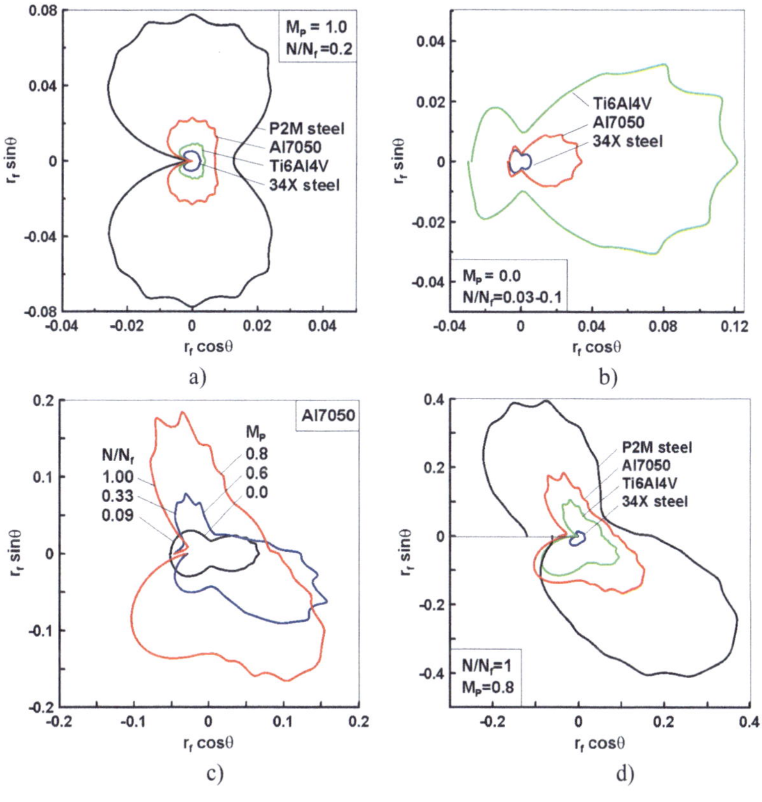

Fig. 3. Fracture process zone contour under (a) pure mode I, (b) pure mode II and (c, d) mixed modes cyclic fracture.

A fatigue damage evolution equation was proposed by the authors [4] through connecting the damage variable with the static toughness exhausting during cyclic loading. This equation has the following form

$$\omega_N = \frac{\omega_{N-1}}{\ln N_f} \ln\left(1 - \frac{N}{N_f}\right); \tag{6}$$

where ω_{N-1} denotes the critical value of the damage variable, that is expressed by

$$\omega_{N-1} = 1 - \frac{\Delta\sigma^2}{2E\sigma_u} \tag{7}$$

An amplitude of cyclic stress at the crack tip can be determined by applying the singular HRR-type stress field

$$\Delta\sigma = \bar{K}_P \bar{r}_c^{\frac{-1}{n'+1}} \tilde{\sigma}(\theta) \tag{8}$$

where K_P is the plastic stress intensity factor and r_c is the critical distance. The damage variable ω_N indicate in Eq. (6) the measure of fatigue damage with $\omega_N = 0$ denoting the undamaged state and $\omega_n = 1$ the fully damaged state.

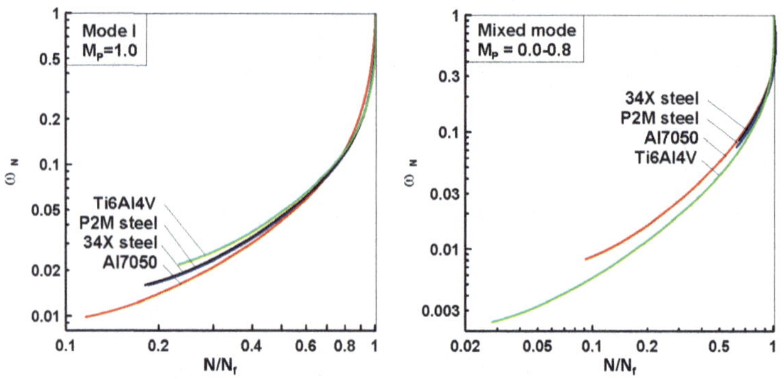

a) b)

Fig. 4. Damage distributions as a function of accumulated number of loading cycles (a-pure mode I, b-mixed mode).

Figure 4 presents a comparison of the predicted values of the fatigue fracture damage evolution for pure mode I (Fig. 4a) and mixed modes (Fig. 4b) according to Eqs. (6–8) for all considered structural metallic materials. It may be seen that the mode mixity changed significantly the processes of fatigue damage accumulation only in the range $N/N_f < 0.5$.

In this work, an expression for the low cycle fatigue fracture process is derived; this includes static and cyclic material properties, as well as mixed mode fracture parameters. The elastic–plastic crack tip singularity fields are employed. The main finding, based on the numerical and experimental results, is that the effect of mixed mode fracture during low-cycle fatigue crack growth in considered steels, aluminum and titanium alloys leads to the conclusion that under pure mode I, pure mode II and mixed mode loading conditions and shape and size of the plastic, fracture process and damage zones do not coincide with each other and have a hierarchical structure.

References

1. Shlyannikov, V.N.: Creep–fatigue crack growth rate prediction based on fracture damage zone models. Eng. Fract. Mech. **214**, 449–463 (2019)
2. Citarella, R., Cricri, G.: Comparison of DBEM and FEM crack path predictions in a notched shaft under torsion. Eng. Fract. Mech. **77**, 1730–1749 (2010)
3. Shlyannikov, V.N.: Mixed-mode static and fatigue crack growth in central notched and compact tension shear specimens. ASTM STP **1359**, 279–294 (1999)
4. Duyi, Y., Zhenlin, W.: A new approach to low-cycle fatigue damage based on exhaustion of static toughness and dissipation of cyclic plastic strain energy during fatigue. Int. J. Fatigue **23**, 679–687 (2001)

Miscellaneous (Computational Mechanics, Dynamics, Nanomaterials, Plasticity, Wave Propagation)

Parametric Resonance of a Micro-beam with Flexible Support Under Fringing Field Excitation

Zhichong Wang(✉)

Hebei University of Engineering, Handan, China
wangzhichong@tju.edu.cn

Abstract. A new resonance structure with flexible support under electrostatic fringing field excitation is designed. This structure can be used in sensor. In this paper, the nonlinear frequency responses of the movable beam in resonance structure are analyzed by the Method of Multiple Scales. The effects of electrode thickness, slit gap and initial displacement on the vibration of the movable beam are investigated in detail. The key results are as follows. A smaller electrode thickness and slit gap can lead to a larger change of the actuated frequency, and they can also lead to a larger vibration amplitude. The initial displacement has an influence on the frequency response, but this influence is not obvious, since the electrostatic force is small when the initial displacement is not large. This resonance structure can avoid superposition of input signal and output signal. Thus, the signal-to-noise ratio and the accuracy of the sensor can be improved.

Keywords: Electrostatic MEMS · Parametric resonance · Fringing field excitation · Flexible support · Nonlinear response

1 Introduction

The sensing mechanisms or actuation mechanisms for Micro Electro Mechanical Systems (MEMS) mainly include electro-thermal mechanism [1, 2], piezoelectric mechanism [3, 4] and electrostatic mechanism [5–10]. Among them, the last one is the most popular because of its fast speed, low power consumption, ease of fabrication and small size. In electrostatic MEMS, the gap between movable structure and stationary electrodes can be changed, once the electric field is applied on the surface of the movable structure. The electrical stiffness acts in conjunction with the mechanical stiffness. By modulating the applied voltage, the overall stiffness of the structure can be controlled. The electric force is non-linearly related to the gap distance between movable structure and stationary electrodes. Thus, it can drive the system into parametric resonance if the voltage is large enough to overcome non-conservative forces and the driving frequency is inside an instability tongue (usually at twice the natural frequency). This parametric excitation system is excited by changing the parameters of the system instead of applying a direct force in the direction of desired motion [11].

E. Gdoutos and M. Konsta-Gdoutos (Eds.): ICTAEM 2020, STIN 16, pp. 127–132, 2020.
https://doi.org/10.1007/978-3-030-47883-4_22

More attractive than other nonlinear system, parametric excitation system, in which resonance occurs at twice the natural frequency, can avoid superposition of input signal and output signal.

In this paper, a new resonance structure with flexible support is designed; an analytical investigation of the parametric vibration actuated by fringing electrostatic fields is provided; the nonlinear behaviors are analyzed.

2 Problem Formulation

A new resonance structure with flexible support is designed. This structure includes movable beam, support beam and electrode, as shown in Fig. 1. In Fig. 1, l_b w_b and t_b denote the length, width and thickness of the movable beam respectively; l_{bO} w_{bO} and t_{bO} denote the length, width and thickness of the support beam respectively; w_s and t_s denote the width and thickness of the stationary electrode respectively; d_g is the slit gap between the movable beam and stationary electrode in the width direction; d_{gO} is the slit gap between the support beam and stationary electrode in the width direction; d is the initial displacement of the movable beam measured from the stationary electrode in the thickness direction. In order to study the vibration behaviors of the beam, the geometric parameters of the structure are taken as l_b = 5 mm, w_b = 0.4 mm, t_b = 0.01 mm, l_{bO} = 1 mm, w_{bO} = 0.4 mm, t_{bO} = 0.01 mm, w_s = 0.5 mm, t_s = 0.3 mm, d_g = 0.1 mm and d_{gO} = 0.5 mm. The range of the initial displacement is less than half of the electrode thickness in this structure.

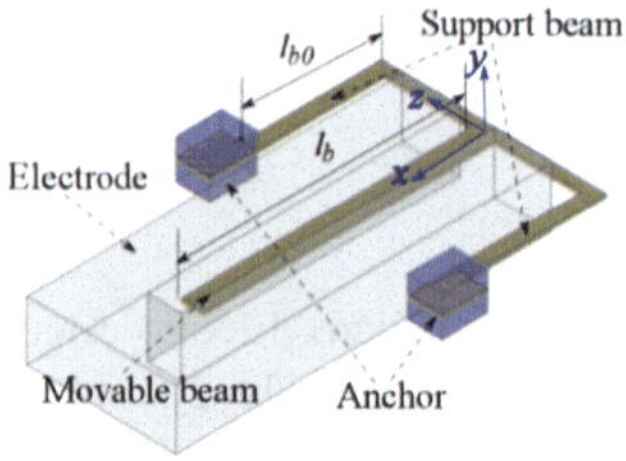

(a)

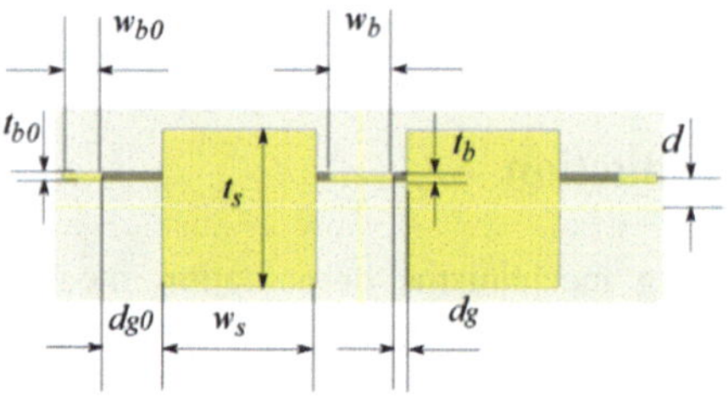

(b)

Fig. 1. Schematic illustration of resonance structure (a) 3D view (b) section view

Make $u(x, t)$ denote the deflection of the movable beam. According to the elastic beam theory, the governing equation and the boundary conditions of the beam are

$$EI\frac{\partial^4 u(x,t)}{\partial x^4} + \rho w_b t_b \frac{\partial^2 u(x,t)}{\partial t^2} = q_a + q_e \tag{1}$$

$$EI \cdot \frac{\partial^2 u(0,t)}{\partial x^2} = k_r \cdot \frac{\partial u(0,t)}{\partial x},\ EI \cdot \frac{\partial^3 u(0,t)}{\partial x^3} = -k_t \cdot u(0,t),\ \frac{\partial^2 u(l_b,t)}{\partial x^2} = 0,\ \frac{\partial^3 u(l_b,t)}{\partial x^3} = 0. \tag{2}$$

In this paper, the coefficients in the above equations are as following. The Young's modulus is E = 108GPa. The inertial moment of the movable beam cross section is $I = w_b \times t_b^3/12$. The mass density of the beam is ρ = 8500 kg/m^3. q_a represents the aerodynamic force per unit length, which is $q_a = -0.5 \times \rho_a \times w_b \times c_a \times (\partial u/\partial t) \times |\partial u/\partial t|$, and the drag coefficient is set as $c_a = 5$, ρ_a is the density of the air. q_e represents the excitation force per unit length. The electrostatic force is generated on the basis of the fringing fields. The distributed electrostatic force is $q_e = f_e \times v_e^2$. f_e is electrostatic force on the movable beam per unit length per square voltage, which is set as $f_e = r_1 \times d \times r_3 \times d^{3,}$ and r_1, r_3 are fitting parameters. Thus, the relationship between electrostatic force and deflection is $f_e = e_p \times (1 + e_{p1} \times u + e_{p2} \times u^2 + e_{p3} \times u^3)$, in which $e_p = r_1 \times d + r_3 \times d^3$, $e_{p1} = (r_1 + 3 \times r_3 \times d^2)/e_p$, $e_{p2} = 3 \times r_3 \times d/e_p$, $e_{p3} = r_3/e_p$. v_e is a combined DC/AC voltage actuating the micro-beam, i.e., $v_e(t) = v_{DC} + v_{AC} \times \cos(\omega_e \times t)$. In the boundary conditions, $k_r = 2 \times E \times I_0/l_{b0}$ and $k_t = -2 \times 3 \times E \times I_0/l_{b0}^3$ are the rotational and translational stiffness of the flexible boundary respectively, in which $I_0 = w_{b0} \times t_{b0}^3/12$ is the inertial moment of the support beam cross section.

The Galerkin discretization method is used, and the steady-state solution of the non-dimensional governing equation is written by $U(X, T) = \Phi(X) \times \Theta(T)$, and the mode function $\Phi(X)$ for movable beam is $\Phi(X) = \xi_1 \times \cos(\lambda_r \times X) + \xi_2 \times \sin(\lambda_r \times X) + \xi_3 \times \mathrm{ch}(\lambda_r \times X) + \xi_4 \times \mathrm{sh}(\lambda_r \times X)$, Substituting this mode function into the non-dimensional boundary conditions yields

$$\begin{bmatrix} -\lambda_r & -K_R & \lambda_r & -K_R \\ K_T & -\lambda_r^3 & K_T & \lambda_r^3 \\ -\cos\lambda_r & -\sin\lambda_r & \mathrm{ch}\lambda_r & \mathrm{sh}\lambda_r \\ \sin\lambda_r & -\cos\lambda_r & \mathrm{sh}\lambda_r & \mathrm{ch}\lambda_r \end{bmatrix} \begin{bmatrix} \xi_1 \\ \xi_2 \\ \xi_3 \\ \xi_4 \end{bmatrix} = 0 \tag{3}$$

For a nontrivial solution, the determinant of the matrix on the left side must be zero, thus the value λ_r is given. Then, the equation of the boundary condition is solved to find ξ_1, ξ_2, ξ_3 and ξ_4. Thus, the mode function can be obtained.

Multiplying the non-dimensional governing equation by the mode function, and integrating the resultant equation from $X = 0$ to 1. The displacement splits into a static displacement and a dynamic displacement, i.e. $\Theta(T) = \Theta_{0p} + \vartheta_p(T)$, and we obtain static equation and dynamic equation.

$$S_0 + S_1\Theta_{0p} + S_2\Theta_{0p}^2 + S_3\Theta_{0p}^3 = 0 \tag{4}$$

$$\begin{aligned} &\vartheta_p'' + K_1\vartheta_p + K_2\vartheta_p^2 + K_3\vartheta_p^3 + \alpha_a\vartheta_p'\left|\vartheta_p'\right| = \\ &\begin{bmatrix} (F_{E1}\cos W_eT + F_{E2}\cos 2W_eT) \\ + \left(K_{E1P1}\vartheta_p + K_{E1P2}\vartheta_p^2 + K_{E1P3}\vartheta_p^3\right)\cos W_eT \\ + \left(K_{E2P1}\vartheta_p + K_{E2P2}\vartheta_p^2 + K_{E2P3}\vartheta_p^3\right)\cos 2W_eT \end{bmatrix} \end{aligned} \tag{5}$$

The Method of Multiple Scales is used to investigate the response of the resonator with small vibration amplitude. We obtain steady-state response as following

$$\left(\frac{W_e^2\alpha_a}{\pi}\right)^2\alpha_0^2\,KEP_1^2+\left(\frac{3K_3}{2\varepsilon}\alpha_0^2-\frac{W_e^2}{2\varepsilon^2}+\frac{2K_1}{\varepsilon^2}-\frac{15K_3^2}{16W_e^2}\alpha_0^4\right)^2KEP_2^2=KEP_2^2\,KEP_1^2 \quad (6)$$

$$\tan 2\beta_0=\left[\left(\frac{W_e^2\alpha_a}{\pi}\right)\alpha_0 KEP_1\right]\Big/\left[\left(\frac{3K_3}{2\varepsilon}\alpha_0^2-\frac{W_e^2}{2\varepsilon^2}+\frac{2K_1}{\varepsilon^2}-\frac{15K_3^2}{16W_e^2}\alpha_0^4\right)KEP_2\right] \quad (7)$$

3 Dynamic Analysis

Based on the steady-state response equation, the thickness of electrode is set as t_s = 0.20 mm and t_s = 0.30 mm, the slit gap between the movable beam and stationary electrode in the width direction is set as d_g = 0.06 mm, d_g = 0.08 mm and d_g = 0.10 mm, the frequency response curves are shown in Fig. 2 to analyze the effect of electrode thickness and slit gap. As shown in Fig. 2, a smaller electrode thickness and slit gap can lead to a larger change of the actuated frequency, and they can also lead to a larger vibration amplitude. What's more, as shown in Fig. 2, the resonance occurs at twice the natural frequency, which can avoid superposition of input signal and output signal; the range of the actuated frequency is related small, which can serve and improve the signal-to-noise ratio. These advantages can improve the accuracy of the sensor.

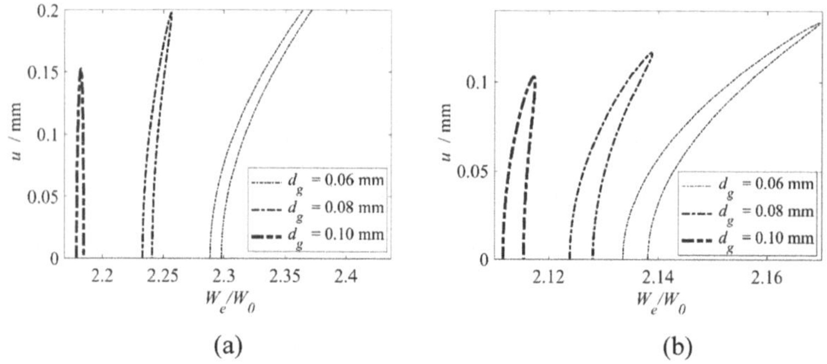

Fig. 2. Frequency response curves showing the effect of electrode thickness and slit gap (a) t_s = 0.20 mm (b) t_s = 0.30 mm

The system parameters are set as t_s = 0.30 mm and d_g = 0.10 mm. The dynamic equation is solved by using the MMS and the Fourth - Order Runge - Kutta Method (RK4). Figure 3 shows the effects of the initial displacement on the frequency response when l_{b0} = 0.01 mm. As shown in the figure, the results of the MMS and the RK4 are in a good agreement when the amplitude is small; the error between the results of the

MMS and the RK4 increases, when the vibration amplitude increases. The initial displacement has an influence on the frequency response, but this influence is not obvious, since the electrostatic force is small when the initial displacement is not large.

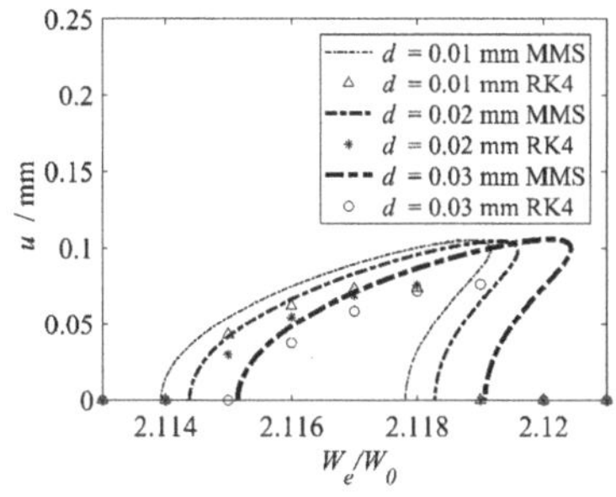

Fig. 3. Frequency response curves of the MMS and the RK4 showing the effect of initial displacement

4 Conclusions

This paper investigates a new resonance structure with flexible support. This resonator is a parametric excitation system, which resonance occurs at twice the natural frequency. The nonlinear behaviors of the movable beam are analyzed by using the MMS to get the frequency response equation. The effects of electrode thickness, slit gap and initial displacement on the vibration of the movable beam are focused. A smaller electrode thickness and slit gap can lead to a larger change of the actuated frequency, and they can also lead to a larger vibration amplitude. The initial displacement has an influence on the frequency response, but this influence is not obvious, since the electrostatic force is small when the initial displacement is not large. This resonance structure can avoid superposition of input signal and output signal. Thus, the signal-to-noise ratio and the accuracy of the sensor can be improved.

References

1. Shi, H., Fan, S., Zhang, Y., Sun, J.: Nonlinear dynamics study based on uncertainty analysis in electro-thermal excited MEMS resonant sensor. Sens. Actuators A **232**, 103–114 (2015)
2. Ramini, A.H., Hajjaj, A.Z., Younis, M.I.: Tunable resonators for nonlinear modal interactions. Sci. Rep. **6**, 9 (2016)
3. Ahmed, S., Ounaies, Z., Arrojado, E.A.F.: Electric field-induced bending and folding of polymer sheets. Sens. Actuators A **260**, 68–80 (2017)
4. Wang, C., Zhang, Q., Wang, W., Feng, J.: A low-frequency, wideband quad-stable energy harvester using combined nonlinearity and frequency up-conversion by cantilever-surface contact. Mech. Syst. Signal Process. **112**, 305–318 (2018)
5. Ilyas, S., Arevalo, A., Bayes, E., Foulds, I.G., Younis, M.I.: Torsion based universal MEMS logic device. Sens. Actuators A **236**, 150–158 (2015)

6. Alqasimi, J.E., Ouakad, H.M.: Vibrational response of initially deformed bistable microbeams under the combined effect of mechanical shock loads and electrostatic forces. J. Vib. Acoust. **140** (2017). 021013-021013-021012
7. Wu, G.Q., Chua, G.L., Gu, Y.D.: A dual-mass fully decoupled MEMS gyroscope with wide bandwidth and high linearity. Sens. Actuators A **259**, 50–56 (2017)
8. Feng, J., Liu, C., Zhang, W., Hao, S.: Static and dynamic mechanical behaviors of electrostatic MEMS resonator with surface processing error. Micromachines **9**, 34 (2018)
9. Han, J., Li, L., Jin, G., Feng, J., Li, B., Jia, H., Ma, W.: Vibration identification of folded-MEMS comb drive resonators. Micromachines **9**, 381 (2018)
10. Uhlig, S., Gaudet, M., Langa, S., Schimmanz, K., Conrad, H., Kaiser, B., Schenk, H.: Electrostatically driven in-plane silicon micropump for modular configuration. Micromachines **9**, 190 (2018)
11. Pallay, M., Towfighian, S.: A parametric electrostatic resonator using repulsive force. Sens. Actuators A **277**, 134–141 (2018)

Measurement of Elastic Nonlinearities Using the Fundamental Edge Wave Mode

James Martin Hughes[1(✉)], Andrei Kotousov[1], and Ching-Tai Ng[2]

[1] School of Mechanical Engineering, The University of Adelaide, Adelaide, SA 5005, Australia
james.m.hughes@adelaide.edu.au
[2] School of Civil, Environmental, and Mining Engineering, The University of Adelaide, Adelaide, SA 5005, Australia

Abstract. Measurement of the nonlinear elastic properties of materials represents a great interest in engineering and materials science. Changes of these properties are often related to mechanical damage and applied stresses, which are paramount for maintaining integrity and safety of structures. Current ultrasonic techniques typically utilise bulk, Lamb, or Rayleigh waves to measure material nonlinearities, however, spatial and velocity dispersion make this a very difficult task, and the use of several (empirical) correction factors is usually required. In this work we suggest using the fundamental edge wave mode – a natural analogue of the classical Rayleigh wave propagating in a finite thickness plate – for the purpose of measuring elastic nonlinearities. Edge waves naturally avoid spatial dispersion as they are guided by the apex of a plate, thus avoiding the need for correction factors. Additionally, the fundamental edge wave mode can propagate long distances without significant attenuation under certain conditions. The outcomes of this study demonstrate that the measurement of material nonlinearities is achievable using the fundamental edge wave mode. A linear trend between the nonlinearity parameter and propagation distance is experimentally observed, which is predicated by theoretical studies. Therefore, potential applications of the fundamental edge wave mode are very promising for the evaluation of mechanical damage and measurement of applied or residual stresses.

Keywords: Harmonic generation · Edge waves · Feature guided waves

1 Introduction

Material nonlinearities can be acquired through mechanical processes such as fatigue, creep, thermal stress, and physical damage [1]. Measurement of the nonlinear properties can therefore provide an insight into the health of the material and structure [2], which explains the growing interest in incorporating nonlinear methods for Structural Health Monitoring (SHM). Typically, bulk, Lamb, or Rayleigh waves are used to measure material nonlinearity through the growth of higher-order harmonics [3, 4], however, these wave modes are subject to spatial dispersion, multimodal generation,

E. Gdoutos and M. Konsta-Gdoutos (Eds.): ICTAEM 2020, STIN 16, pp. 133–139, 2020.
https://doi.org/10.1007/978-3-030-47883-4_23

and velocity dispersion, meaning the matching conditions required for cumulative second harmonic generation are only achievable for certain frequency-thickness products [5]. Using such wave modes therefore requires the application of numerous correction factors [6]. Consequently, there is a growing interest in alternative methods or wave modes which may be used to measure material nonlinearity more directly.

One promising candidate is the fundamental edge wave mode, ES_0, which is a natural analogue of the Rayleigh wave propagating in a finite thickness plate [7]. Similarly to Rayleigh waves, edge waves are surface waves, but are guided by the apex of a plate-structure. From a mathematical standpoint, the main difference between Rayleigh and edge waves is the latter must additionally satisfy traction-free boundary conditions on the side surfaces of the plate, whereas the former propagates under either plane stress or plane strain conditions. The inclusion of the plate side faces in the mathematical formulation naturally results in the absence of spatial dispersion, which presents a major problem facing measurements conducted using Lamb and Rayleigh waves. In addition, edge waves are weakly velocity dispersive, which allows approximate internal resonance conditions to be satisfied over a wide frequency range – a requirement for cumulative harmonic growth and measurement of material nonlinearity.

Although edge waves are not spatially dispersive, they are energy dispersive, and gradually decay through conversion to other wave modes, namely the S_0 Lamb wave mode. This amplitude decay is negligible for low frequency-thickness values ($\omega 2h/c_2 < 6$), however, theoretical studies indicate that at higher frequencies and Poisson's ratios the decay is more significant [8]. The problem of multimodal generation can also be reduced by exciting at a relatively low frequency-thickness product, see Fig. 1, and by utilising the wedge excitation method, which has been used in a previous study to excite the fundamental edge wave mode [9].

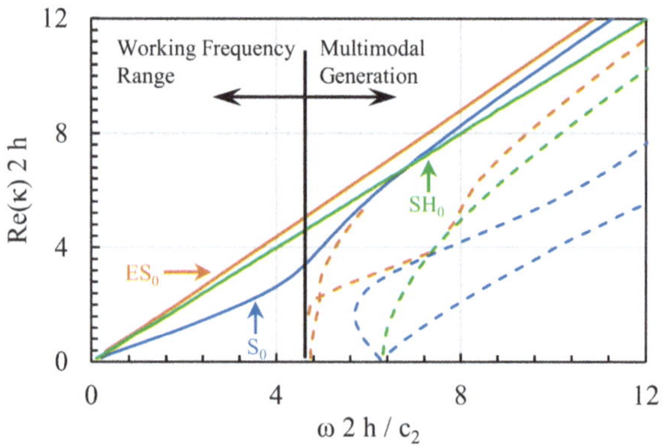

Fig. 1. Dispersion curves showing edge, shear, and Lamb wave modes. Solid lines depict the fundamental modes (ES_0, S_0, and SH_0) and dotted lines represent higher-order modes.

Edge waves were first documented by Shaw in 1956 [10], and later described by Gazis and Mindlin in 1960 [11]. In the following decades several theoretical studies were published which focused on the mathematical formulation of edge waves, higher-order modes, and the velocity dispersion characteristics [7, 8]. However, there have been relatively few studies which experimentally investigate the properties of edge waves for practical purposes. One recent study conducted by Wilde et al. [7] showed that edge waves may be generated using a bonded piezoelectric transducer, although this produced a complicated wave field due to generation of other wave modes such as the fundamental shear-horizontal and fundamental Lamb. This work was extended by the current authors who demonstrated edge wave excitation using the wedge method, which is more commonly applied to Lamb and Rayleigh waves, and produces a cleaner edge wave packet [9].

The purpose of this paper is therefore to demonstrate the excitation of ES_0 mode using the wedge method, and to investigate cumulative harmonic generation at low frequency-thickness values. It is believed that the results of this study will be of interest for researchers and engineers working in SHM and NDT fields.

2 Mechanics of Higher-Harmonic Generation

The excitation of harmonics in an ultrasonic signal occurs as a result of the adoption of nonlinear equations of elasticity. A perturbation technique can be applied to solve these equations, where the resultant relationship for a weakly nonlinear one-dimensional wave is:

$$\frac{A_2}{A_1^2} = \beta' x = \frac{\kappa^2}{8}\beta x \tag{1}$$

where A_1 and A_2 are the amplitudes of the fundamental and second harmonic respectively, x is the distance from the excitation source, κ is the wavenumber, β' is the relative nonlinearity parameter, and β is the acoustic nonlinearity parameter, which is a material property related to the second- and third-order elastic constants. Similar relationships can be derived for higher-order harmonics [12]. It is widely accepted that Eq. (1) holds true for other types of waves when some correction factor is applied [6]. Therefore, through measurement of the fundamental and second harmonic of a propagating elastic wave, it is possible to determine the relative or acoustic nonlinearity parameters, which are an indicator of damage as mentioned in the Introduction.

Measurement of the harmonic amplitudes is typically achieved using a fast Fourier transform (FFT) to decompose the displacement signal produced by the wave into its frequency components. The amplitudes of the fundamental and higher-harmonic components are then extracted, and the relative nonlinearity parameter can be determined if the displacement signal is recorded at a number of propagation distances, x. Typically, corrections related to the source, spatial dispersion, and attenuation must be applied to reveal the linear relationship between A_2/A_1^2 and β', however, the favourable characteristics of edge waves (no spatial dispersion and minimal attenuation at low frequency-thicknesses) allows direct measurements without requiring correction factors. This represents a significant advantage over current ultrasonic techniques which use bulk, Lamb, and Rayleigh waves.

3 Experimental Methodology

Excitation of the ES_0 mode was achieved using the wedge method, which has been well documented for the generation of many types of guided waves [13, 14]. The wedge was constructed from an ultra-high molecular weight polyethylene plastic called 'Polystone' by manufacturer 'Dotmar'. The wedge angle of 52° was chosen to facilitate conversion of the longitudinal wave to the edge wave mode due to Snell's law. The wedge assembly was coupled to a 5083-H116 aluminium plate with dimensions 700 mm × 500 mm and thickness 5 mm using motor oil.

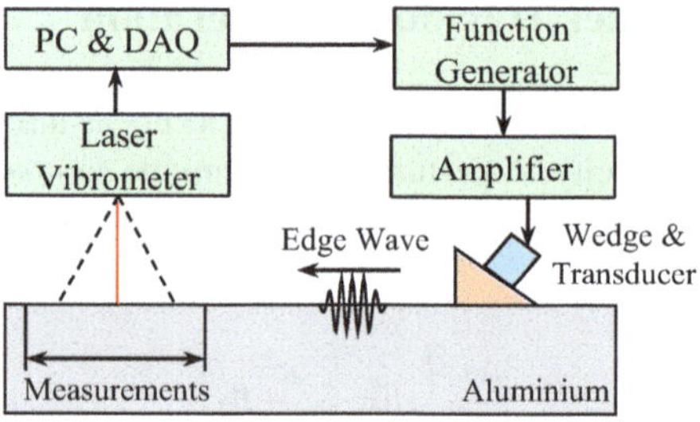

Fig. 2. Signal flow schematic of experimental setup.

A longitudinal wave of frequency f = 300 kHz was excited using an Ultran GC350-D13 contact transducer. The 15 cycle square-windowed excitation signal was generated using an AFG3012B arbitrary function generator and amplified to approximately $\pm 150V_{pp}$ using a Khron-Hite 7500 amplifier. Measurement of the out-of-plane displacement was achieved using a Polytec PSV-400-M2-20 scanning laser vibrometer in 1D mode, which was directed perpendicularly along the centre line of the aluminium plate. The out-of-plane displacement was recorded at 7 locations equispaced between 350 mm and 500 mm from the front of the wedge, and 10 trials were recorded at each point. A schematic of the experimental setup is shown above in Fig. 2.

4 Results

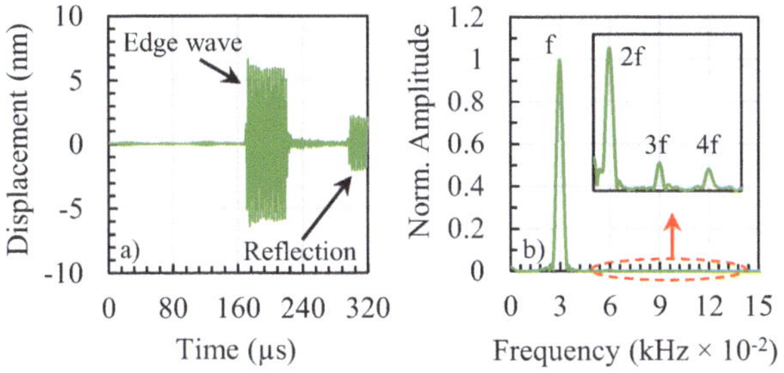

Fig. 3. a) Typical time domain signal showing the edge wave and a reflection from the plate end, and b) FFT of the edge wave signal after applying a Hann-window.

A typical out-of-plane displacement signal recorded by the laser vibrometer is shown above in Fig. 3(a). The region of the signal corresponding to the edge wave is clearly identified, and verification that the signal corresponds to the fundamental edge wave mode was achieved through measurement of the wave speed. A reflection from the plate end is also visible. A Hann-window is applied to the portion of the signal corresponding to the edge wave before performing an FFT, the results of which are presented in Fig. 3(b). There is a clear peak at the second harmonic frequency (600 kHz), as well as peaks which correspond to the third and fourth harmonics, indicating that harmonic generation does indeed occur for edge waves under approximate internal resonance conditions. Although beyond the scope of this paper, the presence of the two higher-order harmonics is promising for future research and could allow measurement of nonlinearities with the third- or fourth-order harmonics.

Figure 4 shows the relationship between the relative nonlinearity parameter, β', and propagation distance, x. The data gathered is well represented by a linear trendline, which is supported by the high R-squared value of 0.87 and theoretically predicted by Eq. (1). Although the fundamental edge wave mode is weakly dispersive, the linearisation error for the current experiment amounts to approximately 10% at the furthest measurement point, and therefore Eq. (1) can be applied. It is difficult to compare the rate of accumulation of the second harmonic found in this study against other studies which were based on Rayleigh or Lamb waves due to various correction and conversion factors, however, the corresponding experimental values of β' are of the same order of magnitude.

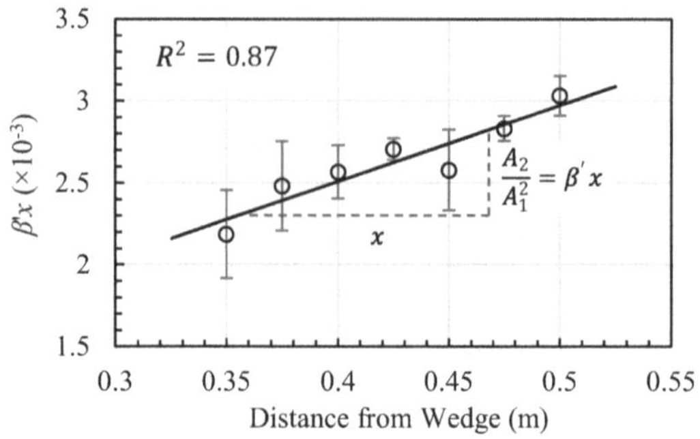

Fig. 4. Growth of the relative nonlinearity parameter using the fundamental edge wave mode, where error bars show the standard deviation. The linear trend seen is supported by both theoretical predictions (Eq. (1)) and previous studies which utilised Rayleigh and Lamb waves.

5 Conclusion

The outcomes of this study demonstrate the potential of the ES_0 mode for measurement of material nonlinearities. A linear growth of the relative nonlinearity parameter with propagation distance was observed as predicated by both theoretical and experimental studies. Though it is difficult to compare the current results to those which use Rayleigh and Lamb waves, the experimental values of β' are of the same order. It is important to highlight that the measurements using the ES_0 mode can be achieved without applying empirical or numerically obtained correction factors. Finally, it is believed that the fundamental edge wave mode is the most promising candidate for ultrasonic measurements of elastic nonlinearities. The applications of this work therefore have great potential for use in the evaluation of mechanical damage and applied or residual stresses.

Acknowledgment. This work was supported by the Australian Research Council through DP160102233, LE170100079, DP200102300 and the Australian Research Training Program Scholarship. Their support is greatly appreciated.

References

1. Matlack, K.H., Kim, J.-Y., Jacobs, L.J., Qu, J.: Experimental characterization of efficient second harmonic generation of Lamb wave modes in a nonlinear elastic isotropic plate. J. Appl. Phys. **109**, 014905 (2011)
2. Matlack, K.H., Kim, J.-Y., Jacobs, L.J., Qu, J.: Review of second harmonic generation measurement techniques for material state determination in metals. J. Nondestr. Eval. **34**, 273 (2014)
3. Zuo, P., Zhou, Y., Fan, Z.: Numerical and experimental investigation of nonlinear ultrasonic Lamb waves at low frequency. Appl. Phys. Lett. **109**(2), 021902 (2016)
4. Pruell, C., Kim, J.-Y., Qu, J., Jacobs, L.J.: Evaluation of plasticity driven material damage using Lamb waves. Appl. Phys. Lett. **92**(23), 231911 (2007)

5. Li, W., Cho, Y., Achenbach, J.D.: Detection of thermal fatigue in composites by second harmonic Lamb waves. Smart Mater. Struct. **21**(8), 085019 (2012)
6. de Lima, W.J.N., Hamilton, M.F.: Finite-amplitude waves in isotropic elastic plates. J. Sound Vib. **265**(4), 819–839 (2003)
7. Wilde, M.V., Golub, M.V., Eremin, A.A.: Experimental and theoretical investigation of transient edge waves excited by a piezoelectric transducer bonded to the edge of a thick elastic plate. J. Sound Vib. **441**, 26–49 (2019)
8. Zvernov, V., Kaplunov, J.: Three-dimensional edge waves in plates. Proc. R. Soc. A **464**, 301–318 (2008)
9. Hughes, J.M., Kotousov, A., Ng, C.-T.: Generation of higher harmonics with the fundamental edge wave mode. Appl. Phys. Lett. **116**, 101904 (2020)
10. Shaw, E.A.G.: On the resonant vibrations of thick barium titanate disks. J. Acoust. Soc. Am. **28**(1), 38–50 (1956)
11. Gazis, D.G., Mindlin, R.D.: Extensional vibrations and waves in a circular disk and a semi-infinite plate. J. Appl. Mech. **27**(3), 541–547 (1960)
12. Kube, C.M., Arguelles, A.P.: Ultrasonic harmonic generation from materials with up to cubic nonlinearity. J. Acoust. Soc. Am. **142**(2), 224–230 (2017)
13. Hughes, J.M., Vidler, J., Ng, C.-T., Khanna, A., Mohabuth, M., Rose, L.F.R., Kotousov, A.: Comparative evaluation of in situ stress monitoring with Rayleigh waves. Struct. Health Monit. **18**(1), 205–215 (2019)
14. Hughes, J.M., Vidler, J., Khanna, A., Mohabuth, M., Kotousov, A., Ng, C.-T.: Measurement of residual stresses in rails using Rayleigh waves. In: SIF18 Conference, Perth (2019)

Unsteady Axisymmetric Electro-Magneto-Elastic Oscillations of a Continuous Cylinder Under the External Displacement Field

Vladimir Vestyak(✉) and Vasily Scherbakov

Moscow Aviation Institute (National Research University),
Volokolamskoe shosse, 4, 125993 Moscow, Russian Federation
v.a.vestyak@mail.ru, vasily.a.scherbakov@gmail.com

Abstract. In the present work is considered an axisymmetric time-dependent waves of an infinite cylindrical body. The body material is taken to be isotropic and electro magneto elastic. Piezoelectric effects are not taken into account. The deformation process is described by a system of equation with respect to radial and angular components of deformation of the body points in cylindrical coordinate system. In additional, it takes into account the effect of current density, surface charges, electric and magnetic fields. All parameters and ratios are reduced to dimensionless form. To solve the problem, are used the Fourier transformation of angles and the Laplace transformation of time. Then, the resulting expressions expansion in series in terms of a small parameter. The small parameter characterizes the relationship between mechanic and electro-magnetic fields. To move into the space of the originals using the inverse Laplace transformation via residue theorem.

Keywords: Electromagnetoelasticity · Axisymmetric waves · Residue theorem · Coupled problems · Time-dependent axisymmetric problems · Green's functions

1 Problem Statement

1.1 Basic Equations

Consider the unsteady oscillations of a continuous cylinder of infinite length with a radius R. The material of a cylindrical body is considered electro magneto elastic and isotropic. Piezoelectric effects will not be taken into account. The flat deformed state of the continuous medium will be investigated, therefore all the required functions of the problem will depend on three parameters: τ – time, r – radial coordinate and θ – angular coordinate. The interaction of mechanical and electromagnetic fields is described by the following system of equations [1–7]:

E. Gdoutos and M. Konsta-Gdoutos (Eds.): ICTAEM 2020, STIN 16, pp. 140–144, 2020.
https://doi.org/10.1007/978-3-030-47883-4_24

– equations of motion respect to displacement

$$\ddot{u}_r = \left(1-\eta^2\right)\frac{\partial I_1}{\partial r} + \eta^2 \Delta u_r - \frac{\eta^2}{r^2}\left(2\frac{\partial u_\theta}{\partial \theta} + u_r\right) + \alpha F_{er}, \tag{1}$$

$$\ddot{u}_\theta = \frac{\left(1-\eta^2\right)}{r}\frac{\partial I_1}{\partial \theta} + \eta^2 \Delta u_\theta + \frac{\eta^2}{r^2}\left(2\frac{\partial u_r}{\partial \theta} - u_\theta\right) + \alpha F_{e\theta}, \tag{2}$$

where $I_1 = \frac{1}{r}\left[\frac{\partial (r u_r)}{\partial r} + \frac{\partial u_\theta}{\partial \theta}\right]$, $\Delta = \frac{1}{r}\frac{\partial}{\partial r}\left(r\frac{\partial}{\partial r}\right) + \frac{1}{r^2}\frac{\partial^2}{\partial \theta^2}$;

– Maxwell's equations

$$\begin{aligned} &\frac{1}{r}\left[\frac{\partial (rE_\theta)}{\partial r} - \frac{\partial E_r}{\partial \theta}\right] = -\dot{H}, -\frac{\partial H}{\partial r} = \eta_e^2\left(\gamma j_\theta + \dot{E}_\theta\right), \\ &\frac{1}{r}\frac{\partial H}{\partial \theta} = \eta_e^2\left(\gamma j_r + \dot{E}_r\right), \frac{1}{r}\left[\frac{\partial (rE_r)}{\partial r} + \frac{\partial E_\theta}{\partial \theta}\right] = \rho_e; \end{aligned} \tag{3}$$

– Ohm's law

$$j_r = E_r + H_0 \dot{u}_\theta + \rho_{e0}\dot{u}_r/\gamma,\ j_\theta = E_\theta - H_0 \dot{u}_r + \rho_{e0}\dot{u}_\theta/\gamma; \tag{4}$$

– Lorentz force

$$F_{er} = \rho_{e0}E_r + \rho_e E_{0r} + \gamma(j_{0\theta}H + j_\theta H_0), F_{e\theta} = \rho_{e0}E_\theta + \rho_e E_{0\theta} - \gamma(j_{0r}H + j_r H_0); \tag{5}$$

– Physical relationships of electromagnetism

$$D_r = E_r,\ D_\vartheta = E_\vartheta,\ B = H. \tag{6}$$

Here u_r, u_θ – components of the displacement vector; $F_{er}, F_{e\theta}$ – components of the Lorentz force; E_r, E_θ – components of electric field; H, B – components of magnetic field; j_r, j_θ – components of current density; ρ_e – volume charge density; D_r, D_θ – components of electric displacement field. All of these parameters are unknown and depend on three variables: τ, r and θ. H_0, ρ_{e0}, E_{0r}, $E_{0\theta}$, j_{0r} and $j_{0\theta}$ are the parameters of the initial electromagnetic field and do not depend on time.

The above equations are dimensionless. In order to make these equations dimensionless, the following substitutions were applied (the "-" sign indicates a dimensional parameter):

$$r = \frac{\bar{r}}{R},\ \tau = \frac{c_1 t}{R},\ u_k = \frac{\bar{u}_k}{R},\ H = \frac{\bar{H}\mu_e c_1}{cE_*}, B = \frac{\bar{B}c_1}{cE_*},\ \rho_e = \frac{4\pi\bar{\rho}_e R}{\varepsilon E_*},$$
$$E_k = \frac{\bar{E}_k}{E_*},\ D_k = \frac{\bar{D}_k}{\varepsilon E_*}, j_k = \frac{\bar{j}_k}{\sigma E_*},\ F_{ek} = \frac{\bar{F}_{ek} R}{\lambda + 2\mu},\ \eta_e = \frac{c_1}{c_e},$$
$$\alpha = \frac{\varepsilon E_*^2}{4\pi(\lambda + 2\mu)},\ \gamma = \frac{\gamma_e R}{c_1} = \frac{4\pi\sigma R}{\varepsilon c_1}, c_1^2 = \frac{\lambda + 2\mu}{\rho},\ c_2^2 = \frac{\mu}{\rho},$$
$$\eta = \frac{c_2}{c_1}, \gamma_e = \frac{4\pi\sigma}{\varepsilon},\ c_e^2 = \frac{c^2}{\mu_e \varepsilon},$$

where c_1 and c_2 - propagation velocities of tension-compression waves and shear waves; E_* - characteristic level of electric field; λ and μ – Lame's parameters; ε and μ_e - dielectric and magnetic permeability coefficients; $k = r, \theta$; c - light speed in vacuum; σ - conductivity coefficients.

1.2 Initial and Boundary Conditions

It is assumed that at the initial moment of time, there are no perturbation in the solid body.

As conditions on the boundary of a cylindrical body, the following are assumed:

$$u_r(\tau, r, \theta)|_{r=1} = U_r(\tau, \theta), u_\theta(\tau, r, \theta)|_{r=1} = 0.$$

At the point at $r = 0$, disturbances in a cylindrical body are considered limited:

$$u_r(\tau, r, \theta)|_{r=0} = O(1), u_\theta(\tau, r, \theta)|_{r=0} = O(1).$$

The system of Eqs. (3)–(6) can be reduced to a system of equations of two equations:

$$\begin{aligned}
&\eta_e^2(\ddot{E}_r + \gamma\dot{E}_r) = N_{11}(E_r) + N_{12}(E_\vartheta) - \eta_e^2(\rho_{e0}\ddot{u}_r + \gamma H_0 \ddot{u}_\theta),\\
&\eta_e^2(\ddot{E}_\vartheta + \gamma\dot{E}_\vartheta) = N_{21}(E_r) + N_{22}(E_\vartheta) + \eta_e^2(\gamma H_0 \ddot{u}_r - \rho_{e0}\ddot{u}_\theta),\\
&N_{11} = \frac{1}{r^2}\frac{\partial^2}{\partial\vartheta^2},\ N_{12} = -\frac{1}{r}\frac{\partial}{\partial\vartheta}\left(\frac{\partial}{\partial r} + \frac{1}{r}\right),\\
&N_{21} = -\frac{1}{r}\frac{\partial}{\partial\vartheta}\left(\frac{\partial}{\partial r} - \frac{1}{r}\right),\ N_{22} = \frac{\partial}{\partial r}\left[\frac{1}{r}\frac{\partial(r)}{\partial r}\right].
\end{aligned} \tag{7}$$

2 Solution Methods

To solve the problem of the field of displacement and the electric field are expressed through the vector and scalar potentials:

$$\begin{aligned}
&\mathbf{u} = grad\varphi + rot\psi,\\
&\mathrm{E} = grad\varphi_e + rot\psi_e,\\
&div\psi = 0,\ div\psi_e = 0,
\end{aligned} \tag{8}$$

where φ and φ_e - scalar potentials, $\boldsymbol{\psi}$ and $\boldsymbol{\psi}_e$ - vector potentials.
The Lorentz force vector can also be represented via scalar and vector potentials:

$$\mathbf{F_e} = grad\Phi + rot\boldsymbol{\Psi}, \tag{9}$$

where

$$\Delta\Phi = divF_e = \frac{1}{r}\frac{\partial(rF_{er})}{\partial r} + \frac{1}{r}\frac{\partial F_{e\theta}}{\partial\theta}, \Delta\Psi = -rotF_e = -\frac{1}{r}\frac{\partial(rF_{e\theta})}{\partial r} + \frac{1}{r}\frac{\partial F_{er}}{\partial\theta}.$$

In order to fulfill the axisymmetric conditions, it is necessary that the defining parameters of the problem, as well as the vector and scalar potentials correspond to the following equations:

$$\begin{aligned}
&u_r = u_r(\tau, r, \theta),\ u_\theta = u_\theta(\tau, r, \theta),\ u_z \equiv 0, F_{ez} \equiv 0, E_r = E_r(r, \theta, t),\\
&E_\theta = E_\theta(r, \theta, t), E_z = E_{0z} \equiv 0, E_{0r} = E_{0r}(r, \theta),\ \ E_{0\theta} = E_{0\theta}(r, \vartheta),\\
&\varphi = \varphi(r, \theta, t), \varphi_e = \varphi_e(r, \theta, t), F_{er} = F_{er}(r, \theta, t),\ \ E_\theta = E_\theta(r, \theta, t),\\
&\psi_r = \psi_\theta \equiv 0, \psi_{er} = \psi_{e\theta} \equiv 0, \psi_z = \psi = \psi(r, \theta, t), \psi_{ez} = \psi_e = \psi_e(r, \theta, t),\\
&H_r = H_{0r} = H_\theta = H_{0\theta} \equiv 0,\ \ H_z = H(r, \theta, t),\ \ H_{0z} = H_0(r, \vartheta).
\end{aligned}$$

By substituting (8) and (9) into systems (1), (2) and (7), we obtain a system of equations for φ, ψ and φ_e, ψ_e.
First, the unknown functions are decomposed into a complex Fourier series:

$$\varphi(\tau, r, \theta) = \sum_{n=-\infty}^{+\infty} \varphi_n^F(\tau, r)e^{i\theta},\ \psi(\tau, r, \theta) = \sum_{n=-\infty}^{+\infty} \psi_n^F(\tau, r)e^{i\theta},$$
$$\varphi_e(\tau, r, \theta) = \sum_{n=-\infty}^{+\infty} \varphi_{en}^F(\tau, r)e^{i\theta},\ \psi_e(\tau, r, \theta) = \sum_{n=-\infty}^{+\infty} \psi_{en}^F(\tau, r)e^{i\theta},$$

then, the Laplace transforms in time are applied to the obtained coefficients $\varphi_n^F(\tau, r)\ \ldots\ \psi_{en}^F(\tau, r)$ (s - conversion parameter):

$$\varphi_n^F(\tau, r) \xrightarrow{L} \varphi_n^{FL}(s, r),$$
$$\cdot$$
$$\cdot$$
$$\psi_{en}^F(\tau, r) \xrightarrow{L} \psi_{en}^{FL}(s, r).$$

The resulting coefficients $\varphi_n^{FL}(s, r)\ldots\psi_{en}^{FL}(s, r)$ are expanded in a power series for a small parameter α:

$$\varphi_n^{FL} = \sum_{m=0}^{\infty} \varphi_{nm}^{FL} \alpha^m, \ldots, \psi_{en}^{FL} = \sum_{m=0}^{\infty} \psi_{enm}^{FL} \alpha^m.$$

After that we get a new recurrent system of differential equations for the coefficient of expansion $\varphi_{nm}^{FL} \ldots \psi_{nm}^{FL}$.

And finally in order to move obtained $\varphi_{nm}^{FL} \ldots \psi_{nm}^{FL}$ parameters into the space of the originals it is using the inverse Laplace transformation via residue theorem.

References

1. He, T., Tian, X., Shen, Y.: A generalized electromagneto-thermoelastic problem for an infinitely long solid cylinder. Eur. J. Mech. A. **24**(2), 349–359 (2005)
2. Wang, X., Dong, K.: Magnetothermodynamic stress and perturbation of magnetic field vector in a non-homogeneous thermoelastic cylinder. Eur. J. Mech. A. **25**(1), 98–109 (2006)
3. Vestyak, V.A., Lemeshev, V.A.: Rasprostranenie nestatsionarnykh radialnykh vozmushcheniy ot tsilindricheskoy polosti v elektromagnitouprugoy srede. In: Mater. XIV Intern. Symp. "Dynamic and technolo-ronmental problems of mechanics of structures and continua" named A.G. Gorshkov. Moscow, vol. 1, pp. 59–60 (2008)
4. Tarlakovskii, D.V., Vestyak, V.A., Zemskov, A.V.: Dynamic processes in thermoelectromagnetoelastic and thermoelastodiffusive media. In: Encyclopedia of Thermal Stresses, vol. 2, pp. 1064–1071. Springer, Dordrecht (2014)
5. Gorshkov, A.G., Medvedsky, A.L., Rabinsky, L.N., Tarlakovsky, D.V.: Waves in Continuous Mediums Textbook for Higher Education Institutions, p. 472. FIZMATLIT, Moscow (2004). (in Russian)
6. Vestyak, V.A., Lemeshev, V.A., Tarlakovsky, D.V.: One-dimensional time-dependent waves in an electromagnetoelastic half-space or in a layer. Doklady Phys. **54**(6), 262–264 (2009)
7. Vestyak, V.A., Lemeshev, V.A., Tarlakovskii, D.V.: The propagation of time-dependent radial perturbations from a spherical cavity in an electromagnetoelastic space. Doklady Phys. **55**(9), 468–470 (2010)

Development of Micro-mechanical Models of Fatigue Damage

James Vidler[1(✉)], Andrei Kotousov[1], and Ching-Tai Ng[2]

[1] School of Mechanical Engineering, The University of Adelaide, Adelaide SA 5005, Australia
james.vidler@adelaide.edu.au
[2] School of Civil, Environmental, and Mining Engineering, The University of Adelaide, Adelaide SA 5005, Australia

Abstract. Accumulation of fatigue damage in high and low cycle regimes is largely associated with nucleation and development of irreversible plastic deformations and voids. The effect of these microstructural changes on conventional (second-order) elastic constants is typically very small, which makes experimental evaluation of the progressive fatigue damage accumulation difficult and the classical damage theories inapplicable. It was demonstrated in the past that the third-order elastic constants are sensitive to fatigue damage and these material constants can be evaluated using various ultrasonic techniques. The ultimate aim of this study is to develop micro-mechanical models, which link the micro-porosity and micro-plastic deformations, to the effective third-order elastic constants of the material. These models could provide a foundation for the evaluation of early fatigue damage, i.e. the damage prior formation of a micro-defect (crack), as well as the remaining fatigue life of structures.

Keywords: Fatigue damage · Third-Order Elastic Constants · Finite elasticity

1 Introduction

Inspired by Eshelby, micromechanics literature abounds with studies of dilatations, inhomogeneities, and inclusions, which are relevant to analysis of many problems of mechanics and material science involving plasticity, thermal expansion and phase transformations [1]. Unfortunately, in the case of finite deformations the analysis is usually prohibitively difficult. The finite deformation analysis requires solutions to the nonlinear equations of equilibrium, which are scarce. The lowest order of nonlinear response of an elastic body is represented by the third order elastic constants (TOECs), which can be formally obtained from the series expansion of strain energy density with respect to the strain components [2].

The TOECs of a material have both an intrinsic and an acquired nature. The intrinsic TOECs, which are related to the anharmonic forces in crystals, play an important role in predictions of mechanical behaviour, the ideal strength and ductility in metals and in the explanation of anharmonic properties of solids such as the thermal expansion. The acquired TOECs are often related to various dislocation, micro-plasticity and phase

E. Gdoutos and M. Konsta-Gdoutos (Eds.): ICTAEM 2020, STIN 16, pp. 145–150, 2020.
https://doi.org/10.1007/978-3-030-47883-4_25

transformation phenomena as well as to accumulation of mechanical damage such as fatigue and radiation damage [3].

Material nonlinearities and TOECs were investigated extensively with ultrasonic techniques in the early 1960s. Over the past two decades, however, the focus has shifted to ultrasonic guided waves, which can be used in non-destructive techniques as well as in on-line damage monitoring systems. Monitoring of accumulated mechanical damage is currently considered as a significant engineering problem, specifically for structures subjected to fatigue. In the absence of manufacturing defects the accumulation of micro-damage (or the early damage stage) normally precedes the crack nucleation and propagation stages. Depending on the material, its microstructure and the presence of pre-existing defects, the early damage stage can occupy from 10% to up to 90% of the total fatigue life in high-cycle fatigue (HCF) regime, and it fully dominates in the ultra-high cycle fatigue (UHCF) regime (up to 10^9 cycles and beyond) [4].

Previous studies have found that fatigue damage in metals and advanced engineering alloys in LCF and HCF regimes increases with applied cycles in a cumulative manner, and material degradation is largely associated with the accumulation of irreversible micro-plastic deformations and nucleation and growth of voids, respectively. Classical damage mechanics links the conventional (or second-order) elastic constants to fatigue damage [5]. However, the conventional elastic constants (e.g. shear and bulk moduli) and the damage parameter(s)/function(s) are not very sensitive to microstructural damage until the final stages of the fatigue life, particularly in the HCF and UHCF regimes. This has been well documented in past experimental studies, e.g. [6, 7] and has recently been confirmed by direct numerical computations using grain-size models [8]. Therefore, it is difficult to validate the damage mechanics models in the early stages of the fatigue life, or quantify the cumulative microstructural damage experimentally based on measurements of the conventional (second-order) elastic constants. At the same time, many recent studies have demonstrated that the change of nonlinear (or third-order) elastic constants is usually much stronger than the change of conventional elastic constants, particularly at the initial (or early) stages of the fatigue life. As discussed above these changes can be detected using various ultrasonic techniques, which are under development since 1960s.

The ultimate aim of this work is to develop micro-mechanical models, which describe the change of the effective third-order elastic constants (TOECs) with accumulation of micro-plastic deformations, as well as nucleation and growth of micro-voids. Previous analytical studies [9–11] have investigated the effective properties of nonlinear elastic material in the theory of micromechanics, but have not combined their research with a model of damage accumulation. The development of such models could provide a theoretical foundation for the development of new early-damage assessment techniques as well as mode adequate fatigue life evaluation procedures.

2 Theoretical Framework

Consider a material body undergoing deformation with material points $\boldsymbol{X}$ mapped to spatial points $\boldsymbol{x}$, so that $\boldsymbol{x} = \chi(\boldsymbol{X})$. The deformation gradient is

$$F = \text{Grad}\,\chi. \tag{1}$$

For a hyperelastic and isotropic material, a strain energy density function exists, and may be represented as a function of the principal invariants of the material strain tensor. In Murnaghan's form [2], the strain energy density function including terms up to third-order is

$$W(I_1, I_2, I_3) = \frac{1}{2}(\lambda + 2\mu)I_1^2 - 2\mu I_2 + \frac{1}{3}(l + 2m)I_1^3 - 2mI_1I_2 + nI_3 \tag{2}$$

where I_1, I_2, I_3 are the principal invariants of the material strain tensor,

$$E = \frac{1}{2}\left(F^T F - I\right)$$

the second-order elastic constants are λ and μ, and the TOECs of the material are l, m, and n. The nominal stress tensor for compressible materials is

$$S = \partial W / \partial F \tag{3}$$

and the equilibrium condition expressed in the referential coordinates, neglecting body forces, is

$$\text{Div}\, S = 0. \tag{4}$$

The governing equations of elasticity presented above are nonlinear, and most often solved using numerical methods. Exact, closed form solutions have been derived only under highly restrictive boundary conditions, or using specific forms of the strain energy density function. However, provided that the displacements and rotations be sufficiently small, second-order corrections to the linear elasticity solution may be derived using a perturbation approach [12].

Consider the expansion of the deformation

$$\chi(X) - X = ku_1 + k^2 u_2$$

where k is a small parameter, the vector u_1 corresponds to the solution for the displacement vector in the linear theory of elasticity, and the vector u_2 is the second-order correction to the linear solution. Substitution of the above expansion into the governing equations of elasticity and the boundary conditions yields a hierarchy of systems of equations, each corresponding to a linear elasticity problem. Each system of equations features a body force and surface tractions dependent on the solution to the lower-order systems of equations.

Applying the perturbation approach to the problem of a medium containing voids allows the derivation of the effective elastic constants of the material. If the medium is assumed to consist of a dilute distribution of spherical voids at a volume concertation of $c \ll 1$, the relevant problem for prediction of effective properties is a spherical shell containing an isolated void at the centre, with outer radius such that the volume concentration of voids is c.

Three loading cases are considered: pure dilatation, simple extension, and simple tension. In each loading case, the linear solution $\boldsymbol{u}_1$ is constructed in the standard manner using spherical harmonic functions; the second-order correction is constructed using spherical harmonic functions, a displacement potential, and a Galerkin vector. After solving the second-order elasticity problem for each of the three loading conditions, the total strain energy stored in the medium with voids is calculated and equated to the strain energy stored in an equivalent, homogeneous medium [11, 13],

$$\int W(I_1, I_2, I_3) dV = \int W_H(I_1, I_2, I_3) dV_H \tag{5}$$

where W_H is the strain energy function of the effective homogeneous medium, and $I_{1,H}$, $I_{2,H}$, $I_{3,H}$ are the principal invariants of the material strain tensor for the homogeneous medium.

The equivalent homogeneous medium has elastic properties which coincide with the effective properties of the porous medium, and is subjected to identical surface loading conditions. By considering three separate loading cases, all three TOEC of an isotropic hyperelastic material may be calculated and related to the volume concentration of voids.

3 Selected Results

The derivations and final expressions for the effective third order elastic constants of a medium containing a dilute distribution of micro-voids are extremely lengthy and cumbersome, which prevents the presentation of explicit expressions in this short

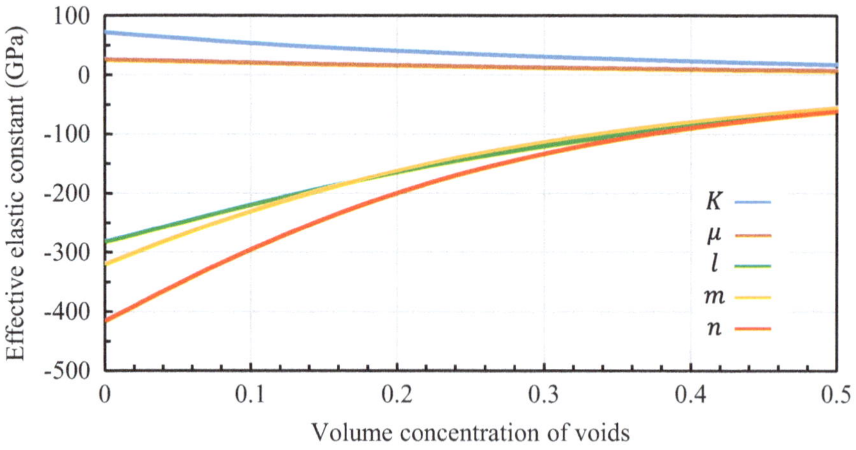

Fig. 1. Change in effective elastic constants for 6061-T6 aluminium due to a dilute distribution of micro-voids.

paper. For the purposes of illustration, the change in the effective elastic constants as the volume concentration of micro-voids increases is presented in Fig. 1, for the elastic constants of 6061-T6 aluminium reported in [14], $\lambda = 54.308\,\text{GPa}$, $\mu = 27.174\,\text{GPa}$, $l = -281.5\,\text{GPa}$, $m = -339.0\,\text{GPa}$, $n = -416.0\,\text{GPa}$.

The five (two second-order and three third-order) effective elastic constants are shown in Fig. 1. The plots are truncated at $c = 0.5$, due to the assumption of a dilute distribution of voids. It is seen that, for 6061-T6 aluminium, the TOECs feature a much higher sensitivity to the volume fraction of micro-voids. These constants have been derived from consideration of three different load cases: pure dilatation, simple extension, and simple tension.

4 Conclusion

A micro-mechanical model for the evaluation of effective conventional and third-order constants has been developed for an isotropic material with a dilute distribution of voids, which can be related to early low-cycle fatigue damage. The numerical calculations for aluminum support previously published experimental results claiming that the non-linear elastic constants (TOECs) are more sensitive to fatigue damage than the conventional (second-order) constants. Future work will be focused on the validation of theoretical results by conducting ultrasonic measurements of the change of TOECs in LCF regime. Similar models will be developed for micro-plastic deformations to describe the similar changes in HCF regime.

Acknowledgement. This work was supported by the Australian Research Council through DP160102233, LE170100079, DP200102300 and the Australian Research Training Program Scholarship. Their support is greatly appreciated.

References

1. Eshelby, J.D.: The determination of the elastic field of an ellipsoidal inclusion, and related problems. Proc. R. Soc. Lond. A **241**, 376–396 (1957)
2. Destrade, M., Ogden, R.W.: On stress-dependent elastic moduli and wave speeds. IMA J. Appl. Math. **78**, 965–997 (2012)
3. Matlack, K.H., Kim, J.-Y., Jacobs, L.J., Qu, J.: Review of second harmonic generation measurement techniques for material state determination in metals. J. Nondestruct. Eval **34**, 273 (2014)
4. Chan, K.S.: Roles of microstructure in fatigue crack initiation. Int. J. Fatigue **32**(9), 1428–1447 (2010)
5. Lemaitre, J.: A continuous damage mechanics model for ductile fracture. ASME J. Eng. Mat. Technol. **107**, 83–89 (1985)
6. Chaboche, J.L., Lesne, P.M.: A non-linear continuous fatigue damage model. Fatigue Fracture Eng. Mater. Struct. **2**, 1–17 (1988)
7. Lissenden, C., Liu, Y., Rose, J.: Use of non-linear ultrasonic guided waves for early damage detection. Insight Non-Destr. Test. Cond. Monitoring **57**, 206–211 (2015)

8. Benedetti, I., Gulizzi, V.: A grain-scale model for high-cycle fatigue degradation in polycrystalline materials. Int. J. of Fatigue **116**, 90–105 (2018)
9. Ogden, R.W.: On the overall moduli of non-linear elastic composite materials. J. Mech. Phys. Solids **22**, 541–553 (1974)
10. Chen, Y., Jiang, X.: Nonlinear elastic properties of particulate composites. J. Mech. Phys. Solids **41**(7), 1177–1190 (1993)
11. Imam, A., Johnson, G.C.: Ferrari, M,: Determination of the overall moduli in second order incompressible elasticity. J. Mech. Phys. Solids **43**(7), 1087–1104 (1995)
12. Clayton, J.D.: Defects in nonlinear elastic crystals: differential geometry, finite kinematics, and second-order analytical solutions. Z. Angew. Math. Mech. **95**(5), 476–510 (2015)
13. Hill, R.: On constitutive macro-variables for heterogeneous solids at finite strain. Proc. R. Soc. Long. A. **326**, 131–147 (1972)
14. Mohabuth, M.: Effect of uniaxial stress on the propagation of higher-order Lamb wave modes. Int. J. Non-Linear Mech. **86**, 104–111 (2016)

Research on Springback of 5A02 Aluminum Alloy Considering Thickness Normal Stress in Hydroforming

Peicheng Jiang, Lihui Lang(✉), and Sergei Alexandrov

Beihang University, No. 37 Xueyuan Road, Haidian District Beijing, People's Republic of China
{jiangpc, lang}@buaa.edu.cn
https://ev.buaa.edu.cn/

Abstract. Springback is an unavoidable problem in sheet metal forming. When the amount of sheet springback exceeds the allowable range, it becomes a forming defect. In the process of hydraulic bulging, the fluid medium acts on the surface of the sheet uniformly instead of the rigid mold, so that the material is simultaneously stressed, which makes the deformation law of the material different from that of the sheet in the traditional rigid bulging. In this paper, the research on the springback of 5A02 aluminum alloy is carried out by taking the active hydroforming process of a typical shallow drawn rotatory body arc surface part as the research object. The theoretical analysis method is used to take the thickness normal stress into the theoretical calculation of the springback amount, and a theoretical springback amount calculation model of the three-dimensional state of stress. Numerical simulations are carried out under different loading paths with hydraulic pressure of 1, 2, 3, 10, 20 and 30 MPa. The influence and law of springback of the parts under different loading paths and different hydraulic conditions are also obtained.

Keywords: Springback · Hydraulic bulging · Theoretical analysis · Three-dimensional state of stress · Numerical simulation

1 Introduction

Springback has been a focus of attention in the field of sheet metal forming manufacturing. Scholars have carried out a large number of fruitful studies on sheet springback since the last century [1–4]. Research on the springback of ordinary steels such as carbon steel, stainless steel, and automotive steel has accumulated a wealth of experimental data and mature research methods. As for the springback research of aluminum alloys and other light alloy materials, it is just in its infancy, especially aluminum alloy materials for advanced manufacturing. The prediction accuracy of the springback numerical simulation depends largely on the mechanical properties of the aluminum alloy sheet. Among them, anisotropy and Bauschinger effect are two important aspects, which directly affect the accuracy of springback prediction. The anisotropy of the material is closely related to the selected yield criterion, while the Bauschinger effect is directly related to the plastic hardening model. Therefore, the

E. Gdoutos and M. Konsta-Gdoutos (Eds.): ICTAEM 2020, STIN 16, pp. 151–157, 2020.
https://doi.org/10.1007/978-3-030-47883-4_26

coupling relationship between the yield criterion and the plastic hardening model is of great significance for springback prediction.

In this paper, the springback research is carried out with the active hydroforming process of a typical shallow drawn rotatory body arc surface part. Considering the influence of key process parameters on the springback during hydroforming, such as the hydraulic pressure loading path, two types of high-pressure and low-pressure loading have been performed. Among them, numerical simulation of hydraulic bulging in 6 cases of high pressure of 10, 20, 30 MPa and low pressure of 1, 2, 3 MPa respectively. Finally, the results of the numerical simulation can be used to verify the accuracy of the springback theoretical analysis Equation for the hydraulic bulging process to predict the springback trend of sheet metal under different hydraulic pressure paths.

2 Theoretical Analysis of Springback in Hydraulic Bulging Process

Taking the contour of the meridian passing through the rotating body as the research object, the deformation of the metal sheet during the hydraulic bulging process can be regarded as the traditional bending deformation. In order to facilitate the study of springback in the process of hydraulic bulging, the model is simplified. Based on the calculation process of the classic bending springback theory, the thickness normal stress is introduced into it. The simplified sheet bending model with thickness normal stress is shown in Fig. 1.

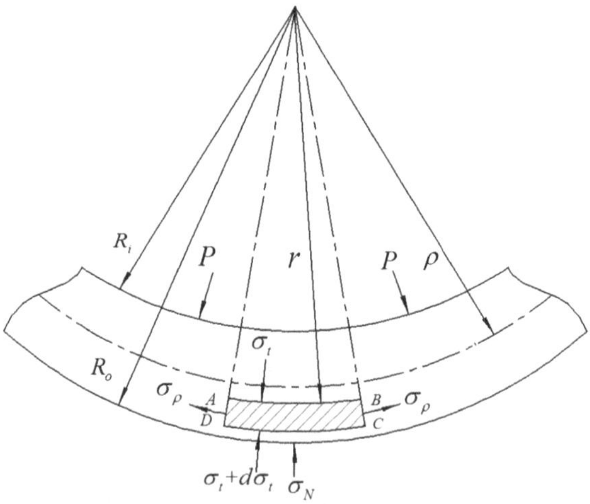

Fig. 1. The theoretical diagram of sheet metal bending process

Here R_i—Inner edge radius of the sheet, R_o—Outer edge radius of the sheet, ρ—Neutral layer radius, r—Micro-unit radius.

During the bending deformation of the sheet, the bending moment effect occurs inside the sheet under the action of external force, which causes the sheet to change from elastic deformation to plastic deformation. When the external force is unloaded, a part of the sheet recovers elastically [5].

We choose a micro-unit and name it ABCD, assuming the material has ideal plasticity, according to the Mises yield criterion [6, 7]

Outside the neutral layer: $\sigma_\rho > \sigma_\theta > \sigma_t$, the σ_ρ and σ_t have opposite signs, we can know that

$$\sigma_\rho + \sigma_t = \beta\sigma_s \tag{1}$$

And inside the neutral layer: $\sigma_t > \sigma_\theta > \sigma_\rho$, the σ_ρ and σ_t have the same signs, so we can get

$$\sigma_\rho - \sigma_t = \beta\sigma_s \tag{2}$$

The width of the micro-unit ABCD is a unit length. At any moment of bending deformation, the micro-unit should be in a state of force equilibrium. The stress σ_θ in the width direction has no effect on the equilibrium of the micro-unit in this plane. In addition, in order to maintain its tangential balance, the tangential stress σ_ρ on both sides of the micro-unit should be equal. Because the micro-unit must satisfy the equilibrium condition of the force, the algebraic sum of the forces on the micro-unit in the radial direction must be zero. Therefore, we can list the equilibrium equation of the micro-unit ABCD according to the equilibrium conditions:

$$\sigma_t \cdot r \cdot d\alpha - (\sigma_t + d\sigma_t) \cdot (r + dr) \cdot d\alpha - \sigma_\rho \cdot dr \cdot d\alpha = 0 \tag{3}$$

Among them, $\sigma_t \cdot r \cdot d\alpha$ is the force on the AB arc surface, $(\sigma_t + d\sigma_t) \cdot (r + dr) \cdot d\alpha$ is the force on the AB arc surface, and $\sigma_\rho \cdot dr \cdot d\alpha$ is the resultant force on the AC and BD surfaces.

Further simplification and derivation

$$\sigma_t = -\beta\sigma_s \cdot \ln r + c \tag{4}$$

Outside the neutral layer, when $r = R_o$, $\sigma_t = \sigma_N$. Among them, σ_N is the reaction force of the die to the sheet. So we can get the value of the constant c: $c = \sigma_N + \beta\sigma_s \cdot \ln R_o$.

From (1) and (4) we can know:

$$\sigma_\rho = \beta\sigma_s \cdot \ln r + \beta\sigma_s - \sigma_N - \beta\sigma_s \cdot \ln R_o \tag{5}$$

As for the inner side of the neutral layer, when $r = R_i$, $\sigma_t = P$. P is the hydraulic pressure. So we can get the value of the constant c: $c = P - \beta\sigma_s \cdot \ln R_i$.

Then

$$\sigma_\rho = \beta\sigma_s \cdot \ln r + \beta\sigma_s + P - \beta\sigma_s \cdot \ln R_i \tag{6}$$

The radius of the neutral layer before unloading is ρ, which increases to ρ_0 after unloading. If we use Δk to represent the amount of curvature reduction, then $\Delta k = \frac{1}{\rho} - \frac{1}{\rho_0}$.

It is assumed that the load bending moment of the sheet in plastic bending deformation is M, and the moment of inertia of the section is J. Using the formula about elastic bending in "Materials Mechanics", it can be known that after the bending moment is removed, the springback Δk of the sheet is:

$$\Delta k = \frac{M}{EJ} \tag{7}$$

Here $J = \frac{Bt^3}{12}$. B is the sheet width, and we take its size as 1. The sheet is compacted in the middle of the mold, and there is no external force, so the load bending moment M is generated by the tangential stress σ_ρ inside the sheet. With the neutral layer as the boundary, it can be divided into the bending moment caused by the inner tangential stress and the outer tangential stress, that is:

$$M = M_i + M_o \tag{8}$$

Here M_i is the bending moment caused by the inner tangential stress, and M_o is the bending moment caused by the outer tangential stress.

$$M_i = \int_{R_i}^{\rho} \sigma_\rho \cdot r \cdot b \cdot dr = \int_{R_i}^{\rho} (\beta\sigma_s \cdot \ln r + \beta\sigma_s + P - \beta\sigma_s \cdot \ln R_i) \cdot r \cdot dr \tag{9}$$

$$M_o = \int_{\rho}^{R_o} \sigma_\rho \cdot r \cdot b \cdot dr = \int_{\rho}^{R_o} (\beta\sigma_s \cdot \ln r + \beta\sigma_s - \sigma_N - \beta\sigma_s \cdot \ln R_o) \cdot r \cdot dr \tag{10}$$

Finally, the equation of sheet springback amount in the hydraulic bulging process considering the thickness normal stress is obtained, as shown in Eq. (11).

$$\begin{aligned} \Delta k = \frac{12}{Et^3} \cdot \Bigg[& \int_{R_i}^{\rho} (\beta\sigma_s \cdot \ln r + \beta\sigma_s + P - \beta\sigma_s \cdot \ln R_i) \cdot r \cdot dr \\ & + \int_{\rho}^{R_o} (\beta\sigma_s \cdot \ln r + \beta\sigma_s - \sigma_N - \beta\sigma_s \cdot \ln R_o) \cdot r \cdot dr \Bigg] \end{aligned} \tag{11}$$

3 Finite Element Simulation Analysis of Hydraulic Bulging

The software Dynaform is used to simulate the hydraulic bulging process. In the process of hydraulic bulging, the pressure value and loading rate of the fluid medium affect the deformation behavior of the sheet forming process. Therefore, in this section, the simulation scheme studies of the hydraulic pressure of 1, 2, 3, 10, 20, and 30 MPa are performed. The loading path of the hydraulic pressure is shown in Fig. 2.

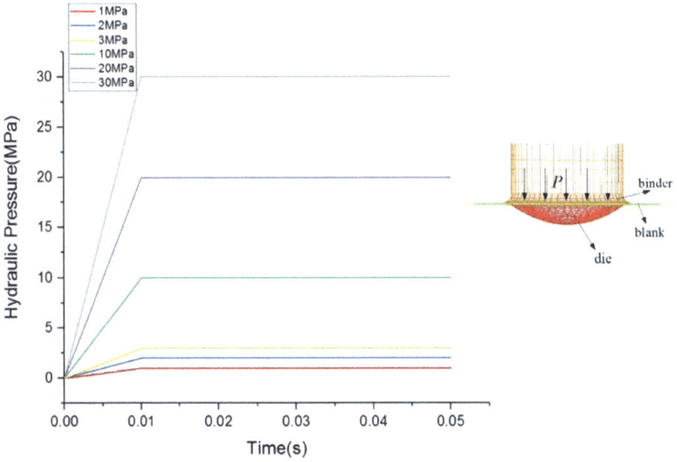

Fig. 2. Hydraulic pressure loading path

In the process of sheet hydraulic bulging, the central area is the first to undergo plastic deformation due to the bending moment effect and then expands to other areas. With the increase of the hydraulic pressure, the deformation of the vertex area gradually increases until the mold is fully fitted (20 MPa has been fully fitted), and the reaction force of die on the sheet is also greater. The maximum bulging height H, the curvature radius ρ, and the wall thickness δ of the bulging apex are measured and calculated, as shown in Table 1.

Table 1. Relevant data obtained from numerical simulation of hydraulic bulging

P (MPa)	H (mm)	δ (mm)	ρ (mm)	σ_N (MPa)
1	10.2	0.983	194.116	5.5
2	13.4	0.970	146.500	9.5
3	16.8	0.944	112.986	13.3
10	19.2	0.938	98.725	40.0
20	20.0	0.937	98.820	80.0
30	20.0	0.940	98.817	120.0

The wall thickness data of the parts under different hydraulic loading paths are measured. The 13 measurement points are located on the symmetry plane of the center of the part and are evenly distributed. The simulation curve of the wall thickness distribution is shown in Fig. 3.

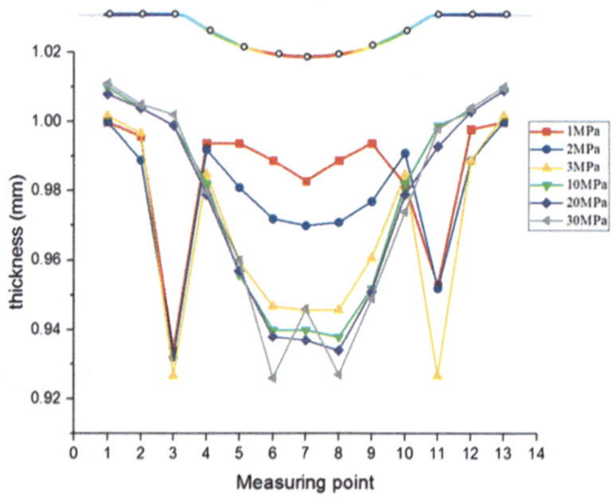

Fig. 3. The distribution of wall thickness

4 Conclusion

Based on 5A02 aluminum alloy, this paper uses the theoretical analysis method to derive the theoretical calculation model of springback amount considering thickness normal stress. In the later stage, ρ value and σ_N value obtained by numerical simulation and experiment can be substituted into Eq. (11) for comparison and verification. Numerical simulations are carried out under hydraulic loading paths with hydraulic pressures of 1, 2, 3, 10, 20, 30 Mpa. Studies have shown that due to strain strengthening or strain rate strengthening, the increase in deformation resistance in the region with the largest amount of deformation is also the largest, thereby making the deformation more dispersed. Macroscopically, the deformation is more uniform, so the wall thickness distribution will be more uniform.

References

1. Lingbeek, R., Huétink, J., Ohnimus, S., et al.: The development of a finite elements based springback compensation tool for sheet metal products. J. Mater. Process. Technol. **169**(1), 115–125 (2005)
2. Kinzel, G.L.: A new model for springback prediction for aluminum sheet forming. J. Eng. Mater. Technol. **127**(3), 279–288 (2005)
3. Vrh, M., Halilovič, M., Starman, B., et al.: Modelling of springback in sheet metal forming. Int. J. Mater. Form. **2**(1), 825–828 (2009)

4. Tamura, S., Sumikawa, S., Uemori, T., et al.: Experimental observation of elasto-plasticity behavior of type 5000 and 6000 aluminum alloy sheets. Mater. Trans. **61**(6), 255–261 (2011)
5. Shiguang, H.: Engineering Analysis of Sheet Metal Cold Stamping. Beijing University of Aeronautics and Astronautics Press, Beijing (2004)
6. Yu, H.Q., Chen, J.D.: Principles of Metal Forming. China Machine Press, Beijing (2010)
7. Zhang, Q.: Research on Composite Hydroforming Process and Key Technology of Aluminum Alloy Sheet. Beihang University, Beijing (2018)

On the Friction Test for Metal Forming Applications

Sergei Alexandrov[1,2(✉)], Marko Vilotic[3], and Dragisa Vilotic[3]

[1] School of Mechanical Engineering and Automation, Beihang University, 37 Xue Yuan Lu, Haidian Qu, Beijing Shi, China
sergei_alexandrov@yahoo.com

[2] Ishlinsky Institute for Problems in Mechanics, Russian Academy of Sciences, 101-1 Prospect Vernadskogo, 119526 Moscow, Russia

[3] Faculty of Technical Sciences, University of Novi Sad, Trg Dositeja Obradovica 6, 21000 Novi Sad, Serbia

Abstract. The ring compression test is a conventional test for identifying the friction law for metal forming applications. This test is very widely used but has a significant disadvantage. In particular, if the friction stress is high, then a sticking region occurs over a part of the friction surface. In this case, the interpretation of experimental results is difficult because the friction law at sliding is not valid over the sticking region. If the friction stress is very high, then this region occupies the entire friction surface. In this case, the process becomes insensitive to the friction factor at sliding at all. In order to overcome this difficulty and keep the conventional procedure for identifying the friction law, it is proposed to carry out the compression test between fat and conical dies. The geometry of the conical die should be chosen such that no sticking region occurs on the friction surface.

Keywords: Friction · Sticking · Metal forming

1 Introduction

The most popular friction law in metal forming applications is Tresca's law, which postulates that the friction stress is equal to a portion of the local shear yield stress. The friction factor involved in this law should be found from experiment. A typical test for this purpose is the ring compression test. The test must be supplemented with a theoretical analysis. The latter is often based on upper bound solutions. Several upper bound solutions for axisymmetric ring forging have been found in [1]. Solutions based on the velocity field proposed in [2] have been derived in [3, 4]. Other upper bound solutions for axisymmetric ring forging have been given in [5–8]. The solutions above are based on kinematically admissible velocity fields with no rigid regions. On the other hand, conventional friction laws used in conjunction with rigid perfectly plastic models demand the existence of such a region in certain cases. For example, in the case of axisymmetric ring forging a rigid region must appear if the friction stress changes its direction. The existence of a rigid zone can affect the interpretation of theoretical

E. Gdoutos and M. Konsta-Gdoutos (Eds.): ICTAEM 2020, STIN 16, pp. 158–161, 2020.
https://doi.org/10.1007/978-3-030-47883-4_27

solutions. For example, the definition for the neutral radius accepted in [2] is not valid in this case.

To overcome the difficulty above, it is proposed to use the compression test between flat and conical dies. In this case, the geometry of the conical die can be chosen such that no sticking region occurs independently of the friction stress.

2 Conventional Ring Compression Test

A schematic diagram of the conventional ring compression test is shown in Fig. 1.

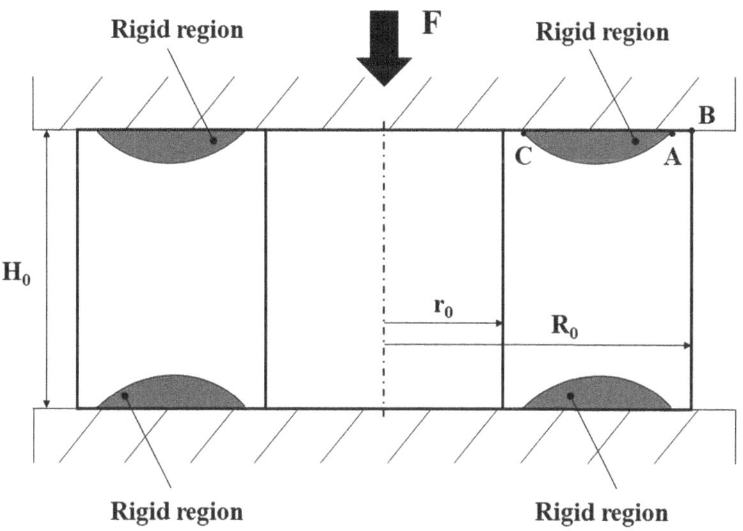

Fig. 1. Schematic diagram of the conventional ring compression test.

The interpretation of this test requires a theoretical solution. It is usually assumed that the regime of sliding occurs over the entire friction surface. This means that the friction law in the form

$$\tau_f = mk \tag{1}$$

is valid over the entire friction surface. In (1) τ_f is the friction stress, m is the friction factor and k is the shear yield stress. However, in reality, a sticking region usually occurs over a portion of the friction surface (Fig. 1). In this case, the friction law (1) is valid over region AB whereas the friction stress is not controlled by the friction law at sliding over region AC. Therefore, the interpretation above is not valid. Moreover, the region of sticking may occupy the entire friction surface. In this case, the solution becomes insensitive to the value of m at all. In the case of constraint forging, it has been found in [9] that the upper bound solution is not affected by the friction law if $m > m_{cr}$. The value of m_{cr} depends on geometric parameters.

3 New Ring Compression Test

In the present paper, it is proposed to use the ring compression test between flat and conical dies (Fig. 2).

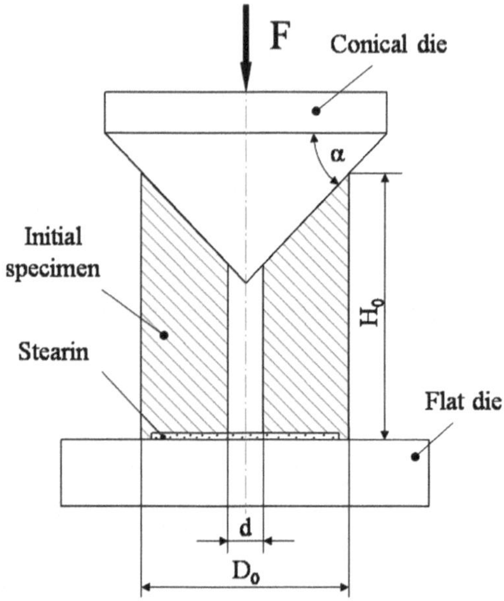

Fig. 2. Schematic diagram of the new ring compression test.

In general, the geometric parameters can be chosen such that no sticking region occurs on the friction surface between the conical die and specimen. Then, the procedure similar to that used in conjunction with the conventional ring compression test can be applied to evaluate the friction factor involved in (1). Several tests have been carried out to show the efficiency of the method proposed (Fig. 3).

Fig. 3. Conical dies used in the new ring compression test (cone angle: 8° 30′, 15° and 30°)

4 Conclusions

A disadvantage of the conventional ring compression test for identifying the friction law (1) is that a rigid region usually occurs over a part of the friction surface. This region may occupy the entire friction surface if the friction stress is high enough. It is proposed to use the compression between flat and conical dies (Fig. 2) to overcome the difficulty above. Several tests have been carried out to confirm the efficiency of the new test.

Acknowledgments. This work was carried out within the framework of the international project supported by grants RFBR-18-51-76001 (Russia) and BULKSURFACE – 359 (Ministry of education, science and technological development, Serbia).

References

1. Liu, J.Y.: Upper-bound solutions of some axisymmetric cold forging problems. ASME J. Eng. Ind. **93**, 1134–1144 (1971)
2. Avitzur, B., Van Tyne, C.J.: Ring formation: an upper bound approach. Part 1: flow pattern and calculation of power. ASME J. Eng. Ind. **104**, 231–237 (1982)
3. Avitzur, B., Van Tyne, C.J.: Ring formation: an upper bound approach. Part 2: process analysis and characteristics. ASME J. Eng. Ind. **104**, 238–247 (1982)
4. Avitzur, B., Van Tyne, C.J.: Ring formation: an upper bound approach. Part 3: constrained forging and deep drawing applications. ASME J. Eng. Ind. **104**, 248–252 (1982)
5. Avitzur, B., Sauerwine, F.: Limit analysis of hollow disk forging. Part 1: upper bound. ASME J. Eng. Ind. **100**, 340–344 (1978)
6. Lee, J.H.: Upper bound analysis of the upsetting of pressure – sensitive polymeric rings. J. Mater. Process. Technol. **30**, 601–612 (1988)
7. Alexandrov, S.: An analysis of the axisymmetric compression of viscous materials. J. Mater. Process. Technol. **105**, 278–283 (2000)
8. Wu, M.-C., Yeh, W.-C.: Effect of natural boundary conditions on the upper-bound analysis of upset forging of ring and disks. Mater. Des. **28**, 1245–1256 (2007)
9. Alexandrov, S., Lyamina, E., Jeng, J.-R.: A general kinematically admissible velocity field for axisymmetric forging and its application to hollow disk forging. Int. J. Adv. Manuf. Technol. **88**, 3113–3122 (2017)

From the Linear Theory of Elasticity to Bending Equations for Beams with Variable Cross-Section

Boris Gusev[1] and Vasily Saurin[2(✉)]

[1] RUT – MIIT, Obrazcova Str. 9b9, 127994 Moscow, Russia
[2] IPM RAS, Vernadskogo Pr-t 101-1, 119626 Moscow, Russia
saurin@ipmnet.ru

Abstract. A formulation of a boundary value problem to find natural frequencies of an inhomogeneous beam in the framework of the linear theory of elasticity is represented. The main attention in the article is paid to the analysis of transverse vibrations of beams with a variable cross-section. A system of two connected ordinary differential equations with variable coefficients is obtained in the framework of the linear theory of elasticity, by using semidiscrete approximations and a projection approach. Various bilateral energy quality estimates for approximate solutions that follow from the method of integro-differential relations are proposed. In the final part of the paper the advantages of the variational technique in problems of free vibrations of inhomogeneous beams are discussed based on a numerical example.

Keywords: Dynamics · Beams · Variable cross-section

1 Statement of the Problem

Consider an isotropic plate occupied in the plane Ox_1x_2 some area symmetric with the axis Ox_1:

$$V = \{x : -L/2 < x_1 < L/2, \quad -h(x_1) < x_2 < h(x_1)\} \quad \text{with} \quad x = (x_1, x_2).$$

It is considered that the stress-strain state of the plate is described by two-dimensional equations of the linear theory of elasticity and, also external plate boundaries are free of loading. Similar problems are considered in [1] and [2].

By using symmetry properties of the plate shape the low dimensional polynomial approximations of the unknown displacements u and stress σ functions with respect to one of the coordinates x_2 describing the lateral vibrations can be given as

$$\sigma_{11} = \sum_{j=0}^{M} \hat{x}^{2j+1} \sigma_{11}^{2j+1}(x_1), \quad \sigma_{12} = \sum_{j=0}^{M} \hat{x}^{2j} \sigma_{12}^{2j}(x_1), \quad \sigma_{22} = \sum_{j=0}^{M} \hat{x}^{2j+1} \sigma_{22}^{2j+1}(x_1),$$
$$u_1 = \sum_{j=0}^{M} \hat{x}^{2j+1} u_1^{2j+1}(x_1), \quad u_2 = \sum_{j=0}^{M} \hat{x}^{2j} u_2^{2j}(x_1), \quad \hat{x} = x_2/h(x_1).$$

E. Gdoutos and M. Konsta-Gdoutos (Eds.): ICTAEM 2020, STIN 16, pp. 162–163, 2020.
https://doi.org/10.1007/978-3-030-47883-4_28

In a projection approach, it is required that the displacements and stresses provide zero values to the integral projections of the constitutive vector and tensor on the selected tensor fields. These conditions give the possibility to construct different systems of ordinary differential equations with variable coefficients to study spectral properties of the beam.

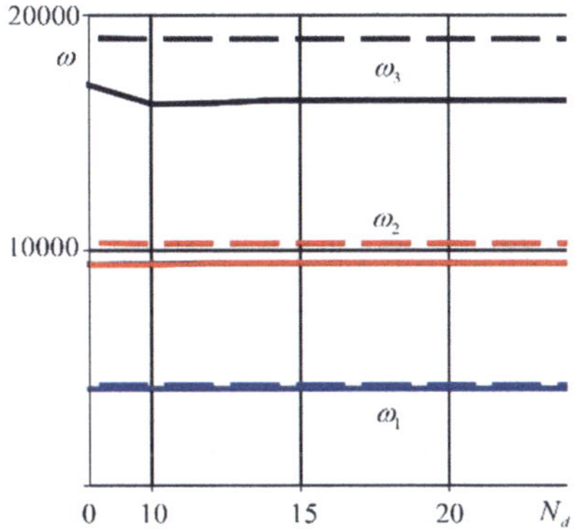

Fig. 1. Convergence of the eigenfrequencies.

Figure 1 shows the behavior of the first three eigenfrequencies versus the number of degrees of freedom N_d for two different mathematical models, describing the natural motion of a beam with variable cross section. The solid lines correspond to the eigenfrequencies that are obtained from the proposed equations with $M = 1$ and the dashed lines show the frequencies obtained from the classical beam equation

$$\rho(x)\frac{\partial^2 v}{\partial t^2} + \frac{\partial^2}{\partial t^2}\left(EJ(x)\frac{\partial^2 v}{\partial x^2}\right) = 0.$$

Here v is the transverse displacements of the beam center line points, ρ is the linear density of the material.

This work is financially supported in part by the State Program № AAAA-A17-117021310380-1 and Russian Foundation for Basic Research, projects no. 18-01-00812, 18-08-01346, 18-29-03228, 19-01-00173.

References

1. Saurin, V.: Analysis of dynamic behavior of beams with variable cross-section. Lobachevskii J. Math. **40**, 364–374 (2019)
2. Kostin, G., Saurin, V.: Dynamics of Solid Structures. Method Using Integrodifferential Relations. De Gruyter, Berlin (2017)

On Different Representations of Equations for Beam with Variable Cross-Section

Vladimir Poliakov[1] and Vasily Saurin[2(✉)]

[1] RUT – MIIT, Obrazcova Str. 9b9, 127994 Moscow, Russia
[2] IPM RAS, Vernadskogo Pr-t 101-1, 119626 Moscow, Russia
saurin@ipmnet.ru

Abstract. The main part of the paper is devoted to the formulations of the boundary value problem to finding natural frequencies of an inhomogeneous beam in the framework of the Euler-Bernoulli hypotheses. Three different formulations of the beam equation in displacements, momentums, and bending moments are discussed. Next, the possibilities to constructing various bilateral energy quality estimates for approximate solutions that follow from the method of integro-differential relations are investigated. In the final part, based on a numerical model example, the advantages of the variational technique in problems of free vibrations of inhomogeneous beams are discussed.

Keywords: Dynamics · Beams · Variable cross-section

1 Beam Equations

Consider a thin beam with a length L, which is described by an equation under the Euler-Bernoulli hypotheses. It is supposed that the cross section of the beam is a rectangle with the width $b = \text{const}$ and the height $h = h(x)$, which can be varied along its length. The beam lateral motions in the framework of this model can be represented by a linear partial differential equation

$$\rho(x)y_{tt} + (EJ(x)y_{xx})_{xx} = q(x) \quad x \in (0, L). \tag{1}$$

Here, y is the function described lateral displacements of the beam midline, $\rho(x)$ is a linear density of the beam material, E is Young's modulus, $q(x)$ is the external distributed load, $J(x)$ is a cross sectional moment of inertia. Similar problems are considered in [1] and [2].

Let us introduce new variables $p(t,x)$ and $m(t,x)$ that characterize the beam dynamics and, at the same time, have a clear physical meaning. The function $p(t,x)$ is the linear momentum density and $m(t,x)$ is the bending moment in a cross section of the beam. Introduce miscellaneous functions η and ξ

$$\eta = \rho y - p = 0, \quad \xi = EJy_{xx} - m = 0 \tag{2}$$

By using relations (2) Eq. (1) can be respectively expressed through momentum and moment functions

E. Gdoutos and M. Konsta-Gdoutos (Eds.): ICTAEM 2020, STIN 16, pp. 164–165, 2020.
https://doi.org/10.1007/978-3-030-47883-4_29

$$p_{tt} + \left(EJ \left(\frac{p}{\rho} \right)_{xx} \right)_{xx} = 0, \quad m_{tt} + EJ \left(\frac{m_{xx} - q}{\rho} \right)_{xx} = 0. \tag{3}$$

It can be seen that these equations differ from each other and this is primarily due to the fact that the variable functions of bending stiffness $EJ(x)$ and linear density $\rho(x)$ enter differentiation with respect to time t and spatial coordinate x in different way, respectively. However, one should not assume that these equations lead to different results, since they are obtained from a unique generating system of equations. But they can have different convergence rates when a numerical solution is finding.

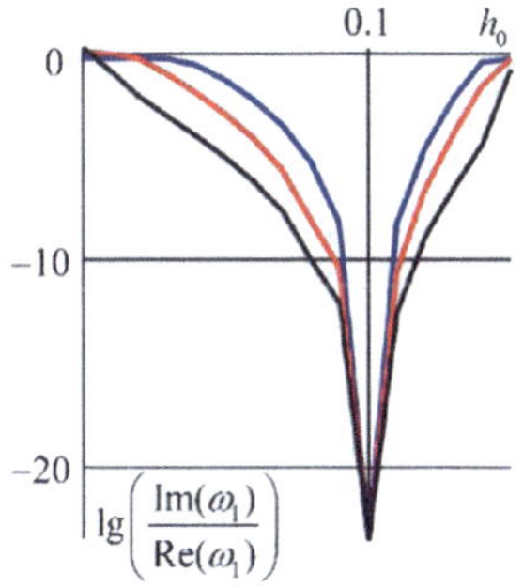

Fig. 1. Dependence the value of accuracy $\lg(\text{Im}(\omega_1)/\lg(\text{Re}(\omega_1))$ with respect to the parameter h_0.

Figure 1 shows the behavior of the accuracy of the first eigenfrequencies ω_1 versus the to the parameter h_0, $h(x) = (-12h_0 + 1.2)x^2 + h_0)$ obtained for three beam formulations discussed in this paper (black curve correspond to the statement in displacements, red – momentums, blue – moments). The value $h_0 = 0.1$ reflects the beam with constant height. It is seen that better accuracy is obtained for moment formulation and difference in accuracies reach several orders.

Acknowledgments. This work is financially supported in part by the State Program № AAAA-A17-117021310380-1 and Russian Foundation for Basic Research, projects no. 18-01-00812, 18-08-01346, 18-29-03228, 19-01-00173.

References

1. Saurin, V.: Analysis of dynamic behavior of beams with variable cross-section. Lobachevskii J. Math. **40**, 364–374 (2019)
2. Kostin, G., Saurin, V.: Dynamics of Solid Structures. Method Using Integrodifferential Relations. De Gruyter, Berlin (2017)

Shear Residual Stresses Induced by Torsional Loading

D. Pérez Gallego[(✉)], J. Ruiz Hervías, and D. A. Cendón Franco

Departamento de Ciencia de Materiales, Universidad Politécnica de Madrid, ETSI Caminos, Canales y Puertos, c/Profesor Aranguren, 3, 28040 Madrid, Spain
d.perez@upm.es

Abstract. Residual stresses can be induced by fabrication processes and service conditions that lead to non-uniform deformations. In the standard test method for residual stress analysis by x-ray diffraction (EN 15305), the validation is done by using reference specimens with known normal residual stress. However, the method also stablishes restrictions to the values of both normal and shear stresses, although the latter can be negligible with respect to the former. Consequently, it would be interesting to have a reference specimen with known shear residual stresses for independent validation. In the present work, torsion tests have been performed in a steel sample with the aim of producing shear residual stresses. Residual stresses were simulated by FEM modelling and measured by X-ray diffraction.

Keywords: Residual shear stresses · X-ray diffraction · Finite element method

1 Motivation

The analysis of residual stresses has an important role in the study of materials science or manufacturing processes. There are several methods to calculate the residual stresses, one of the most extended techniques being the X-ray diffraction. The standard EN 15305 [1] establishes certain criteria for reference specimens to validate the method. Most reference specimens are prepared by shoot peening [2], which produces residual compressive stresses at the surface. These kind of reference specimens have bee qualified by means of round robin tests [3].

With shot peening, a homogenous compressive residual stress can be generated in metallic materials in a controlled way, where shear stresses are negligible. However, the standard imposes restrictions for both normal and shear stresses which could be difficult to fulfill a reference specimen with known normal residual stresses. In summary, the standard limits the value of shear stresses in reference specimens although there are no such specimens with known residual shear stresses.

The main idea of the present work is to develop an experimental procedure to create reference specimens with known shear residual stresses. Torsional loads have been applied to pearlitic steel sample to induce this kind of stresses The residual stresses will to be measured by X-ray diffraction and the results will be compared with numerical simulations to study the residual stress distribution in the sample.

E. Gdoutos and M. Konsta-Gdoutos (Eds.): ICTAEM 2020, STIN 16, pp. 166–167, 2020.
https://doi.org/10.1007/978-3-030-47883-4_30

2 Experimental and Numerical Methods

Round bars of pearlitic steel have been loaded under torsion up to a given value of shear strain (see Fig. 1a). After unloading the bars, their surface shear residual stresses will be measured by X-ray diffraction, and compared with the numerical predictions obtained with the commercial FEM code *ABAQUS* (see Fig. 1b).

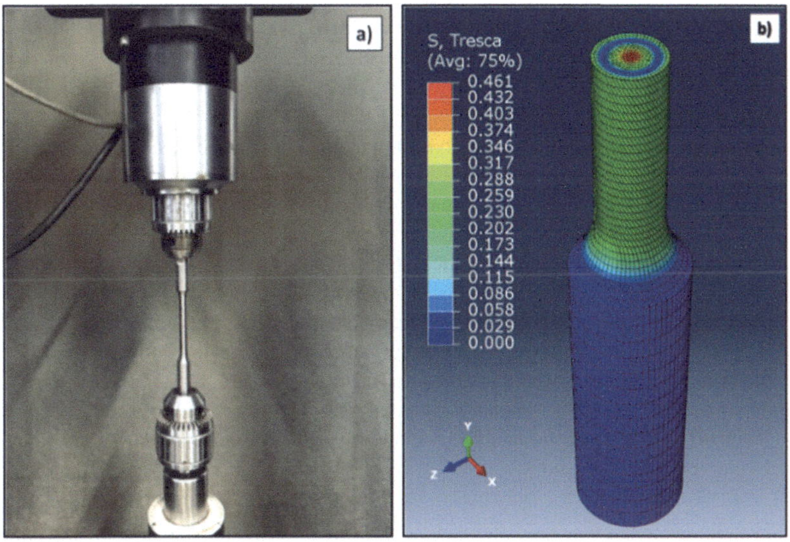

Fig. 1. **a)** Torsion test; **b)** Shear residual stresses obtained by FEM simulation.

References

1. European Standards: Non-destructive testing. Test method for residual stress analysis by X-ray diffraction (EN 15305) (2008)
2. Kobayashi, M., Matsui, T., Murakami, Y.: Mechanism of creation of compressive residual stress by shot peening. Int. J. Fatigue **20**(5), 351–357 (1998). https://doi.org/10.1016/S0142-1123(98)00002-4
3. Lefebvre, F., Wasniewski, E., François, M., Cacot, J., Le-Bec, P., Baumhauer, E., Bouscaud, D., Bergey, T., Blaize, D., Gloaguen, D., Cosson, A., Jégou, S., Cheynet, Y., Leray, S., Meheux, M., Monvoisin, J., Allain, P., Vidal, J.C., Sprauel, J., Hamdi, H.: External reference samples for residual stress analysis by x-ray diffraction. Adv. Mater. Res. **996**, 221–227 (2014). https://doi.org/10.4028/www.scientific.net/AMR.996.221

Elastoplastic Analysis of Re-entrant Cell Formation in Auxetic Foams Under Compression and Heat Treatment

Yun-Che Wang(✉), Tsai-Wen Ko, and Hsiang-Wei Lai

National Cheng Kung University, Tainan 70101, Taiwan, Republic of China
yunche@mail.ncku.edu.tw

Abstract. Negative Poisson's ratio of polymeric foam can be obtained by using the thermal transformation method to convert cell shape from convex to reentrant ones. In this work, the large-deformation finite element analysis is performed to quantify the magnitudes of compression and temperature for subsequent heat treatment to reduce elastic recovery, such that the reentrant shape can be maintained when the compression is removed. The amount of compression and temperature for heat treatment that can produce NPR foam can be used as a guide for experimental investigations, instead of trial-and-error methods to determine manufacturing parameters.

Keywords: Elastoplastic analysis · Finite element method · Auxetic foam · Negative Poisson's ratio

1 Introduction

Classical elasticity allows the Poisson's ratio (PR) of materials in three dimensions be in the range between -1 and 0.5. Through design of material microstructures, effective Poisson's ratio of polymeric foam can be negative when its cell shape is reentrant [1]. Recently, numerical investigations of the negative Poisson's ratio (NPR) properties in the 2D star-shaped lattice have been reported [2–4]. A method to convert the initially convex cell shape, which gives rise to positive Poisson's ratio, to the concave, i.e. reentrant, one is through the thermal transformation technique [1]. The thermal transformation method requires triaxial compression of the foam first and subsequent heat treatment while the compression is maintained. Permanent deformation due to plasticity and release of stress under heat treatment are responsible for minimizing elastic recovery when the compression load is removed. For metallic foam, sequential compression along three orthogonal directions on a cube is sufficient. In this work, we numerically model the deformation processes of the foam changing from positive to negative PR by finite element analysis with the elastic-perfectly-plastic material model and of the Mises flow rule.

E. Gdoutos and M. Konsta-Gdoutos (Eds.): ICTAEM 2020, STIN 16, pp. 168–169, 2020.
https://doi.org/10.1007/978-3-030-47883-4_31

2 Results and Discussion

Starting from the conventional cell shape, as shown in Fig. 1 at time t = 0, we adopt large-deformation plasticity theory to model the deformation of the unit cell under hydrostatic compression in two dimensions. The color bar indicates the Mises stress in units of MPa. The boundary conditions of the left and bottom edges are of the roller type. Uniform compressive stress is applied on the top and right edges. As time progresses, the snapshots of deformation at time t = 1, 1.5, 2 s are shown in Fig. 1. The reentrant cell shape can be clearly observed. The stress in the skeleton is removed by heat treatment, so that the reentrant shape is maintained. Elastic analysis of the reentrant unit cell reveals the effective Poisson's ratio is about -0.3.

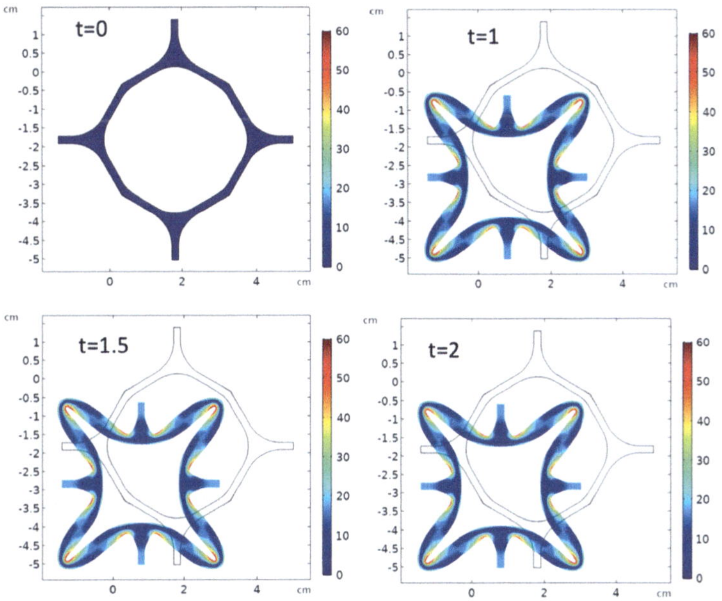

Fig. 1. Snapshots of the deformation processes of the re-entrant unit cell at time t = 0, 1, 1.5, 2 s under biaxial compression.

References

1. Lakes, R.: Foam structures with a negative Poisson's ratio. Science **235**(4792), 1038–1040 (1987)
2. Wang, Y.-C., Shen, M.-W., Liao, S.M.: Microstructural effects on the Poisson's ratio of star-shaped two-dimensional systems. Phys. Status Solidi B **254**, 1700024 (2017)
3. Wang, Y.-C., Lai, H.-W., Ren, X.J.: Enhanced auxetic and viscoelastic properties of filled reentrant honeycomb. Phys. Status Solidi B, 1900184 (2019)
4. Wang, Y.-C.: Negative Poisson's ratio in plane elasticity. In: Altenbach, H., Öchsner, A. (eds.) Encyclopedia of Continuum Mechanics. Springer, Heidelberg (2019)

Analytical Modeling of the Thermo-Mechanical Behavior of a Friction-Pendulum Seismic Isolator

Todor Zhelyazov(✉)

Technical University of Sofia, 1000 Sofia, Bulgaria
elovar@yahoo.com

Abstract. The contribution outlines the modeling of the thermo-mechanical behavior of a curved surface slider employed as a device for passive seismic isolation of buildings and structures. The Lagrange formalism is employed to derive the equation of motion of the slider on the spherical surface of the lower sliding pad. Besides the equations of motion, accurate modeling requires the formulation of the response of the friction interface. Previous studies have shown the dependence of the friction coefficient on the sliding velocity. Reportedly, the rise in temperature during sliding also affects the magnitude of the friction coefficient. The model is being developed in order to be further integrated into a finite element model of a base-isolated structure.

Keywords: Seismic isolation · Friction-Pendulum system · Analytical model

1 Introduction

Bearing devices for passive seismic isolation are increasingly used for buildings, bridges, and other types of structures. By dissipating energy, seismic isolation reduces the demands of the superstructure in terms of ductility and load-carrying capacity.

Among other types of devices, the most frequently used are high-damping rubber bearings, lead-core rubber bearings, and friction-pendulum systems. The latter type is the object of the research.

The simplest version of the friction-pendulum isolator (sometimes also referred to as a curved surface slider) consists of a slider and two sliding pads- upper and lower. Generally, contact surfaces have a spherical form.

2 Equations of Motion

The equations of motion are derived from the Lagrangian of the slider constrained to move on the spherical surface of the inferior sliding pad of the bearing device [1]. The resulting system of differential equations of motion is solved numerically. Mechanical behavior of the bearing device, subjected to strong ground motion, is investigated (see Fig. 1). The analysis assumes a constant value of the friction coefficient and an effective unilateral response of the sliding interface.

E. Gdoutos and M. Konsta-Gdoutos (Eds.): ICTAEM 2020, STIN 16, pp. 170–171, 2020.
https://doi.org/10.1007/978-3-030-47883-4_32

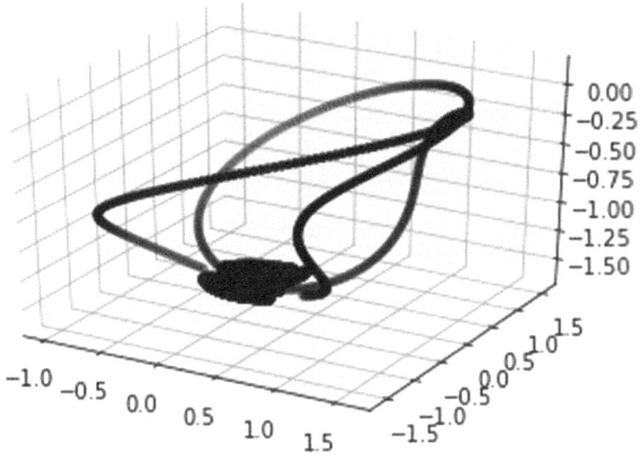

Fig. 1. Trajectory of the slider's motion (on the spherical surface of the lower sliding pad) in response to strong ground motion.

3 Response of the Sliding Interface

A model proposed by [2] is used to model the frictional behavior at the slider/spherical surface interface. The analytical model includes empirical relationships defined to take into account the influence of the normal pressure acting on the interface [3], as well as of the temperature rise during sliding [4, 5].

The outlined model is developed in order to be implemented into a finite element analysis of a base-isolated structure. Thus, in a more general context, the analytical model is employed in the assessment of the robustness of base-isolated structures.

References

1. Monti, G., Petrone, F.: Analytical thermo-mechanics 3D model of friction pendulum bearings. Earthquake Eng. Struct. Dyn. **45**, 957–977 (2016)
2. Park, Y.J., Wen, Y.K., Ang, A.H.S.: Random vibration of hysteretic system under a bi-directional ground motions. Earthquake Eng. Struct. Dyn. **14**(4), 543–557 (1985)
3. Mokha, A.S., Amin, N., Constantinou, M.C., Zayas, V.: Seismic isolation retrofit of large historic building. J. Struct. Eng. **122**(3), 298–308 (1996)
4. Kumar, M., Whittaker, A.S., Constantinou, M.C.: Characterizing friction in sliding isolation bearings. Earthquake Eng. Struct. Dynam. **44**, 1409–1425 (2015)
5. Quaglini, V., Bocciarelli, M., Gandelli, E., Dubini, P.: Numerical assessment of frictional heating in sliding bearings for seismic isolation. J. Earthquake Eng. **18**, 1198–1216 (2014)

Symposium on Theoretical and Experimental Approaches in Mechanics of Solids with Nonhomogeneities and Defects, by Roman Kushnir

Damage Accumulation and Limit State of Welded Pipelines with Corrosion-Erosion Metal Loss Due to Ultra-Low-Cycle Fatigue

Alexey Milenin(✉), Elena Velikoivanenko, Galina Rozynka, and Nina Pivtorak

E.O. Paton Electric Welding Institute of NAS of Ukraine, 11, K. Malevich Street, Kiev 03150, Ukraine
asmilenin@ukr.net

Abstract. Analytical prediction of the reliability of pipelines under ultra-low-cycle loading demands the consideration of the several interrelated damaging factors, those determine the limit state of specific structure. For that in this work it was developed the complex numerical technique for finite-element assessment of damage accumulation and limit state of welded pipelines with corrosion-erosion metal loss. The ductile mechanism of subcritical damage was considered as the main one that assumes certain material softening due to voids nucleation and accumulation up to limit state. Additionally, it was necessary to take into account the material hardening and softening in plastic strain that was described by isotropic and kinematic mixed hardening laws. This complex approach allowed revealing the major physical-mechanical mechanisms of structure failure depending on the type of external loading, size of the local metal loss and the features of welding on the limit state of typical pipeline element.

Keywords: Pipeline · Corrosion-erosion metal loss · Limit state · Welding · Isotropic hardening · Kinematic hardening · Ductile fracture

1 Introduction

Numerical assessment of residual strength of pipeline elements (PE) with revealed corrosion-erosion metal losses is the typical problem for expert analysis of reliability of various industrial systems. The common approach for solving this problem is the determination of the limit state of specific structure under rated loading. As the most of pipelines are welded ones the influence of residual state of stresses and strains should be taken into account if the metal loss belongs to welding area. This problem is studied well for static loading (e.g. by internal pressure) whereas for cyclic loading complexity of physical-mechanical processes, those cause the subcritical damage and failure, demands sufficient conservativeness of corresponding analytical techniques. Ultra-low-cycle fatigue is notable for the sufficient plastic flow of material which demands taking into account the material hardening and softening as well as nucleation and accumulation of distributed voids of ductile fracture of certain volume concentration f. Presence of welded joint and local geometry anomalies of structure (operational defects of

E. Gdoutos and M. Konsta-Gdoutos (Eds.): ICTAEM 2020, STIN 16, pp. 175–180, 2020.
https://doi.org/10.1007/978-3-030-47883-4_33

corrosion-erosion metal loss) leads to peculiar state of stresses and strains of PE under internal pressure and/or bending moment that influences on the susceptibility of specific welded structure to fracture.

Within the limits of this work it is proposed the complex numerical technique for finite-element prediction of damage accumulation and limit state of welded pipelines with corrosion-erosion metal losses under ultra-low-cycle loading.

2 Numerical Procedure for Prediction of Limit State of Welded Pipeline with Corrosion-Erosion Metal Loss Under Ultra-Low-Cycle Loading

2.1 Main Assumptions and Material Model

Ultimate limit state of corroded (eroded) PE depends on a number of physical-mechanical processes initiated by operational and technological influence. In case of large cyclic strain loading (e.g. earthquake, landslide, inaccurate overloading, buckling, etc.) limit state of pipeline is determined by the evolution of plastic strains and corresponding accumulation of subcritical damage up to nucleation of macroflaw. Local corrosion-erosion metal losses and corresponding stress concentrations lead to a certain decrease of bearing capacity of pressurized PE. Their admissibility is determined by relevant standard specifications for design operational loads, but ultra-low-cycle loading considers exceptional conditions of pipeline operation that complicates expert analysis of their compliance with design demands. The main complexity consists in nonlinear response of material to cyclic plastic deformation and evolution of its properties. Thus, along with characteristic strain hardening, two possible mechanisms of material softening could be assigned [1]: Bauschinger effect due to change of the direction of plastic deformation, and the accumulation of porosity ductile damage which is caused by plastic strain and lead to the reduction of true cross-section of the structure. As the typical technological and main pipelines are the welded ones, the presence of assembly girth welds causes the spatial inhomogeneity of stress-strain state and complex interaction of operational and residual stresses. This also should be taken into account, especially in case of close location of the metal loss and weld.

2.2 Finite-Element Description of Physical-Mechanical Processes in Ultra-Low-Cycle Loading of Welded Pipeline with Local Corrosion-Erosion Metal Loss

The physical phenomena, those determine the ultimate limit state of typical welded PE with local geometry anomaly, are spatially nonuniform and the conventional approach for their description is the finite-element (FE) modeling. For considered problem the defective pipeline of diameter D and wall thickness t was modeled within the limits of the cylindrical set of coordinates (r, β, z), local metal loss was approximated as semi-elliptical wall thinning of the size of 2 s, $2u$ and a (see Fig. 1). In welding and further low-cycle loading of PE by internal pressure P and bending moment M the current and limit state of stresses and strains is determined by several mechanisms of deformation, those corresponds to respective summand of total strain tensor ε_{ij} ($i, j = r, \beta, z$) namely: elastic ε_{ij}^{e}, plastic ε_{ij}^{p}, thermal $\delta_{ij}\varepsilon_{T}$ and ductile damage $\delta_{ij}(f/3)$.

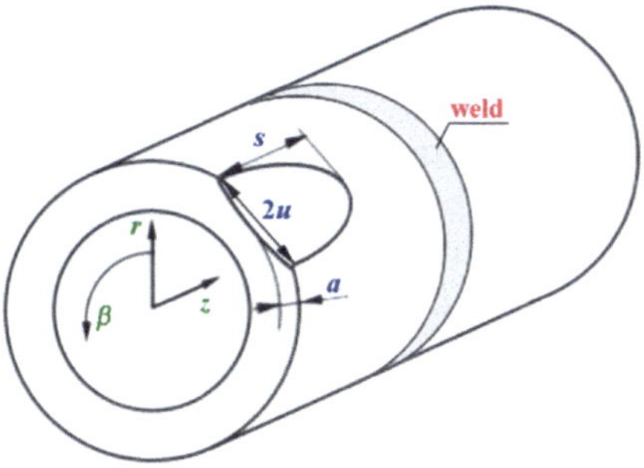

Fig. 1. Scheme of welded pipeline element with defect of semi-elliptic shape in cylindrical coordinate system (r, β, z).

Elastic strains correlate with current stresses as follows:

$$\varepsilon_{ij}^{e} = \frac{\sigma_{ij} - \delta_{ij}\sigma_{m}}{2G} + \delta_{ij}(K\sigma_{m} + \varphi) \tag{1}$$

where $\sigma_m = (\sigma_{rr} + \sigma_{\beta\beta} + \sigma_{zz})/3$; $K = (1 - 2\nu)/E$, $G = 0.5E/(1 + \nu)$, E is Young modulus, ν is Poison's ratio, φ is volumetric changes (thermal or caused by structural changes).

The increment of plastic strains for a specific FE linearly depends on the scalar function Λ and deviator component of the stress tensor, namely:

$$\mathrm{d}\varepsilon_{ij}^{p} = \mathrm{d}\Lambda(\sigma_{ij} - \delta_{ij}\sigma_{m}) \tag{2}$$

In turn, the specific value of the function Λ depends on the stress state in the considered area of the structure, as well as on the shape of the yield surface Φ [2].

For numerical prediction of porosity nucleation in material of welded pipeline with local metal loss the strain-based criterion was used [3], further growth of voids depends on stress triaxiality (ratio of the hydrostatic σ_m to equivalent σ_{eq} stress) and the intensity of the plastic strain ε_{i}^{p} according to Rice-Tracey law [4].

Thus, the increment of strain tensor components could be presented as follows [5]:

$$\mathrm{d}\varepsilon_{ij} = \mathrm{d}\varepsilon_{ij}^{e} + \mathrm{d}\varepsilon_{ij}^{p} + \delta_{ij}(\mathrm{d}\varepsilon_{T} + \mathrm{d}f/3) \tag{3}$$

Equation (3) in differential form has the next expression:

$$\Delta\varepsilon_{ij} = \Psi(\sigma_{ij} - \delta_{ij}\sigma_{m}) + \delta_{ij}(K\sigma_{m} + \Delta\varepsilon_{T} + \Delta f/3) - \frac{1}{2G}(\sigma_{ij} - \delta_{ij}\sigma_{m})^{*} + (K\sigma_{m})^{*} \tag{4}$$

where symbol "*" refers to the variable of the previous tracing step, Ψ is the state function of the material, which is determined by plastic flow surface Φ, that was considered according to Gurson-Tvergaard-Needleman model [6]:

$$\Phi = \left(\frac{\sigma_{eq}}{\sigma_Y}\right)^2 - (q_3 f')^2 + 2q_1 f' \cosh\left(q_2 \frac{3\sigma_m}{2\sigma_Y}\right) \tag{5}$$

where q_1, q_2, q_3 are constants, f' is equivalent concentration of voids, σ_Y is yield stress.

The solvable system of equations in the variables of the vector of displacement increments ΔU_i in FE nodes at each step of tracing and Ψ-iterations was determined by the minimization of the functional L (Lagrangian variational principle), i.e. [2]

$$\frac{\partial L}{\partial \Delta U_i} = 0, \tag{6}$$

where $L = -\frac{1}{2}\sum_V (\sigma_{ij} + J_{ij})\Delta\varepsilon_{ij} V_{m,n,r} + \sum_{S_P} F_i \Delta U_i \Delta S_P^{m,n,r}$, $\sum_V$ is the sum operator by internal FEs, $\sum_{S_P}$ is the sum operator on surface FEs on which the components of the force vector F_i are specified.

It was considered that limit state of specific structure is reached in case of fulfilling the one of three conditions of numerical criterion of brittle-ductile fracture [2]:

$$\begin{aligned} &\Psi - \tfrac{1}{2G} \geq \frac{\varepsilon_f - (\varepsilon_i^p)^*}{1.5\sigma_Y(\varepsilon_i^p, f')}; \\ &f' \geq \tfrac{1}{q_1}\exp\left(-\tfrac{3q_2\sigma_m}{2\sigma_Y}\right); \\ &\tfrac{\sigma_1}{1-2f/3} > S_K, \end{aligned} \tag{7}$$

where S_K is the microcleavage stress, ε_f is the critical strain (Mackenzie rule).

The mixed hardening rule was used for prediction of material properties evolution under ultra-low-cycle fatigue loading. It combines the linear kinematic hardening rule in Von Mises form [7] and isotropic hardening effect for taking into account the of yield surface transformation due to plastic deformation and accumulation of ductile:

$$\sqrt{\frac{3}{2}(\sigma_{ij} - \delta_{ij}\sigma_m - \bar{X})(\sigma_{ij} - \delta_{ij}\sigma_m - \bar{X})} - \sigma'_Y(f')[1 + (\varepsilon^p/\varepsilon_0)]^m \leq 0 \tag{8}$$

where $\bar{X}$ is shift tensor of kinematic hardening, $\sigma'_Y(f')$ is material state according to (5), m, ε_0 are material constants.

3 Results and Discussion

Residual state of stresses and strains in the weld region, kinematic hardening and ductile damage have an influence on limit state of eroded PE under ultra-low-cycle loading with internal pressure and bending moment. One of the fundamental problems, that could be solved with the described above FE approach, is the determination of influence of this interrelated phenomena on bearing capacity of specific welded structure. As a case of study it was considered the PE of $D \times t = 315 \times 10$ mm size made of 316L stainless steel (E = 193 GPa, ν = 0.3, σ_Y = 170 MPa). Individual erosion metal loss of semi-elliptical shape on inner surface of the pipe ($2s \times 2u \times a$ = $40 \times 20 \times 5$ mm). Examples of equivalent stresses distributions over cross-section of the pipe after welding and under operational conditions are given in Fig. 2.

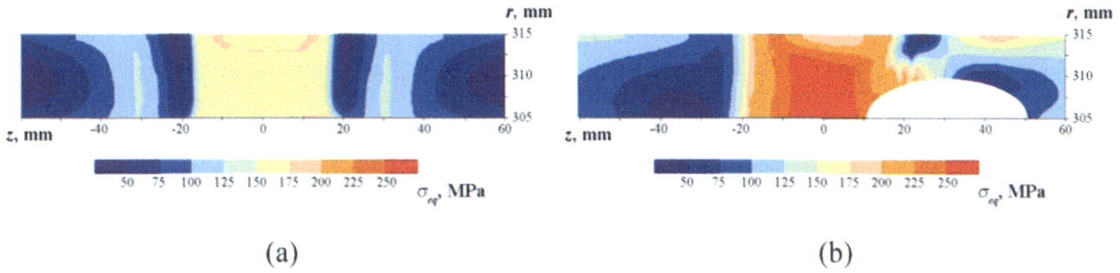

Fig. 2. Numerically assessed distributions of equivalent stresses σ_{eq} in the pipeline ($D \times t$ = 315×10 mm, 316L stainless steel): (a) – residual state in the region of girth weld; (b) – with internal erosion defect ($2s \times 2u \times a$=$40 \times 20 \times 5$ mm) under operational pressure 8.0 MPa.

Figure 3a shows the dependencies of local stresses $\sigma_{\beta\beta}$ on strains $\varepsilon_{\beta\beta}$ near the internal defect of erosion thinning with and without taking into account the ductile damage of material caused both with internal pressure P = 10 MPa and cycling bending moment M from −85 to 85 kN•m (that corresponds to the range of maximum axial stress from −120 to 120 MPa). As it could be seen, accumulation of porosity in plastic deformation of pipe steel leads to the total shifting of stress-strain hysteresis loops to higher strains because of porous material softening and decrease of the structure bearing cross-section.

Intensity of ductile damage accumulation in ultra-low-cycle loading (i.e. growth rate of porosity volume concentration f per number of cycles N) has three main stages: plastic deformation before nucleation of ductile damage; nucleation of porosity and redistribution of the fields of strains and stresses; stable growth plastic strains and porosity volume concentration up to limiting state according (7). First two stages correspond to the static loading and take place on the first cycles whereas the third one is connected with fatigue fracture of plastically deformed material. Figure 3b shows the results of numerical assessment of ductile damage accumulation for considered example of eroded PE in stable growth stage. As it could be seen, ductile porosity concentration f increases quasilinearly starting the second cycle of loading with bending moment under the same internal pressure P. That means, that growth rate of f mostly depends on applied range of cycling load, but not the plastic (or total) strain path.

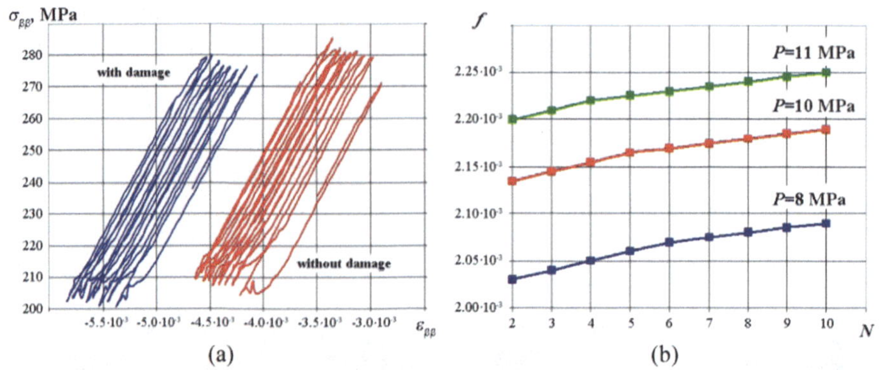

Fig. 3. Dependencies of local stresses $\sigma_{\beta\beta}$ on strains $\varepsilon_{\beta\beta}$ near the internal defect of erosion thinning ($2s \times 2u \times a$ = 40 × 20 × 5 mm) of pipe ($D \times t$ = 315 × 10 mm, 316L) with and without taking into account the ductile damage of material (internal pressure P = 10 MPa and bending moment $M = -85 \div 85$ kN•m) – (a) and growth rate of porosity volume concentration f per number of cycles N – (b).

References

1. Jia, L.-J., Kuwamura, H.: Ductile fracture model for structural steel under cyclic large strain loading. J. Constr. Steel Res. **106**, 110–121 (2015). https://doi.org/10.1016/j.jcsr.2014.12.002
2. Makhnenko, V.: Problems of examination of modern critical welded structures. Paton Weld. J. **5**, 21–28 (2013)
3. Milenin, A., Velikoivanenko, E., Rozynka, G., Pivtorak, N.: Probabilistic procedure for numerical assessment of corroded pipeline strength and operability. Int. J. Press. Vessels Pip. **171C**, 60–68 (2019). https://doi.org/10.1016/j.ijpvp.2019.02.003
4. Chen, Z., Butcher, C.: Micromechanics Modelling of Ductile Fracture. Springer, Dordrecht (2013). https://doi.org/10.1007/978-94-007-6098-1
5. Velikoivanenko, E.A., Rozynka, G.F., Milenin, A.S., Pivtorak, N.I.: Modelling of processes of nucleation and development of ductile fracture pores in welded structures. Paton Weld. J. **9**, 24–29 (2013). https://doi.org/10.1016/j.ijpvp.2019.02.003
6. Xue, L.: A unified expression for low cycle fatigue and extremely low cycle fatigue and its implication for monotonic loading. Int. J. Fatigue **30**, 1691–1698 (2008). https://doi.org/10.1016/j.ijfatigue.2008.03.004
7. Lemaitre, J., Chaboche, J.-L.: Mechanics of Solid Materials. Cambridge University Press, Cambridge (1990)

Unsteady Elastic-Diffusion Oscillations of a Simply Supported Kirchhoff Plate Under the Distributed Transverse Load Action

O. A. Afanasieva[1] and A. V. Zemskov[1,2(✉)]

[1] Moscow Aviation Institute (National Research University), Moscow, Russia
azemskov1975@mail.ru
[2] Institute of Mechanics, Lomonosov Moscow State University, Moscow, Russia

Abstract. We study unsteady vibrations of a isotropic Kirchhoff plate considering mass transfer. In the general case, the plate is subjected to tensile and shear forces as well as bending moments and torque. Densities of diffusion fluxes are also defined. For the problem formulation, we use the coupled elastic diffusion continuum model in a rectangular Cartesian coordinate system. Further, the unsteady model of an elastodiffusive Kirchhoff plate is obtained using the d'Alembert variational principle. The solution is sought in integral form. To find the Green's functions, we use the Laplace integral transform and Fourier series expansion.

Keywords: Elastic diffusion · Coupled problem · Unsteady problem · Integral transformation · Multicomponent continuum · Kirchhoff plate · Green's function

1 Problem Formulation

We consider the unsteady vibrations problem of a rectangular isotropic N - component Kirchhoff plate under action of mechanical and diffusion perturbations. Figure 1 shows the orientation of Cartesian axes as well as how the forces and the moments are applied.

When moment density and transverse load density are absent, the transverse vibrations equation of the plate has the form (the dots denote the time derivative) [1]:

$$\Delta\ddot{w} - \frac{12}{h^2}\ddot{w} = \Delta\Delta w + \sum_{q=1}^{N}\alpha_q\Delta H_q, \quad \dot{H}_q = D_q\Delta H_q + \Lambda_q\Delta\Delta w. \tag{1}$$

All quantities in (1) are dimensionless. We accepted the following notation:

$$x_i = \frac{x_i^*}{l},\ w = \frac{w^*}{l},\ \tau = \frac{Ct}{l},\ \lambda = \frac{\lambda^*}{\lambda^* + 2\mu^*},\ \mu = \frac{\lambda^*}{\lambda^* + 2\mu^*},\ l_m = \frac{l_m^*}{l},$$

$$\alpha_q = \frac{\alpha_q^*}{\lambda^* + 2\mu^*},\ D_q = \frac{D_q^*}{Cl},\ \Lambda_q = \frac{m^{(q)}D_q^*\alpha_q^* n_0^{(q)}}{\rho RT_0 Cl},\ C^2 = \frac{\lambda^* + 2\mu^*}{\rho},\ h = \frac{h^*}{l},$$

E. Gdoutos and M. Konsta-Gdoutos (Eds.): ICTAEM 2020, STIN 16, pp. 181–186, 2020.
https://doi.org/10.1007/978-3-030-47883-4_34

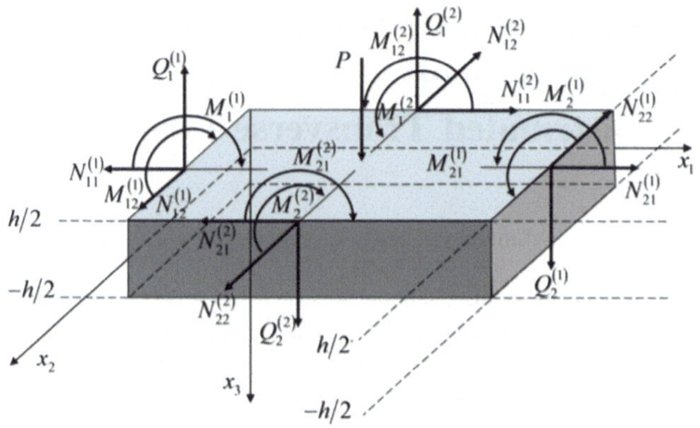

Fig. 1. Forces and moments acting upon the plate.

where t is time; x_i^* are Cartesian coordinates; w^* is a plate transversal displacement; l is the characteristic linear size of the problem (in this case, the plate diagonal, which has dimensions $l_1^* \times l_2^*$ and thickness h^*); H_q are concentration increments for the q-th medium component of an N-component continuum; $n_0^{(q)}$ are initial concentrations; λ^* and μ^* are Lame coefficients; ρ is density; α_q^* are coefficients characterizing the medium volumetric changes due to diffusion; D_q^* are the self-diffusion coefficients; R is the universal gas constant; T_0 is the initial temperature; $m^{(q)}$ is the molar mass.

The Eq. (1) are supplemented by the boundary conditions. In the case of pure bending under the action of bending moments $M_k^{(l)}$, boundary conditions has the form:

$$\begin{aligned}
&\left.\left(\frac{\partial^2 w}{\partial x_1^2} + \lambda \frac{\partial^2 w}{\partial x_2^2} + \sum_{q=1}^{N} \alpha_q H_q\right)\right|_{x_1=0} = f_{111}(x_2, \tau),\\
&\left.\left(\frac{\partial^2 w}{\partial x_1^2} + \lambda \frac{\partial^2 w}{\partial x_2^2} + \sum_{q=1}^{N} \alpha_q H_q\right)\right|_{x_1=l_1} = f_{112}(x_2, \tau),\\
&\left.\left(\lambda \frac{\partial^2 w}{\partial x_1^2} + \frac{\partial^2 w}{\partial x_2^2} + \sum_{q=1}^{N} \alpha_q H_q\right)\right|_{x_2=0} = f_{121}(x_1, \tau),\\
&\left.\left(\lambda \frac{\partial^2 w}{\partial x_1^2} + \frac{\partial^2 w}{\partial x_2^2} + \sum_{q=1}^{N} \alpha_q H_q\right)\right|_{x_2=l_2} = f_{122}(x_1, \tau);
\end{aligned} \tag{2}$$

$$w|_{x_1=0} = f_{211}(x_2, \tau),\ w|_{x_1=l_1} = f_{212}(x_2, \tau),\ w|_{x_2=0} = f_{221}(x_1, \tau),\ w|_{x_2=l_2} = f_{222}(x_1, \tau),$$

$$\begin{aligned}
&H_q|_{x_1=0} = f_{q+2,11}(x_2, \tau), \quad H_q|_{x_1=l_1} = f_{q+2,12}(x_2, \tau),\\
&H_q|_{x_2=0} = f_{q+2,21}(x_1, \tau), \quad H_q|_{x_2=l_2} = f_{q+2,22}(x_1, \tau),
\end{aligned}$$

$$f_{111}(x_2,\tau) = -\frac{12}{h^3}M_1^{(1)}(x_2,\tau), \quad f_{112}(x_2,\tau) = -\frac{12}{h^3}M_2^{(1)}(x_2,\tau),$$
$$f_{121}(x_1,\tau) = -\frac{12}{h^3}M_1^{(2)}(x_1,\tau), \quad f_{122}(x_1,\tau) = -\frac{12}{h^3}M_2^{(2)}(x_1,\tau).$$

The initial conditions are assumed to be zero.

2 Solution Method

The solutions of the problem (1), (2) are sought in integral form

$$\begin{aligned} w(x_1,x_2,\tau) &= \\ &= \sum_{k=1}^{N+2}\int_0^{l_2}[G_{1k1}(x_1,x_2,\zeta,\tau)*f_{k11}(\zeta,\tau)+G_{1k1}(l_1-x_1,x_2,\zeta,\tau)*f_{k12}(\zeta,\tau)]d\zeta \\ &+ \sum_{k=1}^{N+2}\int_0^{l_1}[G_{1k2}(x_1,x_2,\xi,\tau)*f_{k21}(\xi,\tau)+G_{1k2}(x_1,l_2-x_2,\xi,\tau)*f_{k22}(\xi,\tau)]d\xi, \end{aligned}$$

$$\begin{aligned} H_q(x_1,x_2,\tau) &= \\ &= \sum_{k=1}^{N+2}\int_0^{l_2}\left[G_{q+1,k1}(x_1,x_2,\zeta,\tau)*f_{k11}(\zeta,\tau)+G_{q+1,k1}(l_1-x_1,x_2,\zeta,\tau)*f_{k12}(\zeta,\tau)\right]d\zeta \\ &+ \sum_{k=1}^{N+2}\int_0^{l_1}\left[G_{q+1,k2}(x_1,x_2,\xi,\tau)*f_{k21}(\xi,\tau)+G_{q+1,k2}(x_1,l_2-x_2,\xi,\tau)*f_{k22}(\xi,\tau)\right]d\xi, \end{aligned}$$

where G_{ikl} are the surface Green's functions, which satisfy the equations:

$$\Delta\ddot{G}_{1kl} - \frac{12}{h^2}\ddot{G}_{1kl} = \Delta\Delta G_{1kl} + \sum_{q=1}^{N}\alpha_q\Delta G_{q+1,kl}, \quad \dot{G}_{q+1,kl} = D_q\Delta G_{q+1,kl} + \Lambda_q\Delta\Delta G_{1kl}, \tag{3}$$

and the following boundary conditions:

$$
\begin{gathered}
\left.\left(\frac{\partial^2 G_{1kl}}{\partial x_1^2}+\lambda\frac{\partial^2 G_{1kl}}{\partial x_2^2}+\sum_{q=1}^{N}\alpha_q G_{q+1,kl}\right)\right|_{x_1=0}=\delta_{1k}\delta_{1l}\delta(\tau)\delta(x_2-\zeta),\\
\left.\left(\lambda\frac{\partial^2 G_{1kl}}{\partial x_1^2}+\frac{\partial^2 G_{1kl}}{\partial x_2^2}+\sum_{q=1}^{N}\alpha_q G_{q+1,kl}\right)\right|_{x_2=0}=\delta_{1k}\delta_{2l}\delta(\tau)\delta(x_1-\xi),\\
\left.\left(\frac{\partial^2 G_{1kl}}{\partial x_1^2}+\lambda\frac{\partial^2 G_{1kl}}{\partial x_2^2}+\sum_{q=1}^{N}\alpha_q G_{q+1,kl}\right)\right|_{x_1=l_1}=0,\\
\left.\left(\lambda\frac{\partial^2 G_{1kl}}{\partial x_1^2}+\frac{\partial^2 G_{1kl}}{\partial x_2^2}+\sum_{q=1}^{N}\alpha_q G_{q+1,kl}\right)\right|_{x_2=l_2}=0,\\
G_{1kl}|_{x_1=0}=\delta_{2k}\delta_{1l}\delta(\tau)\delta(x_2-\lambda),\ G_{1kl}|_{x_2=0}=\delta_{2k}\delta_{2l}\delta(\tau)\delta(x_1-\lambda),\\
G_{q+1,kl}\big|_{x_1=0}=\delta_{q+2,k}\delta_{1l}\delta(\tau)\delta(x_2-\lambda),\ G_{q+1,kl}\big|_{x_2=0}=\delta_{q+2,k}\delta_{2l}\delta(\tau)\delta(x_1-\lambda),\\
G_{1kl}|_{x_1=l_1}=0,\ G_{1kl}|_{x_2=l_2}=0,\ G_{q+1,kl}\big|_{x_1=l_1}=0,\ G_{q+1,kl}\big|_{x_2=l_2}=0.
\end{gathered}
\tag{4}
$$

To find the Green functions G_{ikl}, the expanding into double trigonometric Fourier series in spatial coordinate and the Laplace transform in time are use. As a result, problem (3), (4) is reduced to the following system of linear algebraic equations (s is Laplace transform parameter):

$$
\begin{aligned}
&k_1(\nu_{nm},s)G_{1kl}^{Lss}(\lambda_n,\mu_m,\xi,\zeta,s)-\sum_{q=1}^{N}\alpha_q\nu_{nm}^2 G_{q+1,kl}^{Lss}(\lambda_n,\mu_m,\xi,\zeta,s)\\
&=F_{1kl}(\lambda_n,\mu_m,\xi,\zeta,s),\\
&-\Lambda_q\nu_{nm}^4 G_{1kl}^{Lss}(\lambda_n,\mu_m,\xi,\zeta,s)+k_{q+1}(\nu_{nm},s)G_{q+1,kl}^{Lss}(\lambda_n,\mu_m,\xi,\zeta,s)\\
&=F_{q+1,kl}(\lambda_n,\mu_m,\xi,\zeta,s),
\end{aligned}
\tag{5}
$$

where $\lambda_n=\pi n/l_1$, $\mu_m=\pi m/l_2$. Other quantities are defined as:

$$
\begin{aligned}
&k_1(\nu_{nm},s)=s^2\left(\nu_{nm}^2+\frac{12}{h^2}\right)+\nu_{nm}^4,\ k_{q+1}(\nu_{nm},s)=s+D_q\nu_{nm}^2,\ \nu_{nm}^2=\lambda_n^2+\mu_m^2,\\
&F_{1kl}(\lambda_n,\mu_m,\xi,\zeta,s)=-\frac{4\lambda_n}{l_1l_2}\delta_{1l}\left\{\delta_{1k}-\delta_{2k}\left[\lambda_n^2+(2-\lambda)\mu_m^2+s^2\right]\right\}\sin\mu_m\zeta\\
&-\frac{4\mu_m}{l_1l_2}\delta_{2l}\left\{\delta_{1k}-\delta_{2k}\left[\mu_m^2+(2-\lambda)\lambda_n^2+s^2\right]\right\}\sin\lambda_n\xi,
\end{aligned}
\tag{6}
$$

$$F_{q+1,kl}(\lambda_n, \mu_m, \xi, \zeta, s) = -\frac{4}{l_1 l_2} \Lambda_q \lambda_n \delta_{2k} \delta_{1l} \left[\lambda_n^2 + (2-\lambda)\mu_m^2\right] \sin \mu_m \zeta$$
$$-\frac{4}{l_1 l_2} \Lambda_q \mu_m \delta_{2k} \delta_{2l} \left[(2-\lambda)\lambda_n^2 + \mu_m^2\right] \sin \lambda_n \xi$$
$$+\frac{4}{l_1 l_2} \left[\Lambda_q \delta_{1k} + D_q \delta_{q+2,k} - \Lambda_q \sum_{j=1}^{N} \alpha_j \delta_{j+2,k}\right] (\mu_m \delta_{2l} \sin \lambda_n \xi + \lambda_n \delta_{1l} \sin \mu_m \zeta), \tag{7}$$
$$G_{ikl}^{Lss}(\lambda_n, \mu_m, \xi, \zeta, s) = \frac{4}{l_1 l_2} \int_0^{l_1} \int_0^{l_2} G_{ikl}^{L}(x_1, x_2, \xi, \zeta, s) \sin \lambda_n x_1 \sin \mu_m x_2 dx_2 dx_1,$$
$$G_{ikl}^{L}(x_1, x_2, \xi, \zeta, s) = \sum_{n=1}^{\infty} \sum_{m=1}^{\infty} G_{ikl}^{Lss}(\lambda_n, \mu_m, \xi, \zeta, s) \sin \lambda_n x_1 \sin \mu_m x_2$$

The solution of the system (5) has the form $\left(q, p = \overline{1, N}, \quad l = 1, 2\right)$

$$G_{1kl}^{Lss}(\lambda_n, \mu_m, \xi, \zeta, s) = \frac{P_{1kl}(\lambda_n, \mu_m, \xi, \zeta, s)}{P(\nu_{nm}, s)},$$
$$G_{q+1,1l}^{Ls}(\lambda_n, \mu_m, \xi, \zeta, s) = \frac{4\Lambda_q}{l_1 l_2} \frac{(\delta_{1l}\lambda_n \sin \mu_m \zeta + \delta_{2l} \mu_m \sin \lambda_n \xi)}{k_{q+1}(\nu_{nm}, s)} + \frac{P_{q+2,1l}(\lambda_n, \mu_m, \xi, \zeta, s)}{Q_q(\nu_{nm}, s)},$$
$$G_{q+1,2l}^{Ls}(\lambda_n, \mu_m, \xi, \zeta, s) = -\frac{4}{l_1 l_2} \frac{\Lambda_q \lambda_n \delta_{1l} \left[\lambda_n^2 + (2-\lambda)\mu_m^2\right] \sin \mu_m \zeta}{k_{q+1}(\nu_{nm}, s)}$$
$$-\frac{4}{l_1 l_2} \frac{\Lambda_q \mu_m \delta_{2l} \left[(2-\lambda)\lambda_n^2 + \mu_m^2\right] \sin \lambda_n \xi}{k_{q+1}(\nu_{nm}, s)} + \frac{P_{q+2,2l}(\lambda_n, \mu_m, \xi, \zeta, s)}{Q_q(\nu_{nm}, s)} \tag{8}$$
$$G_{q+1,p+2,l}^{Ls}(\lambda_n, \mu_m, \xi, \zeta, s) = \frac{4\Lambda_q}{l_1 l_2} \frac{\delta_{1l} \lambda_n \left(D_q \delta_{qp} - \Lambda_q \alpha_p\right) \sin \mu_m \zeta}{k_{q+1}(\nu_{nm}, s)}$$
$$+\frac{4\Lambda_q}{l_1 l_2} \frac{\delta_{2l} \mu_m \left(D_q \delta_{qp} - \Lambda_q \alpha_p\right) \sin \lambda_n \xi}{k_{q+1}(\nu_{nm}, s)} + \frac{P_{q+2,p+2,l}(\lambda_n, \mu_m, \xi, \zeta, s)}{Q_q(\nu_{nm}, s)},$$

where

$$P(\nu_{nm}, s) = k_1 \Pi - \nu_{nm}^6 \sum_{j=1}^{N} \alpha_j \Lambda_j \Pi_j, \quad Q_q(\nu_{nm}, s) = k_q(\nu_{nm}, s) P(\nu_{nm}, s).$$

$$P_{11l}(\lambda_n, \mu_m, \xi, \zeta, s) = -\frac{4}{l_1 l_2} \left(\Pi - \nu_{nm}^2 \sum_{j=1}^{N} \alpha_j \Lambda_j \Pi_j\right) (\lambda_n \delta_{1l} \sin \mu_m \zeta + \mu_m \delta_{2l} \sin \lambda_n \xi),$$

$$P_{12l}(\lambda_n, \mu_m, \xi, \zeta, s)$$
$$= \frac{4\lambda_n}{l_1 l_2} \delta_{1l} \left\{ \left[\lambda_n^2 + (2-\lambda)\mu_m^2 + s^2\right] \Pi - \nu_{nm}^2 \sum_{j=1}^{N} \alpha_j \Lambda_j \left[\lambda_n^2 + (2-\lambda)\mu_m^2\right] \Pi_j \right\} \sin \mu_m \zeta$$
$$+ \frac{4\mu_m}{l_1 l_2} \delta_{2l} \left\{ \left[\mu_m^2 + (2-\lambda)\lambda_n^2 + s^2\right] \Pi - \nu_{nm}^2 \sum_{j=1}^{N} \alpha_j \Lambda_j \left[\mu_m^2 + (2-\lambda)\lambda_n^2\right] \Pi_j \right\} \sin \lambda_n \xi,$$

$$P_{1,q+2,l}(\lambda_n, \mu_m, \xi, \zeta, s) = \frac{4\alpha_q v_{nm}^2}{l_1 l_2}\left(D_q\Pi_q - \sum_{j=1}^{N}\alpha_j\Lambda_j\Pi_j\right)(\lambda_n\delta_{1l}\sin\mu_m\zeta + \mu_m\delta_{2l}\sin\lambda_n\xi),$$

$$P_{q+1,kl}(\lambda_n, \mu_m, \xi, \zeta, s) = \Lambda_q v_{nm}^4 P_{1kl}(\lambda_n, \mu_m, \xi, \zeta, s),$$

$$\Pi(v_{nm}, s) = \prod_{j=1}^{N} k_{j+1}(v_{nm}, s), \quad \Pi_j(v_{nm}, s) = \prod_{r=1, r\neq j}^{N} k_{r+1}(v_{nm}, s).$$

Since $P_{rkl}(\lambda_n, \mu_m, \xi, \zeta, s)$ are rational functions of the transformation parameter s, the originals in equalities (8) are calculated on the base of residues and the tables of operational calculus.

This work was funded by the subsidy from RFBR (Project №20-08-00589 A).

Reference

1. Zemskov, A.V., Tarlakovskii, D.V.: Model nestatsionarnykh uprugodiffuzionnykh kolebaniy plastiny Kirchhoffa. XII Vserossiyskiy syezd po fundamentalnym problemam teoreticheskoy i prikladnoy mekhaniki: sbornik trudov v 4 tomakh. T. 3: Mekhanika deformiruyemogo tverdogo tela, pp. 906–909, Ufa, RITs BashGU (2019). In Russian). https://doi.org/10.22226/2410-3535-2019-congress-v3

A Dynamic Contact Problem of Torsion that Reduces to the Singular Integral Equation with Two Fixed Singularities

V. Popov(✉) and O. Kyrylova

National University «Odessa Maritime Academy», Odessa, Ukraine
dr.vg.popov@gmail.com, olga.i.kyrylova@gmail.com

Abstract. An elastic cylinder of finite length, one of the ends of which is perfectly coupled to the surface of the elastic half-space is considered. A round rigid plate of the same radius is coupled to the other end of the cylinder, and is loaded the torsion moment that is harmonic depend of time. The surface of the half-space outside the contact area with the cylinder and the side surface of the cylinder are been unload. The formulated boundary problem is reduced to a singular integral equation for a function related to stresses in the contact area of the cylinder and half-space. Since the kernel of this integral equation contains fixed singularities, a numerical method for solving this equation the is main result. After solving the integral equation, approximate formulas for calculating the contact stresses.

Keywords: Torsion oscillation · Contact problem · Singular integral equation

1 Introduction

At nowadays, one of the effective methods for solving the boundary value problems of the mechanics of a deformable body is to reduce them to solving integral equations, most often singular ones. Since the exact solution of these equations is rarely possible, the actual problem is the creation of numerical methods for their solution. The presence in the singular part of kernels with fixed singularities makes it difficult to solve r integral equations. In the monograph [1], as well as in articles [2–4] where exact solutions of singular integral equations are found, it is proved that the presence of fixed singularities affects the asymptotic behavior of the solution near the ends of the integration segments. Despite this, in many cases the real asymptotic of the unknown functions is either not taken into account Therefore, the convergence of these numerical methods is quite slow. Articles [5–8] show that the methods based on the use of special quadrature formulas for singular integrals and taking into account the real asymptotic of the solution are most effective in the sense of convergence. Such method for an integral equation with two fixed singularities, to which the contact problem of torsional vibrations of a cylinder on an elastic half space reduces is proposed in this article.

E. Gdoutos and M. Konsta-Gdoutos (Eds.): ICTAEM 2020, STIN 16, pp. 187–192, 2020.
https://doi.org/10.1007/978-3-030-47883-4_35

2 Statement of the Problem and Its Reduction to a Singular Integral Equation

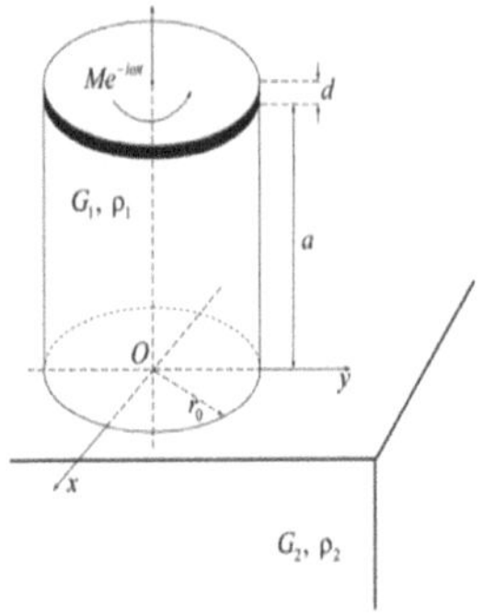

Fig. 1. The elastic cylinder coupled with half space

Let an elastic cylinder $0 \leq r < r_0, 0 \leq z < a, 0 \leq \varphi < 2\pi$ locate on the elastic half-space $0 \leq r < +\infty$, $-\infty < z \leq 0$, $0 \leq \varphi < 2\pi$ (Fig. 1) and coupled to it. A round rigid plate of the same radius as the cylinder and the thickness d is connected to the upper end of the cylinder. A torsion moment $Me^{-i\omega t}$, harmoniously dependent on time, is exerted to the plate. The factor $e^{-i\omega t}$ that determines the dependence on time is omitted farther. Under such conditions, an axisymmetric torsional deformation is realized in the cylinder and half-space and only angular displacements $w_j(r,z)$, $j = 1, 2$ is nonzero. They are determined from the equations

$$\frac{\partial^2 w_j}{\partial r^2} + \frac{1}{r}\frac{\partial w_j}{\partial r} - \frac{w_j^2}{r^2} + \frac{\partial^2 w_j}{\partial z^2} + \kappa_{2j}^2 w_j = 0; \; \kappa_{2j}^2 = \frac{\rho_j^2 \omega^2}{G_j^2}, \; j = 1, 2; \tag{1}$$

where $w_1(r,z)$ is displacement in cylinder, $w_2(r,z)$ is displacement in half space, ρ_1, G_1 are shear modulus and density of the cylinder, ρ_2, G_2 are shear modulus and density of the half space. In the area of contact between the cylinder and the half-space, the following equations are satisfied:

$$\tau_{\varphi z}^{(1)}(r, +0) = q(r), \; \tau_{\varphi z}^{(2)}(r, -0) = q(r), \; 0 \leq r < r_0 \tag{2}$$

where $q(r)$ is unknown stresses in the contact area. Also in the contact area, the condition of continuity of displacements is fulfilled

$$w_1(r, +0) = w_2(r, -0), \;\; 0 \leq r < r_0. \tag{3}$$

The surface of the half-space outside the contact area is considered unloaded

$$\tau_{z\varphi}^{(2)}(r, -0) = 0, \;\; r > r_0. \tag{4}$$

On the upper end of the cylinder, the conditions of couple to the plate is satisfied

$$w_1(r, a) = \theta_0 r, \; 0 \leq r \leq r_0. \tag{5}$$

where θ_0 is unknown plate rotation angle, It is determined from the equation

$$-\omega^2 j_0 \theta_0 = M - M_R, \; j_0 = \pi r_0^4 d \rho_0 / 2 \tag{6}$$

where j_0 is the moment of inertia of the plate, M_R is the moment of the reaction forces. There ρ_0 is density of the plate. The lateral surface of the cylinder is not loaded:

$$\tau_{r\varphi}^{(1)}(r_0, z) = 0,\ 0<z<a. \tag{7}$$

Angular displacement in a cylinder is the solution of the boundary value problem (1), (2), (5), (7) and it equal

$$w_1(r,z) = -\int_0^{r_0} \eta \frac{q(\eta)}{G_1} F_1(\eta, r, z) d\eta + \theta_0 r \frac{\cos \kappa_{21} z}{\cos a\, \kappa_{21}}. \tag{8}$$

$$F_1(\eta, r, z) = \frac{2}{a}\sum_{k=1}^{\infty}\left[g_k^1(\eta, r) + \frac{K_2(q_{k1} r_0)}{I_2(q_{k1} r_0)} I_1(q_{k1}\eta) I_1(q_{k1} r)\right] \cdot \cos \lambda_k z,$$

$$g_k^1(\eta, r) = \begin{cases} I_1(q_{k1} r) K_1(q_{k1}\eta), & r<\eta; \\ I_1(q_{k1}\eta) K_1(q_{k1} r), & r>\eta, \end{cases} \quad \lambda_k = \frac{\pi(2k-1)}{2a};\ q_{k1} = \sqrt{\lambda_k^2 - \eta_{21}^2}.$$

Angular displacement in half-space is the solution of boundary value problem (1), (2) (4), and it equal to

$$w_2(\eta, r) = \int_0^{r_0} \eta \frac{q(\eta)}{G_2} \int_0^{+\infty} \frac{\beta e^{q_2(\beta) z}}{q_2(\beta)} J_1(\beta r) J_1(\beta\eta) d\beta\, d\eta,\ q_2(\beta) a = \sqrt{\beta^2 - \kappa_{22}^2} \tag{9}$$

Now it is necessary to find the unknown contact stresses $q(r)$ for the final determination of the displacement in the cylinder and the half-space For this purpose, the integral equation was obtained by substitution (8), (9) in (3). This integral equation can be transformed into a second-kind integral equation [9] for a new unknown function. This function is related to contact stresses by the next formulas

$$\varphi(x) = \int_x^{r_0} \eta \frac{q(\eta)}{\sqrt{\eta^2 - x^2}} d\eta,\ q(\eta) = -\frac{2}{\pi}\frac{\partial}{\partial \eta}\int_\eta^{r_0} \frac{\varphi(x) dx}{\sqrt{x^2 - \eta^2}}. \tag{10}$$

As a result of the transformations detailed in [9] and the extraction of the singular component, this equation takes the following form:

$$\begin{gathered}(1+c) g(\zeta) + \frac{1}{\pi}\int_{-1}^{1} g(\tau)\left[\frac{1}{2-\tau-\zeta} + \frac{1}{2+\tau+\zeta} - \frac{15}{4}\ln(2+\tau+\zeta)\right. \\ \left. - \frac{15}{4}\ln(2+\tau+\zeta) + R(\tau,\zeta)\right] d\tau = \frac{2\theta_0 \zeta}{\cos(\gamma\,\kappa_0)},\quad -1<\zeta<1.\end{gathered} \tag{11}$$

When deriving Eq. (11) the next notation was taken: $x = r_0\tau$, $y = r_0\zeta$, $\varphi(r_0\tau) = r_0 G_1 g(\tau)$, $\gamma = a/r_0$, $\kappa_0 = \kappa_{21} r_0$, $c = G_1/G_2$.

3 The Numerical Solution of Integral Equation

For construction an efficient numerical method of solving Eq. (11), we need to find out the asymptotic of the unknown function $g(\tau)$ for $\tau \to \pm 1$. This defines the following form of solution:

$$g(\tau) = \left(1 - \tau^2\right)^{\sigma} \psi(\tau), \ \sigma = \frac{1}{\pi} \arcsin \frac{1}{c+1}, \tag{12}$$

where $\psi(\tau)$ is a function that satisfies the Holder conditions.

Quadrature formulas for singular integrals are based on the approximation of the function by the following interpolation polynomial:

$$\psi(\tau) \approx \Psi_n(\tau) = \sum_{m=1}^{n} \psi_m \frac{P_n^{\sigma,\sigma}(\tau)}{[P_n^{\sigma,\sigma}(\tau_m)]'(\tau - \tau_m)}, \tag{13}$$

where $\psi_m = \psi(\tau_m)$, $P_n^{\sigma,\sigma}(\tau)$ are Jacobi polynomials and τ_m are roots of these polynomials. As a result, we obtain the quadrature formulas by the method described in detail in the article [5]

$$\int_{-1}^{1} \frac{(1-\tau^2)^{\sigma} \psi(\tau)}{2 \pm \tau \pm \zeta} d\tau = \sum_{m=1}^{n} \psi_m \frac{A_m^{\sigma\sigma} \left[P_n^{\sigma\sigma}(\tau_m)\right]' + (\mp 1)^{n+1} B_n\left(\frac{1 \pm \varsigma}{2}\right)}{\left(P_n^{\sigma\sigma}(\tau_m)\right)' (2 \pm \tau_m \pm \varsigma)}, \tag{14}$$

$$B_n(Y) = \frac{\Gamma(1+\sigma+n)}{2^{-2\sigma} n!} \left[\sum_{j=0}^{+\infty} C_{nj} Y^{\sigma+j} + \sum_{j=0}^{+\infty} B_{nj} Y^j\right],$$

$$C_{nj} = \frac{(-1)^j \Gamma(-\sigma-n) \Gamma(1+\sigma+n+j)}{j! \Gamma(1+\sigma+n+j)}, B_{nj} = \frac{(-1)^j \Gamma(\sigma-j) \Gamma(1+n+j)}{j! \Gamma(1+2\sigma+n-j)},$$

where $A_m^{\sigma\sigma}$ are coefficients of the Gauss – Jacobi quadrature formula [10].

Similar formulas for integrals with a logarithmic singularity are obtained by the same method [5]:

$$\int_{-1}^{1} (1-\tau^2)^{\sigma} \psi(\tau) \ln(2 \pm \tau \pm \zeta) d\tau. = \sum_{m=1}^{n} A_m^{\sigma\sigma} \psi_m \sum_{j=0}^{n-1} \frac{P_j^{\sigma\sigma}(\tau_m)}{(\mp 1)^j \sigma_j^2} h_j\left(\frac{1 \pm \zeta}{2}\right). \tag{15}$$

The functions $h_j(Y)$ are represented by series similar to these in formula (17).

Then in Eq. (11) the quadrature formulas (15), (17) are applied to the singular integrals, and to the regular ones the Gauss - Jacobi quadrature formulas [12] and as collocation points are taken $\zeta = \tau_k,\ k = 1, 2, \ldots, n$. The result of these actions is a system of linear algebraic equations for $\psi_m, m = 1, 2, \ldots, n$. To the resulting system it is necessary to add the equation of motion of the plate (6). After solving the system, from formulas (10), (12), (13) we obtain formulas for the approximate calculation of the contact stresses.

$$s(\zeta) = \frac{q(r_0\zeta)}{G_1} = \zeta(1-\zeta^2)^{\sigma-\frac{1}{2}} \sum_{m=1}^{n} \Psi_m A_m^{\sigma\sigma} W_m(\zeta) \tag{16}$$

Here functions $W_m(\zeta)$ are represented through hypergeometric functions very cumbersomely.

4 Results of Numerical Analysis and Conclusions

The cylinder of radius $r_0 = 0{,}2\ m$ of aluminum coupled to a cast iron base is considered, as an example. The plate adhered to the upper end of the cylinder is steel and has a thickness of $d = 0{,}02\ m$. The plate is loaded by the moment with amplitude of $M = 1000\ n \cdot m$. The frequency of oscillation changes so that the dimensionless wave number $\kappa_0 = \kappa_2 r_0 = \omega r_0 \sqrt{\rho_1/G_1}$ is in the range $0 \le \kappa_0 \le 10$ (Fig. 2).

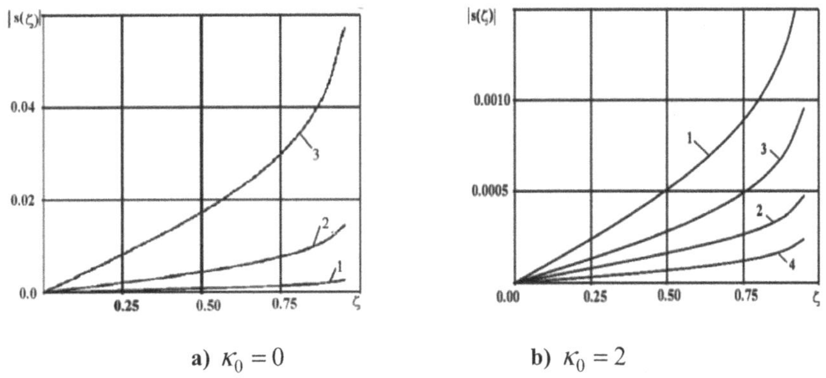

Fig. 2. The contact stresses for other values of frequency and relative height of cylinder

The calculations showed that to obtain the values of contact stresses and the angle of rotation with a relative error of less than for the numerical solution of the integral Eq. (11) in the formula (13), it suffices to use 10–15 interpolation points.

Using formula (18), the influence of the frequency and the ratio of the dimensions of the cylinder on the values of contact stresses is studied. The results of these studies are shown in the figures. The graphs in these figures correspond to the following values of the ratio of the height of the cylinder to its radius $1{-}\gamma = 1,\ 2{-}\gamma = 2,\ 3{-}\gamma = 3,\ 4{-}\gamma = 4$. The following conclusions can be drawn based on the analysis of the results of the calculations. The proposed method for numerically solving a singular integral equation with a fixed singularity with a small volume of calculations allows us to obtain results with high accuracy. This is explained by the fact that the solution takes into account the real asymptotic of unknown functions, and use special quadrature formulas are derived for singular integrals. At torsion of a cylinder which is coupled with an elastic foundation, the highest values of contact stress are observed under static loading $\kappa_0 = 0$. In the same case, with an increase in the relative length of the cylinder, the absolute values of the contact stress increase. Monotonic increase in the absolute values of contact stresses is observed when approaching the boundary of the contact region in the considered frequency range $0 \leq \kappa_0 \leq 10$.

References

1. Duduchava, R.: Integral Equations of Convolution with Discontinuous Presymbols, Singular Integral Equations with Fixed Singularities, and Their Applications to Problems of Mechanics. Razmadze Institute of Mathematics, Academy of Sciences of Gruz. SSR. Tbilisi (1979). (in Russian)
2. Stullybrass, M.: A crack perpe-ndicular to an elastic half-plane. Int. J. Eng. Sci. **8**(5), 351–362 (1970)
3. Hrapkov, A.: The problems of elastic equilibrium of an infinite wedge with an asymmetric incision at the apex, solved in a closed form. Prikladnaya matematika i mehanika. **36**(6), 1062–1069 (1971)
4. Afyan, B.: On the integral equations with fixed singularities in the theory of branched cracks. Dokl. AN Arm. SSR **79**(4), 60–65 (1984)
5. Popov, V.: A dynamic contact problem which reduces to a singular integral equation with two fixed singularities. J. Appl. Math. Mech. **76**(3), 348–357 (2012)
6. Popov, V.: A crack in the form of a three-link broken line under the action of longitudinal shear waves. J. Math. Sci. **222**(2), 143–154 (2017)
7. Popov, V.: Problems of interaction longitudinal shear waves with V-shape tunnels defect. J. Phys: Conf. Ser. **991**(1), 012066 (2018)
8. Lytvyn, O., Popov, V.: Interaction of harmonic longitudinal shear waves with v-shaped. Incl. J. Math. Sci. **240**, 1–16 (2019)
9. Popov, V.: Stress state of finite elastic cylinder with a circular crack undergoing torsion vibration. Int. Appl. Mech. **48**(4), 430–437 (2012)
10. Krylov, V.: Approximate Calculation of Integrals. Dover, New York (2006)

Some Dynamic Problems for Layered Composites

Sergey Pshenichnov(✉)

Institute of Mechanics, Lomonosov Moscow State University, Michurinsky Prospect 1, Moscow 119192, Russia
serp56@yandex.ru

Abstract. The problems of constructing analytical solutions of non-stationary dynamic problems for layered composites in the case when all the layers are linearly viscoelastic, but their hereditary properties are determined by only one kernel, the same for all layers, are considered. In this case, the Poisson's ratio of the material of each layer becomes time-independent. A special case is considered when all the layers are linearly elastic. The perturbation propagation region is assumed to be limited. The integral Laplace transform in time is applied, followed by inversion. Sufficient conditions, under which all the poles of the solution in the transforms are simple, are formulated. The question of the location of these poles is considered. An example is given of constructing an analytical solution to a two-dimensional non-stationary dynamic problem for a cylindrical body consisting of elastic or viscoelastic layers under the above assumption. The results of a study of two-dimensional transient wave processes in a layered cylindrical elastic body, based on the corresponding analytical solution, are presented.

Keywords: Dynamics of layered composites · Piecewise homogeneous bodies · Wave processes

1 Introduction

In the study of transient wave processes in layered solids, one of the important directions is to conduct research based on analytical solutions of the corresponding initial-boundary-value problems. Known analytical results in this area for the case when the layers are linearly viscoelastic can be found, for example, in [1–7]. At the same time, many unexplored issues remain within the framework of this topic. This primarily concerns the study of the influence of the properties of the material on non-one-dimensional transient wave processes in bodies with an arbitrary number of interfaces between homogeneous components. Known analytical solutions often turn out to be valid only in a limited range of time changes, or with significant restrictions on the material properties of homogeneous components (layers), or the solution is difficult to analyze. Earlier, the author of this paper examined theoretical issues related to the construction of solutions to non-stationary dynamic problems for composites with an arbitrary number of homogeneous linear viscoelastic layers by the Laplace transform method with subsequent inversion [8]. At the interfaces between the layers, it

E. Gdoutos and M. Konsta-Gdoutos (Eds.): ICTAEM 2020, STIN 16, pp. 193–197, 2020.
https://doi.org/10.1007/978-3-030-47883-4_36

was assumed that the continuity conditions for the displacement and stress vectors were fulfilled, and the disturbance propagation region was considered limited. Statements about important properties of solutions in images were formulated and proved, which facilitate the construction of originals. These statements reveal the connection between the branch points and the poles of the solution of the problem in images with the spectrum of the corresponding problem of free vibrations of the piecewise-homogeneous body under consideration. This work is devoted to non-stationary dynamic problems for layered composites in the case when all the layers are linearly viscoelastic, but their hereditary properties are determined by only one kernel, the same for all the layers. Moreover, the Poisson's ratio of the material of each layer does not depend on time. A special case when all layers are linearly elastic is considered.

2 Mathematical Formulation of the Problem

Let us consider the non-stationary dynamic problem for a layered composite occupying a domain Ω with a boundary Σ and consisting of N homogeneous isotropic linear-viscoelastic components (layers): $\Omega = \Omega_1 \cup \Omega_2 \cup \ldots \cup \Omega_N$ (Ω_i and Ω_j do not intersect at internal points if $i \neq j$). On the contact surfaces between these components the conditions of continuity of displacement and stresses vectors are satisfied. Consider the case when volumetric and shear relaxations kernels of the material of every component are identical: $T_V^{(n)} \equiv T_s^{(n)}(t) \equiv T(t),\ n = 1, 2, \ldots, N$. In this case Poisson's ratio in every component is time-independent: $\nu^{(n)} \equiv \nu_0^{(n)}$ $(const)$. For each component of the body we write the equations of dynamics ($n = 1, 2, \ldots, N$):

$$(1 - \hat{T})\,\hat{L}^{(n)}\mathbf{u}^{(n)}(\mathbf{x}, t) + \mathbf{f}^{(n)}(\mathbf{x}, t) = \rho_n \ddot{\mathbf{u}}^{(n)}(\mathbf{x}, t) \tag{1}$$

and the constitutive relations

$$\tilde{\boldsymbol{\sigma}}^{(n)}(\mathbf{x}, t) = (1 - \hat{T})\,\hat{l}^{(n)}\,\mathbf{u}^{(n)}(\mathbf{x}, t), \qquad \mathbf{x}(x_1, x_2, x_3) \in \Omega_n \tag{2}$$

For components with numbers $1 \leq m \leq N$, having the common points with a boundary $\Sigma = \Sigma_1 \cup \Sigma_2$, we present the boundary conditions

$$\begin{aligned} \tilde{\boldsymbol{\sigma}}^{(m)}(\mathbf{x}, t)\mathbf{n} = \mathbf{p}^{(m)}(\mathbf{x}, t), \quad & \mathbf{x} \in \Sigma_1 \\ \mathbf{u}^{(m)}(\mathbf{x}, t) = \mathbf{q}^{(m)}(\mathbf{x}, t), \quad & \mathbf{x} \in \Sigma_2, \quad t > 0 \end{aligned} \tag{3}$$

On the contact surface of adjacent components with numbers p and q let us write down the relationships

$$\mathbf{u}^{(p)}(\mathbf{x}, t) = \mathbf{u}^{(q)}(\mathbf{x}, t), \quad \tilde{\boldsymbol{\sigma}}^{(p)}(\mathbf{x}, t)\,\mathbf{n} = \tilde{\boldsymbol{\sigma}}^{(q)}(\mathbf{x}, t)\mathbf{n}, \quad \mathbf{x} \in \Sigma_{pq} \tag{4}$$

For each component, we set the initial conditions

$$\mathbf{u}^{(n)}(\mathbf{x},0) = \mathbf{b}_1^{(n)}(\mathbf{x}), \quad \dot{\mathbf{u}}^{(n)}(\mathbf{x},0) = \mathbf{b}_2^{(n)}(\mathbf{x}), \quad \mathbf{x} \in \Omega_n \tag{5}$$

Here $\hat{L}^{(n)}$ and $\hat{l}^{(n)}$ are differential operators, $\hat{T}$ is integral operator:

$$\hat{L}^{(n)}\mathbf{u}^{(n)}(\mathbf{x},t) = (\lambda_0^{(n)} + \mu_0^{(n)})\operatorname{grad}\operatorname{div}\mathbf{u}^{(n)}(\mathbf{x},t) + \mu_0^{(n)}\Delta\mathbf{u}^{(n)}(\mathbf{x},t), \tag{6}$$

$$\begin{gathered}\hat{l}^{(n)}\mathbf{u}^{(n)}(\mathbf{x},t) = 2\mu_0^{(n)}\operatorname{def}\mathbf{u}^{(n)}(\mathbf{x},t) + \lambda_0^{(n)}\operatorname{div}\mathbf{u}^{(n)}(\mathbf{x},t)\,\tilde{\mathbf{I}}, \\ \hat{T}\xi(t) = \int_0^t T(t-\tau)\xi(\tau)\,d\tau\hat{T}\xi(t),\end{gathered} \tag{7}$$

A dot above a letter means a time derivative; $\tilde{\boldsymbol{\sigma}}^{(n)}$ is the stress tensor; $\mathbf{u}^{(n)}$, $\mathbf{p}^{(m)}$, $\mathbf{q}^{(m)}$ $\mathbf{f}^{(n)}$, $\mathbf{b}_1^{(n)}$, $\mathbf{b}_2^{(n)}$ are the vectors of displacements, the vectors of boundary actions, the volume forces, and the initial displacements and velocities related to the layer with the corresponding number; ρ_n is density; $\mathbf{n}$ is the unit outward normal to the corresponding boundary; Δ is the Laplace operator; $\tilde{\mathbf{I}}$ is the unit tensor; $\lambda_0^{(n)}$, $\mu_0^{(n)}$ are Lame elastic constants. It is assumed that the area of disturbance is bounded, the displacement of the body as a rigid whole is excluded, and the creep of the material is limited.

3 The Theoretical Results

To equations and relations (1)–(4) the Laplace transform in time is applied. Taking into account (6), (7) and the initial conditions (5), we obtain the problem for the transforms. Sufficient conditions under which all the poles of the solution in the transforms $\mathbf{U}^{(n)}(\mathbf{x},s)$ are simple are formulated (s is the complex parameter of the Laplace transform, $\mathbf{U}^{(n)}(\mathbf{x},s)$ is the transforms of variables $\mathbf{u}^{(n)}(\mathbf{x},t)$). The question of the location of these poles is considered.

The case when the initial conditions are zero and the volume forces are absent:

$$\mathbf{b}_1^{(n)} \equiv \mathbf{0}, \quad \mathbf{b}_2^{(n)} \equiv \mathbf{0}, \quad \mathbf{f}^{(n)}(\mathbf{x},t) \equiv \mathbf{0}, \tag{8}$$

vectors of external influences are represented in the form:

$$\mathbf{p}^{(m)}(\mathbf{x},t) = \mathbf{p}_0^{(m)}(\mathbf{x})\,\varphi(t), \quad \mathbf{q}^{(m)}(\mathbf{x},t) = \mathbf{q}_0^{(m)}(\mathbf{x})\,\psi(t) \tag{9}$$

and the hereditary properties of the layers are characterized by a kernel:

$$T(t) \equiv ae^{-bt}, \quad 0<a<b/2, \tag{10}$$

is considered separately.

It is shown that if $\varphi(t) = h(t)$, $\psi(t) = h(t)$, where $h(t)$ is the Heaviside function, then the poles of the transforms $\mathbf{U}^{(n)}(\mathbf{x}, s)$ will obviously be simple. They are located to the left of the imaginary axis (except $s = 0$) and are easily found. If, under these conditions, all homogeneous layers of the body are linearly elastic ($T(t) \equiv 0$), then all the poles of the transforms $\mathbf{U}^{(n)}(\mathbf{x}, s)$, being simple, are located on the imaginary axis.

Here, it is assumed that the disturbance propagation region is limited, therefore, according to the general statement proved in [8], under the assumptions made, the branch points of the transforms $\mathbf{U}^{(n)}(\mathbf{x}, s)$ are absent.

This greatly simplifies the construction of the original $\mathbf{u}^{(n)}(\mathbf{x}, t)$ using the theory of residues, which, under the assumptions made, are not difficult to find.

We note that, having built a solution for the case $\varphi(t) = h(t)$, $\psi(t) = h(t)$, it is possible to obtain solutions using other $\varphi(t)$, $\psi(t)$ as well.

4 An Example of Applying Theoretical Results

Let us consider the process of two-dimensional non-stationary wave propagation in cross-section of a hollow infinite cylinder $R_0 < R < R_N$, $-\pi < \theta \leq \pi$ (R, θ – polar coordinates in the plane of the cross section). The cylinder consists of N homogeneous coaxial layers. On the contact surfaces between these layers $R = R_m, m = 1, 2, \ldots, N - 1$ the conditions of continuity of displacement and stresses vectors are satisfied. We consider the material of layers to be linearly elastic or viscoelastic with the condition $T_v^{(n)} \equiv T_s^{(n)}(t) \equiv T(t)$ when the hereditary kernel has the form (10). Cylinder surface $R = R_0$ is free. Its external surface $R = R_N$ is subjected to the constant along the cylinder element radial load $f_R(\theta, t)\mathbf{n}$, $f_R(\theta, t) = P(t)\,\Theta(\theta)$ ($\mathbf{n}$ is the unit outward normal), starting from time point $t = 0$. Here the function $\Theta(\theta)$ has such a form that the load is self-balanced.

The solution of this problem for the case of elastic layers was constructed in the work [9]. In this paper, for the case when all layers are elastic, the solution is slightly modified and made more convenient for numerical implementation. In addition, a solution is constructed for the case when all the layers are viscoelastic. Fourier series expansion in θ and time Laplace transform with the following inversion were applied. In the represented work we have carried out research of transient wave processes for an elastic layered cylinder with different initial data. In particular, cases where the physicomechanical properties of the layers differ quite significantly are considered. For example, shear moduli differ by two orders of magnitude.

5 Conclusions

Non-stationary dynamic problems of class (1)–(7) are considered for layered composites in the case when the region of propagation of perturbations is bounded. At the same time, all the layers are linearly viscoelastic and their hereditary properties are determined by only one kernel, the same for all the layers. It is shown that in the case of a two-parametrical exponential kernel (10) the solution of a non-stationary dynamic problem for viscoelastic layered composite will not be hard to obtain if a solution of the corresponding problem for elastic layered composite is already known. In the above stated example of a two-dimensional non-stationary dynamic problem, the solution remains correct for the entire period of time changing.

References

1. Sabodash, P.F.: Propagation of longitudinal viscoelastic waves in a three-layer medium. Polymer Mech. **7**(1), 124–128 (1971). [in Russian]
2. Kozlov, V.I., Kucher, N.K.: Dynamic behavior of multilayer cylindrical structures with transient loads. Strength Mater. **12**(5), 639–648 (1980). [in Russian]
3. Nuriev, B.R.: Impact on a Viscoelastic Layered Composite. Izv. Akad. Nauk AzSSR. Ser. Fiz-Tekh. Mat. Nauk **4**, 35–41 (1985). [in Russian]
4. Songnan, L., Ping, G.: Dynamic response of layered viscoelastic half-space and its application to dynamic foundation problems. Hubnan Daxue Xuebao. J. Hunan Univ. **20**(1), 57–64 (1993)
5. Lokshin, A.A.: The head wave at the boundary of two hereditary-elastic half-spaces. The case of a linear source. J. Appl. Math. Mech. **58**(1), 171–176 (1994)
6. Suk Lee, H.: Viscowave – a new solution for viscoelastic wave propagation of layered structures subjected to an impact load. Int. J. Pavement Eng. **15**(6), 542–557 (2014)
7. Hosseini Hashemi, S., Bakhshi Khaniki, H.: Dynamic behavior of multi-layered viscoelastic nanobeam system embedded in a viscoelastic medium with a moving nanoparticle. J. Mech. **33**(5), 559–575 (2017)
8. Pshenichnov, S.G.: Dynamic linear viscoelasticity problems for piecewise homogeneous bodies. Mech. Solids **51**(1), 65–74 (2016)
9. Bulychev, G.G., Pshenichnov, S.G.: Propagation of elastic waves in layered cylinder. Doklady AN USSR **303**(5), 1074–1078 (1988). [in Russian]

The Influence of Geometric Parameters on the Bearing Capacity of Transformable-Volume Structure

L. M. Lobanov(✉), V. S. Volkov, O. V. Makhnenko, S. M. Kandala, and Y. V. Borovyk

E.O. Paton Electric Welding Institute of the NAS of Ukraine, Kiev, Ukraine
valentinvolkov@ukr.net, makhnenko@paton.kiev.ua

Abstract. The metal transformable-volume structure (TVS) for aerospace purposes, which consist of corrugated conical sections, was investigated. The design scheme of the TVS with a minimum mass should provide stability to the action of inertial loads in operating conditions in space orbit. The influence of the main geometric parameters of TVS made of stainless steel on the bearing capacity are determined. The effect of corrugation of the conical sections on natural frequency of TVS was considered. The results of calculating estimation of stress-strain state of the TVS under specified for space orbital station inertial loads were showed.

Keywords: Transformable-volume structures · Stainless steel · Load-carrying shells · Foldable shells · Deployable structures · Inflatable structures · Surfaces of zero total curvature

Light multi-cone TVSs with a large transformation coefficient are promising for use in space apparatus engineering as load-bearing elements for the distancing of an equipment to several meters from the base orbital station. The advantages of multi-cone TVS in comparison with other transformable structures of equal mass and length are, first of all, significantly higher compactness and the ability to compensate for large load values without loss of stability [1].

1 Features of the Formation of Geometry TVS

In the manufacture of multi-cone TVS the truncated conical section is shaped to a compact state using the proposed in E.O.Paton Electric Welding Institute the method of successive mirror image of surface sections $Q_1 \ldots Q_n$ with respect to equidistant parallel planes $\gamma_1 \ldots \gamma_n$ (see Fig. 2a), in which the transformation is close to isometric and occurs with the minimum possible stretching and compression of the middle surface of a thin shell. The sequence of transformed sections of the surface $Q_1 \ldots Q_n$ when using sheet metals takes the form (Fig. 2b), which is explained by restrictions on the minimum allowable bending radius of the material. This restriction is one of the main factors determining the fold step b, the number of annular folds n for a given cone angle α, generatrix length S and the compactness K_T of the shell of a multi-cone TVS:

E. Gdoutos and M. Konsta-Gdoutos (Eds.): ICTAEM 2020, STIN 16, pp. 198–203, 2020.
https://doi.org/10.1007/978-3-030-47883-4_37

$$K_T = 2 \times n = \frac{2 \times S \times \sin \alpha}{b}, \tag{1}$$

at the same time, n increases with decreasing thickness of the material of the conical section, and sections with a smaller thickness at a constant K_T have a lower height H_K.

2 Determination of the Influence of TVS Geometric Parameters on the Bearing Capacity

TVS consists of a certain number of tapered sections hermetically welded together, the geometric parameters of which are determined by the technical requirements for the design. In the process of investigating the influence of TVS parameters on the bearing capacity, such parameters as the diameter, number of cones, and also the thickness and degree of flatness of the corrugation of the metal shell were varied, while its physical and mechanical characteristics were unchanged. The main characteristic characterizing the rigidity of the structure was the first form of natural vibrations, which was determined by numerical calculation methods. The equation for determining the values of the own oscillations has the form [2]:

$$\left(-\omega^2[M] + [K]\right)\{\varphi\} = 0, \tag{2}$$

where [M] is the mass matrix, [K] is the stiffness matrix, $\{\varphi\}$ is the eigenvector.

2.1 Natural Frequencies of Smooth Conical Sections

The investigation was conducted on finite element models TVS 3000 mm long, consisting of a different number of smooth conical shell elements (from 25 to 68), a diameter of 426 mm to 1000 mm and a wall thickness in the range of 0.15…0.4 mm Material - AISI321 stainless steel. The model consisted of smooth conical elements, which eliminates the influence of the degree of corrugation of the surface on the results of a investigation of the influence of the main geometric parameters of TVS (Fig. 3).

The results obtained (Fig. 4) within the investigated range of TVS parameters showed that the natural frequency of TVS vibrations increases with increasing wall thickness (T) and diameter of conical elements (D), as well as reducing their number (N).

Nevertheless, an increase in the wall thickness and the diameter of the conical elements to increase the rigidity of the TVS leads to an increase in the mass of the shell, the limitation of which is critically important in the design of aerospace structures. In addition, an increase in thickness leads to a sharp increase in the force of forming and internal pressure for subsequent disclosure, i.e. to increase the complexity, the required reliability and weight and size characteristics of auxiliary technological equipment.

According to results Fig. 4c, we can conclude that the number of conjugation sections of conical sections (especially, smaller diameter) that have the smallest relative rigidity and in which the strains during the oscillations are maximal contributes to the change in the natural frequency of vibrations of the TVS. Thus, one of the solutions to increase the frequency can be a change in the possible limits of the geometric parameters of the conical sections with their constant thickness, in particular, a slight decrease in the taper angle and an increase in the height of the section. The limits for decreasing the angle α require experimental confirmation, since a concomitant increase in the crease depth a (see Fig. 4) can significantly worsen the quality of the TVS disclosure.

2.2 Natural Frequencies of Corrugated Conical Elements

After the TVS is disclosure by internal excess pressure, the conical sections acquire concentric annular features corresponding to the vertices of the folds in which the front surfaces of the metal shell undergo plastic deformation. The initial generatrix of a smooth conical shell after folding and subsequent disclosure has a wave-like profile, which can be approximately described by a sinusoidal function. Each of the profiles is described by a function of the form

$$y = c \times \sin(px - t), \tag{3}$$

where c = 1..2, p = 0.25 and t = 1.57. The degree of corrugation of the TVS surface is determined primarily by the parameter c, the amplitude of the sinusoidal shape of the corrugation. The surface flatness increases with decreasing parameter c.

To take into account the influence of the degree of corrugation of the TVS surface on the results of the investigation, FE models consisting of conical elements whose surface had a wavy profile were used (Fig. 1a). The results of numerical solutions (Fig. 5) have shown a significant decrease in the natural frequency (f = 2,8…5,1 Hz) of oscillations of a TVS with a wave-like profile compared with a flat design (f = 7,8 Hz). According to Fig. 5b, with an increase in the depth of the corrugation, a decrease in the frequency of natural vibrations is observed. This is due to an increase in the mass of the structure and a decrease in its stiffness.

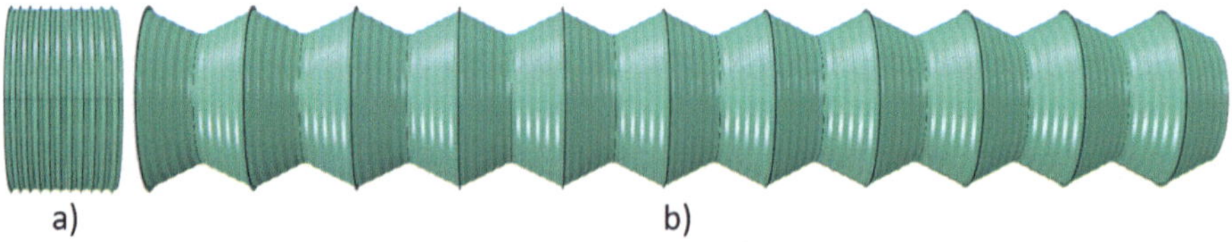

Fig. 1. The geometric scheme of TVS in the folded (a) decomposed (b) state.

Reduction the degree of corrugation can reduce the mass of TVS, but requires more internal pressure to disclosure of structure. Thus, a rational way to optimizing the TVS design scheme is to reduce the number of conical sections while reducing their taper in the approaching a more rigid cylindrical structure.

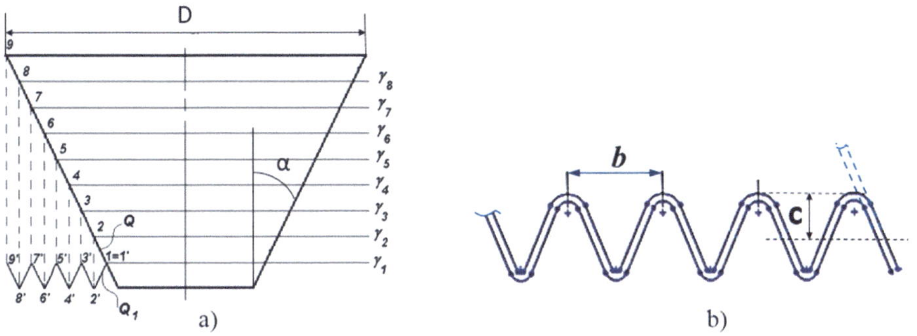

Fig. 2. Theoretical scheme of transformation of the middle surface of a truncated conical shell (a), a sequence of transformed sections of the surface $Q_1 \ldots Q_n$ when using sheet metals (b)

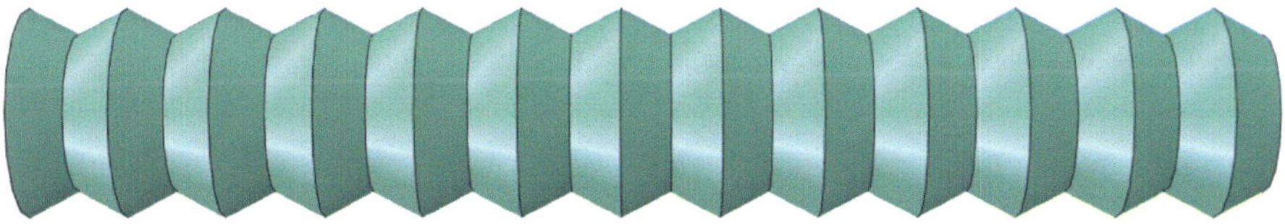

Fig. 3. TVS geometric model for numerical analysis

2.3 TVS Stress Strain State Determination

According to [3], one of the main loads acting on equipment installed inside and outside the ISS modules is inertial loads, which are specified in the form of angular and linear accelerations: $a_X = 12$ m/s^2, $a_Y = 12$ m/s^2, $a_Z = 9$ m/s^2, $e_X = 1.4$ rad/s^2, $e_Y = 1.4$ rad/s^2, $e_Z = 0.4$ rad/s^2, where the Z axis corresponds to the axial direction of the structure, and the X and Y axes correspond to orthogonal radial directions. [4]. Additionally, as a payload for one of the design cases, a load of 3 kg was secured at the top of the TVS.

As a criterion for the strength of TVS, the condition

$$\sigma_{0.2} \geq \sigma^{eqv} \tag{4}$$

where $\sigma_{0.2} = 205$ MPa is the yield strength of the material; σ^{eqv} – equivalent stresses.

The results of calculating the stress-strain state of the TVS (f = 5,1 Hz) under specified inertial loads showed that the maximum equivalent stresses do not exceed 148 MPa, stress concentrators are located in the area of welded joints at the junction of small diameters and adjacent corrugated folds with a predominance of axial tensile stresses. For the loading with 3 kg payload rather substantially higher equivalent stresses up to 200 MPa and maximum displacement 27 mm (Fig. 6) were obtained.

Besides the optimization of geometric parameters the bearing capacity of TVS would be increased by the introduction of additional structural elements (stiffeners).

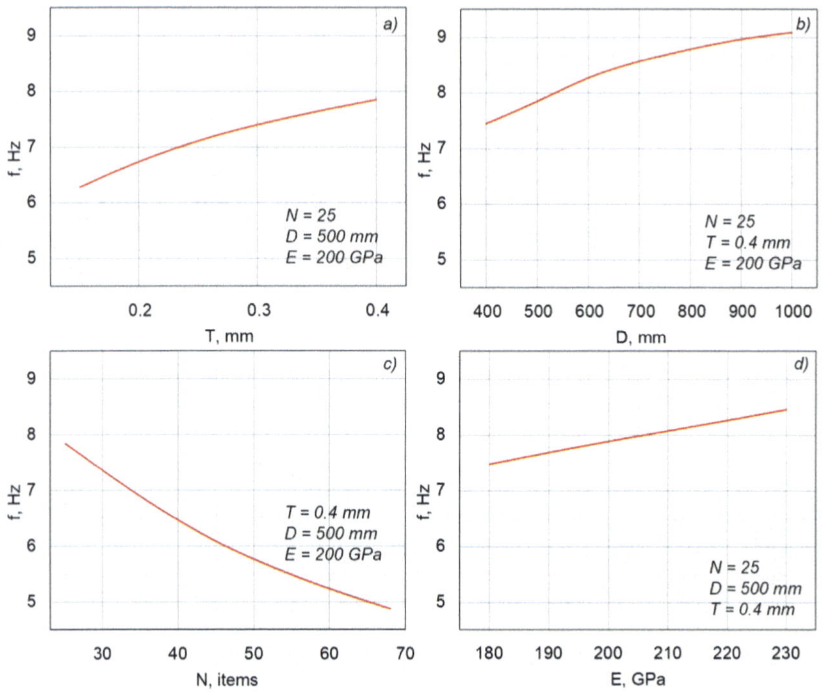

Fig. 4. The dependence of the natural frequencies of vibration of the TVS (L = 3000 mm) on the thickness (a) of the structural material, the diameter of the conical section (b), the number of cones (c), the elastic modulus of the material (d)

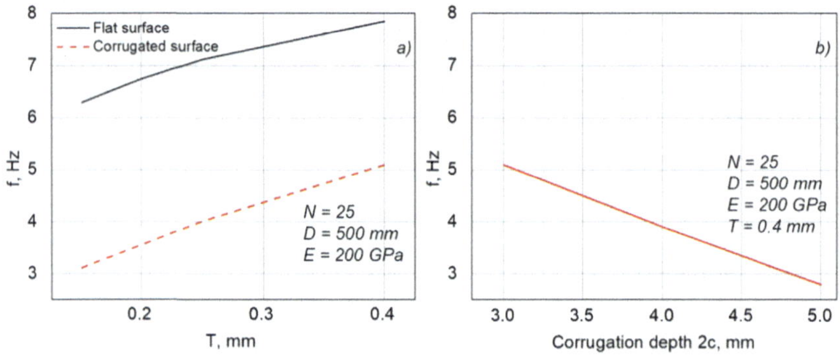

Fig. 5. Natural frequency of TVS: corrugated and flat surface depending on the thickness of the material (a), corrugated surface depending on the depth of the corrugation (b)

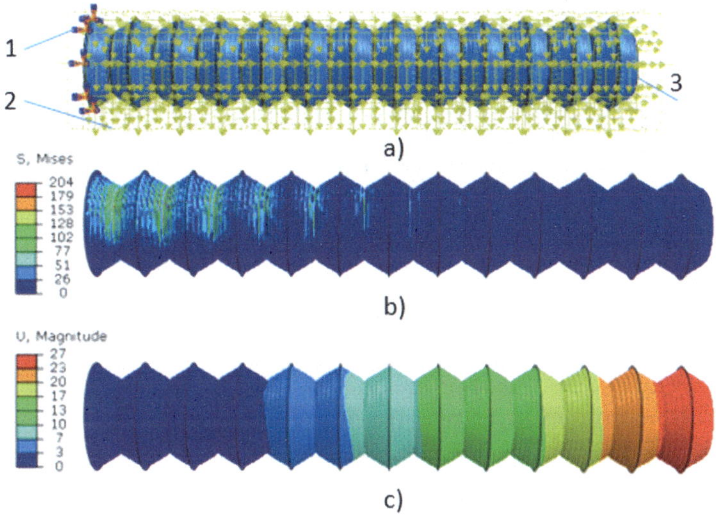

Fig. 6. Calculation model (a), where 1 – rigid fixing, 2 – inertial loads [2], 3 – payload; Mises stresses, MPa (b); equivalent displacements, mm (c).

3 Conclusions

- an increase in the bearing capacity of TVS due to an increase in diameter and wall thickness is advisable but within the mass-dimension requirements of the structure;
- the approximation of TVS to a cylindrical shape by reducing the number of conical elements is an effective way to increase the load-bearing capacity of the structure, but it has strict technological limitations in the manufacturing;
- the corrugation of a conical surface greatly reduces rigidity of TVS, thus the load-bearing capacity of the structure can be significantly increased by reduction the degree of corrugation of the conical sections.

References

1. Lobanov, L., Volkov, V., Yakimkin, A.: Construction methods and comparative evaluation of metal deployable load-carrying shell structures. J. Aerosp. Technol. Manag. **10**, e3818 (2018). https://doi.org/10.5028/jatm.v10.959
2. Abaqus 6.13 Online Documentation. http://dsk.ippt.pan.pl/docs/abaqus/v6.13/
3. ISS User Manual (ISS RS Guide). http://knts.tsniimash.ru/ru/src/CenterInfRes/iss_rs_guide.pdf
4. Lobanov, L.M., Volkov, V.S., Yakimkin, A.V., Savitsky, V.V.: Functional characteristics improvement of metal transformable-volume structures for space applications. J. Aerosp. Technol. Manage. **8**(1), 55–62 (2016). https://doi.org/10.5028/jatm.v8i1.529

The Influence of Mixed Conditions on the Stress Concentration in the Neighborhood of Interfacial Inclusions in an Inhomogeneous Transversely Isotropic Space

Oleksandr Kryvyi[1(✉)] and Yurii Morozov[2]

[1] National University "Odessa Maritime Academy", Odessa, Ukraine
krivoy-odessa@ukr.net
[2] Odessa National Polytechnic University, Odessa, Ukraine

Abstract. Two mixed non-axisymmetric problems on an absolutely rigid circular interphase inclusion in a piecewise-homogeneous trans-versal-isotropic space are considered. One face of inclusion is in conditions of smooth contact, and on the other, conditions of full adhesion to the medium are realized or there is no contact with the medium. Using exact solutions to these problems, the influence of the boundary conditions on the stress concentration in the neighborhood of the inclusion is analyzed. In particular, it has been established that under mixed conditions the stresses have a power-law singularity, the indices of which depend on the elastic constant transversally isotropic half-spaces. In the case of detachment, the power-law singularity is amplified by oscillatory multiplier.

Keywords: Interphase circular inclusion · Singular integral equations · Piecewise-homogeneous transversely isotropic space

1 Formulation of the Problem

Let an absolutely rigid interfacial inclusion occupying a circular region $\Omega = \{\sqrt{x^2+y^2} \le a\}$, in the plane $z = 0$ of connection of two different transversally isotropic half-spaces. An arbitrary load is applied to the inclusion, the action of which leads to the resultant force $\mathbf{P} = (P_1, P_2, P_3)$ and the main moment $\mathbf{M} = (M_1, M_2, M_3)$. The positions of the inclusion faces will be described by the functions

$$\begin{aligned}
&\zeta_6^{\pm} = \zeta_6^0 + \vartheta_0^{\pm}(x,y),\ \zeta_k^{\pm} = \zeta_k^0,\ k = 4,5,\ (x,y) \in \Omega \\
&\zeta_4^0 = \delta_1 - \varphi_3 x_3,\ \zeta_5^0 = \delta_2 + \varphi_3 x_1,\ \zeta_6^0 = \delta_3 + \varphi_2 x_2 + \varphi_1 x_2, \\
&\{\zeta_k^{\pm}\}_{k=1}^{8} = \{\sigma_3(x,y,\pm 0), \sigma_4(x,y,\pm 0), \sigma_5(x,y,\pm 0), u_1(x,y,\pm 0), \\
&\qquad u_2(x_1,x_2,\pm 0), u_3(x_1,x_2,\pm 0)\}, \\
&\boldsymbol{\sigma} = \{\sigma_k\}_{k=1}^{6} = \{\sigma_x, \sigma_y, \sigma_z, \tau_{yz}, \tau_{xz}, \tau_{xy}\}, \quad \mathbf{u} = \{u_k\}_{k=1}^{3} = \{u, v, w\},
\end{aligned} \tag{1}$$

E. Gdoutos and M. Konsta-Gdoutos (Eds.): ICTAEM 2020, STIN 16, pp. 204–209, 2020.
https://doi.org/10.1007/978-3-030-47883-4_38

$\vartheta_0^{\pm}(x,y)$ functions, that describe the inclusion form, respectively, when $z = \pm 0$; δ_k, φ_k, $k = 1,2,3$ – translational displacements and turning angles of the inclusion around the corresponding axes.

The mixed conditions for the interaction of the inclusion with the medium (task **A**) is considered: the upper face is linked to the upper half-space $z = +0$, the lower side is in smooth contact with another half-space $z = -0$. The following functions will be known in the area:

$$\begin{gathered}\chi_6^{\pm}(x,y) = \vartheta^{\pm}(x,y) + (1 \pm 1)\zeta_6^0,\ \vartheta^{\pm} = \vartheta_0^+ \pm \vartheta_0^-,\\ \xi_4^+(x,y) = \zeta_4^0,\ \xi_5^+(x,y) = \zeta_5^0, \xi_2^-(x,y) = \xi_3^-(x,y) = 0, (x,y) \in \Omega \\ \chi_k^{\pm} = \langle \zeta_k(x,y) \rangle^{\pm} = \zeta_k^+(x,y) \pm \zeta_k^-(x,y),\ k = \overline{1,8},\ (x,y) \in \Omega\end{gathered} \tag{2}$$

The case of a delaminated inclusion is also considered (task **B**): one inclusion face is linked to the upper half-space $z = +0$, and there is no inclusion contact with the lower half-space $z = -0$. The following functions will be known in this area:

$$\begin{gathered}\xi_6^+(x,y) = \vartheta_0^+(x,y) + \zeta_6^0,\ \xi_4^+(x,y) = \zeta_4^0, \xi_5^+(x,y) = \zeta_5^0\\ \xi_j^-(x,y) = f_{4-j}^-(x,y), j = 1,2,3, (x,y) \in \Omega,\end{gathered} \tag{3}$$

were $f_j^-(x,y)$– load applied to the lower half-space $z = -0$.

Using the approach described in [1–8], both problems are reduced to systems of two-dimensional singular integral equations. For task **A**, relatively unknown

$$h_1,\ \tau = h_3 + ih_2,\ u = h_4 + ih_5,\ \{h_k(x,y)\}_{k=1}^6 = \{\chi_1^-, \xi_2^+, \xi_3^+, \xi_4^+, \xi_5^+\} \tag{4}$$

this system has the form

$$\begin{gathered}(q_{22}^+ - 2)\tau + 2q_{21}\mathrm{DK}[h_1] + q_{22}^-\mathrm{D}\bar{\mathrm{K}}_0[\bar{\tau}] + q_{23}^+\mathrm{D}\bar{\mathrm{D}}\mathrm{K}[u] + q_{23}^-\mathrm{D}^2\mathrm{K}[\bar{u}] = g_1,\\ (q_{22}^+ - 2)\bar{\tau} + 2q_{21}\bar{D}\mathrm{K}[h_1] + q_{22}^-\bar{D}\mathrm{K}_0[\tau] + q_{23}^+\bar{D}\mathrm{D}\mathrm{K}[\bar{u}] + q_{23}^-\bar{D}^2\mathrm{K}[\bar{u}] = \bar{g}_1,\\ (q_{33}^+ + 2)u - 2q_{31}\bar{\mathrm{K}}_0[h_1] + q_{32}^+\mathrm{K}[\tau] + q_{32}^-\mathrm{DK}[\vartheta\bar{\tau}] - q_{33}^-\mathrm{DK}_0[\bar{u}] = g_2,\\ (q_{33}^+ + 2)u - 2q_{31}\bar{\mathrm{K}}_0[h_1] + q_{32}^+\mathrm{K}[\tau] + q_{32}^-\mathrm{DK}[\omega\bar{\tau}] - q_{33}^-\mathrm{DK}_0[\bar{u}] = g_2,\end{gathered} \tag{5}$$

were

$$\mathrm{K}\left[\chi_j^-\right] = \frac{1}{2\pi}\iint\limits_{\Omega} \frac{\chi_j^-(t,\tau)}{r_0}dtd\tau,\ \mathrm{K}_0\left[\chi_j^-\right] = \frac{1}{2\pi}\iint\limits_{\Omega} \frac{\chi_j^-(t,\tau)}{\vartheta}dtd\tau.$$

$$r_0 = \sqrt{(x-t)^2 + (y-\tau)^2},\ \vartheta = (x-t) + i(y-\tau).$$

For the task **B**, relatively unknown functions

$$h_1,\ \tau = h_3 + ih_2,\ u = h_4 + ih_5, h_6,\ \{h_k(x,y)\}_{k=\overline{1,6}} = \{\chi_k^-(x,y)\}_{k=\overline{1,6}} \tag{6}$$

the task is reduced to a system of six two-dimensional SIE:

$$\begin{gathered}
(q_{11}-1)h_1 + \frac{q_{13}}{2}(\mathrm{D}\,\bar{\mathrm{u}} + \bar{\mathrm{D}}\,\mathrm{u}) - \frac{q_{12}}{2}(\bar{\mathrm{D}}\,\mathrm{K}[\tau] + \mathrm{D}\,\mathrm{K}[\bar{\tau}]) - q_{14}\mathrm{D}\,\bar{\mathrm{D}}\,\mathrm{K}[h_6] = h_1^*,\\
(\frac{\overset{+}{q}_{22}}{2} - 1)\tau - q_{24}\,\mathrm{D}\,h_6 + q_{21}\mathrm{D}\,\mathrm{K}[h_1] + \frac{\overset{-}{q}_{22}}{2}\mathrm{D}\,\overline{\mathrm{K}}_0[\bar{\tau}] - \frac{\overset{+}{q}_{23}}{2}\mathrm{D}\,\bar{\mathrm{D}}\,\mathrm{K}[u] - \frac{\overset{-}{q}_{23}}{2}\mathrm{D}^2\,\mathrm{K}[\bar{u}] = \tau_*,\\
(\frac{\overset{+}{q}_{22}}{2} - 1)\bar{\tau} - q_{24}\mathrm{D}\,h_6 + q_{21}\bar{\mathrm{D}}\,\mathrm{K}[h_1] + \frac{\overset{-}{q}_{22}}{2}\bar{\mathrm{D}}\overline{\mathrm{K}}_0[\bar{\tau}] - \frac{\overset{+}{q}_{23}}{2}\bar{\mathrm{D}}\,\mathrm{D}\,\mathrm{K}[\bar{u}] - \frac{\overset{-}{q}_{23}}{2}\bar{D}^2\mathrm{K}[u] = \bar{\tau}_*,\\
(\frac{\overset{+}{q}_{33}}{2} - 1)u - q_{31}\bar{K}_0[h_1] + \frac{\overset{+}{q}_{32}}{2}\mathrm{K}[\tau] + \frac{\overset{-}{q}_{32}}{2}\mathrm{D}\,\mathrm{K}[\vartheta\bar{\tau}] + \frac{\overset{-}{q}_{33}}{2}\mathrm{D}\,\mathrm{K}_0[\bar{u}] + q_{34}\mathrm{D}\,\mathrm{K}[h_6] = u_*,\\
(\frac{\overset{+}{q}_{33}}{2} - 1)u - q_{31}\bar{\mathrm{K}}_0[h_1] + \frac{\overset{+}{q}_{32}}{2}\mathrm{K}[\tau] + \frac{\overset{-}{q}_{32}}{2}\mathrm{D}\,\mathrm{K}[\vartheta\bar{\tau}] + \frac{\overset{-}{q}_{33}}{2}\mathrm{D}\,\mathrm{K}_0[\bar{u}] + q_{34}\mathrm{D}\,\mathrm{K}[h_6] = u_*,
\end{gathered} \tag{7}$$

For determine of unknown functions $\delta_x, \delta_y, \delta_7, \varphi_x, \varphi_y, \varphi_z$ are use force and moment equilibrium equations

$$\iint\limits_{\Omega} \chi_k dxdy = P_{4-k}, k = 1,2,3,\ \iint\limits_{\Omega} \begin{pmatrix} x \\ y \end{pmatrix} \chi_1 dxdy = \begin{pmatrix} M_2 \\ M_1 \end{pmatrix},\ \iint\limits_{\Omega} (x\chi_2 - y\chi_3)dxdy = M_3. \tag{8}$$

2 The Solution of the SIE

Following the works [1–11], after the transition to polar coordinates (ρ, φ, z), the solution of problems is constructed in the form

$$\begin{gathered}
v_j^{\pm}(\rho,\varphi) = \sum_{n=-\infty}^{\infty} \mathrm{V}_n^{j,\pm}(\rho)e^{in\varphi}, \mathrm{V}_n^{j,-}(\rho) = \frac{1}{2\pi}\int_{-\pi}^{\pi} v_j^-(\rho,\varphi)e^{-in\varphi}d\varphi, j = 1,3,5,\\
v_1(r,\varphi) = h_1(\rho\ \cos\ \varphi, \rho\ \sin\ \varphi), v_2(r,\varphi) = e^{-i\varphi}\tau(\rho\ \cos\ \varphi, \rho\ \sin\ \varphi),\\
v_3(r,\varphi) = e^{-i\varphi}u(\rho\ \cos\ \varphi, \rho\ \sin\ \varphi), v_4(r,\varphi) = h_6(\rho\ \cos\ \varphi, \rho\ \sin\ \varphi).
\end{gathered} \tag{9}$$

Regarding functions $\mathrm{V}_n^{j,\pm}(\rho)$, we obtain systems of integral equations with Weber-Sonin kernels, The solution of which [1–11] made it possible to obtain explicit asymptotic representations for stresses, for the task **A:**

$$\begin{gathered}\sigma_z\big|_{z=0}\simeq k_{11}(\rho-a)^{-\frac{1}{2}-\alpha_0}+k_{12}(\rho-a)^{-\frac{1}{2}}+k_{13}(\rho-a)^{-\frac{1}{2}+\alpha_0},\ \rho\to a+0,\\ \tau_{z\rho}\big|_{z=0}\simeq k_{21}(\rho-a)^{-\alpha_0-\frac{1}{2}}+k_{22}(\rho-a)^{-\frac{1}{2}}+k_{23}(\rho-a)^{\alpha_0-\frac{1}{2}+},\\ \tau_{z\varphi}\big|_{z=0}\simeq k_{31}(\rho-a)^{-\tilde{\alpha}_0-\frac{1}{2}}+k_{32}(\rho-a)^{-\frac{1}{2}}+k_{33}(\rho-a)^{\tilde{\alpha}_0-\frac{1}{2}}.\end{gathered}\tag{10}$$

For the task **B:**

$$\begin{gathered}\begin{pmatrix}\sigma_z\\ \tau_{z\rho}\end{pmatrix}\Bigg|_{z=0}\simeq\frac{1}{\sqrt{\rho-a}}\left\{\sum_{\pm}(\rho-a)^{\pm\alpha_0}\left[\begin{pmatrix}k_{11}^{\pm}\\ k_{21}^{\pm}\end{pmatrix}\cos\left(\beta_0\ln\tfrac{\rho-a}{2a}\right)+\begin{pmatrix}k_{12}^{\pm}\\ k_{22}^{\pm}\end{pmatrix}\sin\left(\beta_0\ln\tfrac{\rho-a}{2a}\right)\right]+\begin{pmatrix}k_{10}\\ k_{20}\end{pmatrix}\right\}\\ \tau_{z\varphi}\big|_{z=0}\simeq\frac{1}{\sqrt{\rho-a}}\left\{k_{31}(\rho-a)^{-\tilde{\alpha}_0}+k_{33}(\rho-a)^{\tilde{\alpha}_0}+k_{32}\right\}.\end{gathered}\tag{11}$$

The expressions for translational displacements for task **A** have the form

$$\begin{gathered}\delta_z=c_z\frac{P_3}{a},\ \varphi_z=s_z\frac{M_3}{a^3},\ \delta_x+i\delta_y=c_x\frac{P_1+iP_2}{a}+c_y\frac{M_2+iM_1}{a^2},\\ \varphi_y+i\varphi_x=s_x\frac{P_1+iP_2}{a^2}+s_y\frac{M_2+iM_1}{a^3}\end{gathered}\tag{12}$$

For task **B** expressions take place.

$$\begin{gathered}\begin{pmatrix}\delta_x\\ \delta_y\end{pmatrix}=\frac{c_{xy}}{a}\begin{pmatrix}P_1\\ P_2\end{pmatrix}+c_{xy}^0a\begin{pmatrix}P_1^0\\ P_2^0\end{pmatrix}+\frac{c_{xy}^m}{a^2}\begin{pmatrix}M_2\\ M_1\end{pmatrix},\ \delta_z=c_z\frac{P_3}{a}-c_z^0aP_3^0,\\ \begin{pmatrix}\varphi_x\\ \varphi_y\end{pmatrix}=\frac{s_{xy}^m}{a}\begin{pmatrix}M_1\\ M_2\end{pmatrix}-\frac{s_{xy}}{a^2}\begin{pmatrix}P_2\\ P_1\end{pmatrix}-s_{xy}^0\begin{pmatrix}P_1^0\\ P_2^0\end{pmatrix},\ \varphi_z=s_z\frac{M_3}{a^3}.\end{gathered}\tag{13}$$

Tables 1 (task **A**) and 2 (task **B**) are given the values of bielastic constants α_0 and $\tilde{\alpha}_0$ for some combinations of materials and opposite combinations: Ceramics A ($BaTiO_3$) (material m1), Ceramics B ($BaTiO_3$+5%$CaTiO_3$) (material m2), Yttrium (material m3), Magnesium (material m4), Beryl (material m5), Cobalt (material m6). From tables it is visible that the replacement of half-spaces significantly affects the behavior of stresses in the neighborhood of the inclusion. This effect is especially noticeable if the elastic characteristics of half-spaces are significantly different and less noticeable in the opposite case.

So, the asymptotic representations for the stresses in the neighborhood of the inclusion have three terms and depend on the bielastic constants α_0 and $\tilde{\alpha}_0$ satisfying the conditions $0<\alpha_0<1/2$, $0<\tilde{\alpha}_0<1/2$. For a homogeneous space $\alpha_0=\tilde{\alpha}_0=1/4$, $\beta=0$. The presence of delamination, which leads to the appearance of oscillations at the neighborhood of the inclusion, $\rho\to a$, has a significant effect on the concentration of stresses.

Table 1. The values α_0 and $\tilde{\alpha}_0$ for the combination of some transversely isotropic materials
(task **A**)

Комбинация материалов	α_0	$\tilde{\alpha}_0$	Комбинация материалов	α_0	$\tilde{\alpha}_0$
m1 – m2	0,25650	0,25644	m2 – m1	0,24367	0,24383
m3 – m2	0.30914	0.30957	m2 – m3	0,20403	0,20655
m3 – m4	0.21835	0.22259	m4 – m3	0,28283	0,28311
m6 – m2	0.20062	0.21931	m2 – m6	0,30662	0,28795
m6 – m3	0.16679	0.18731	m3 – m6	0,36701	0,35202
m6 – m5	0,24159	0,25374	m5 – m6	0,25874	0,24635
m5 – m2	0,20640	0,21652	m2 – m5	0,29729	0,29236
m5 – m4	0,15173	0,17411	m4 – m5	0,38770	0,39060

Table 2. The values α_0 and $\tilde{\alpha}_0$ for the combination of some transversely isotropic materials (task **B**)

Комбинация материалов	α_0	$\tilde{\alpha}_0$	Комбинация материалов	α_0	$\tilde{\alpha}_0$
m1 – m2	0,2553	0,2626	m2 – m1	0,2446	0,2374
m3 – m2	0,2994	0,3481	m2 – m3	0,2006	0,0434
m3 – m4	0,2250	0,1905	m4 – m3	0,2745	0,3095
m6 – m2	0,2000	0,1829	m2 – m6	0,2990	0,3171
m6 – m3	0,1523	0,1028	m3 – m6	0,3437	0,3972
m6 – m5	0,2367	0,2574	m5 – m6	0,2634	0,2426
m5 – m2	0,2134	0,1762	m2 – m5	0,2857	0,3238
m5 – m4	0,1394	0,0683	m4 – m5	0,3532	0,4317

References

1. Efimov, V.V., Krivoi, A.F., Popov, G.Y.: Problems on the stress concentration near a circular imperfection in a composite elastic medium. Mech. Solids **33**(2), 35–49 (1998). Springer ISSN: 0025-6544
2. Kryvyy, O.: The discontinuous solution for the piece-homogeneous transversal iso-tropic medium. Oper. Theory Adv. Appl. **191**, 387–398 (2009). https://doi.org/10.1007/978-3-7643-9921-4_25
3. Kryvyy, O.F.: Singular integral relations and equations for a piecewise homogeneous transversally isotropic space with interphase defects. J. Math. Sci. **176**(4), 515–531 (2011). https://doi.org/10.1007/s10958-011-0419-2
4. Kryvyi, O.F.: Interface crack in the inhomogeneous transversely isotropic space. Mater. Sci. **47**(6), 726–736 (2012). https://doi.org/10.1007/s11003-012-9450-9
5. Kryvy, O.F.: Interface circular inclusion under mixed conditions of interaction with a piecewise homogeneous transversally isotropic space. J. Math. Sci. **184**(1), 101–119 (2012). https://doi.org/10.1007/s10958-012-0856-6

6. Kryvyi, O.F.: Delaminated interface inclusion in a piecewise homogeneous transversely isotropic space. Mater. Sci. **50**(2), 245–253 (2014). https://doi.org/10.1007/s11003-014-9714-7
7. Kryvyi, O.F.: Mutual influence of an interface tunnel crack and an interface tunnel inclusion in a piecewise homogeneous anisotropic space. J. Math. Sci. **208**(4), 406–416 (2015). https://doi.org/10.1007/s10958-015-2455-9
8. Kryvyi, O.F.: The solution method nonaxisymmetric problems of stress concentration for an inhomogeneous transversal isotropic space with interfacial cracks and inclusions. In: ICF2017-14th International Conference on Fracture, Conference Paper, pp. 560–561 (2017)
9. Kryvyi, O., Morozov, Yu.: Interphase circular inclusion in a piecewise-homogeneous transversely isotropic space under the action of a heat flux. In: Gdoutos, E. (eds.) Proceedings of the First International Conference on Theoretical, Applied and Experimental Mechanics. ICTAEM 2018. Structural Integrity, vol. 5. Springer, Cham (2019). https://doi.org/10.1007/978-3-319-91989-8_94
10. Kryvyi, O., Morozov, Yu.: Solution of the problem of heat conduction for the transversely isotropic piecewise-homogeneous space with two circular inclusions. J. Math. Sci. **243**(1), 162–182 (2019). https://doi.org/10.1007/s10958-019-04533-1
11. Kryvyi, O., Morozov, Yu.: Heat-active circular interphase inclusion in the conditions of smooth contact with half-spaces. In: Gdoutos E. (eds) Proceedings of the Second International Conference on Theoretical, Applied and Experimental Mechanics. ICTAEM 2019. Structural Integrity, vol 8. Springer, Cham (2019). https://doi.org/10.1007/978-3-030-21894-2_36

On an Approach to the Thermoelastic Analysis of Nonhomogeneous Solids

Roman Kushnir(✉) and Yuriy Tokovyy

Pidstryhach Institute for Applied Problems of Mechanics and Mathematics, National Academy of Sciences of Ukraine, Lviv 79060, Ukraine
dyrector@iapmm.lviv.ua

Abstract. A general approach to the analysis of thermoelastic response of solids exhibiting spatial or thermal variation in their material properties is addressed in this study. The approach is based on the idea of the direct integration method, which allows for the construction of analytical solutions to the relevant thermoelasticity problems in explicit functional form satisfying the thermal and force boundary conditions on the entire surface of a solid.

Keywords: Nonhomogeneous solids · Thermoelasticity · Explicit solutions

1 Introduction

Construction and improvement of structural elements of nowadays technique along with the development and implementation of new materials with advanced properties necessitate the fundamental analysis of heat- and mass-transfer along with the stress-strain state in multi-phase materials under the simultaneous action of different physical fields with account for wide ranges of the operating and constructional features (e.g., the material anisotropy and non homogeneity, the distributions of fibers and capillaries, thermo sensitivity of the phase components, stratification, etc.), as well as the interaction of the structural elements of different geometry. The exhaustive analysis of the impact caused by mentioned factors on the thermal and mechanical behavior of the structure members allows for controlling the technological processes for composite materials in order to achieve the required operational performance of the objective work pieces. This is vital, in particular, for the development of technologies involved into the fabrication of functionally-graded materials (FGM), whose phase-property profiles can be controlled on the design stage so that the composed materials can have predicted thermal conductivity, deformability and durability under extreme operation conditions. This feature makes FGM to be a perfect choice for various implementations when designing elements of electronic and space techniques, hypersonic aviation, protective systems of nuclear and chemical energetic, technical and engineering hardware of wide-range application. The foresaid makes it clear that the problem on the establishment of by-the-target-values optimal criteria for the implementation of FGM in view of achieving required regimes of the operational performance of the nowadays machinery, devices, and structures becomes extremely important.

E. Gdoutos and M. Konsta-Gdoutos (Eds.): ICTAEM 2020, STIN 16, pp. 210–215, 2020.
https://doi.org/10.1007/978-3-030-47883-4_39

This paper briefly introduces a general approach developed for the thermoelastic analysis of nonhomogeneous and thermosensitive (FGM) solids with arbitrary dependence of the material properties on the spatial coordinates or non-uniform distribution of the transient temperature. The approach is based on implementation of the direct integration method developed for the solution of governing equations of the relevant thermoelasticity problems [1], which is adopted for the case of nonhomogeneous solids with concern to the establishing the fundamental relationships between the stress-tensor, strain-tensor, and displacement-vector components, as well as the integral equilibrium and compatibility conditions. In the final count, this method implies the reduction of the original problems to solution of second-kind integral equations with corresponding boundary and integral conditions [2].

2 Analysis of Thermal Stresses in Nonhomogeneous Solids

2.1 The Idea of the Method

The algorithm of the direct integration method, whose idea was originated from the solution of specific optimization problems [3], implies the following principal stages [1]:

- Select one or several of the stress-tensor components (or their combinations, e.g., the total stress, the difference of normal stresses, etc.) to be the key functions;
- Basing on integration of the equilibrium equations, derive the relations expressing the stress-tensor components through the key functions;
- By making use of these relations and the original boundary conditions, derive local boundary conditions and integral equilibrium conditions for the key functions;
- By making use of the compatibility equations in terms of stresses along with the relations obtained in the second stage, derive the governing equations for the key functions;
- Using the derived integral and boundary conditions for the key stresses, calculate the key stresses from the governing equations;
- Finally, calculate all the stresses from the relations obtained in the second stage.

In order to illustrate the ability of this method, consider its application to solving one-, two-, and three-dimensional problems of thermoelasticity for a radially-non homogeneous cylinder.

Consider a long cylinder C related to the dimensionless cylindrical-polar coordinate system (r, θ, z) with inner and outer radii r_i and $r_o = 1$, respectively. If $r_i = 0$, then the cylinder can be regarded as a solid one. Let the material properties of the cylinder be arbitrary functions of the radial coordinate r. The inner and outer surfaces of the cylinder are exposed to steady-state thermal and force loadings.

2.2 A One-Dimensional Problem

First, assume the applied loadings to be uniform with respect to the angular and axial coordinates θ and z. In this case, the elastic equilibrium of the cylinder is governed by the following equations

$$\frac{d}{dr}(r^2\sigma_{rr}) = r\sigma, \qquad r\frac{d\varepsilon_{\theta\theta}}{dr} = \varepsilon_{rr} - \varepsilon_{\theta\theta} \tag{1}$$

where the total stress σ is given by the formula

$$\sigma = \sigma_{rr} + \sigma_{\theta\theta} \tag{2}$$

and the normal stress-tensor, σ_{rr}, $\sigma_{\theta\theta}$, and strain-tensor, ε_{rr}, $\varepsilon_{\theta\theta}$, components are related through the following constitutive equations

$$\begin{aligned} E\varepsilon_{rr} &= (1-\nu^2)\sigma_{rr} - \nu(1+\nu)\sigma_{\theta\theta} + \alpha(1+\nu)ET, \\ E\varepsilon_{\theta\theta} &= (1-\nu^2)\sigma_{\theta\theta} - \nu(1+\nu)\sigma_{rr} + \alpha(1+\nu)ET. \end{aligned} \tag{3}$$

Here, $E = E(r)$ is the Young modulus, $\nu = \nu(r)$ is the Poisson ratio, $\alpha = \alpha(r)$ is the linear thermal expansion coefficient, and $T = T(r)$ is the temperature field in the cylinder $\mathcal{C}$ due to the applied thermal loadings on the inner and outer lateral surfaces.

The Eqs. (1)–(3) are to be solved under the following boundary conditions

$$\sigma_{rr}(r_i) = -p_i, \qquad \sigma_{rr}(1) = -p_1 \tag{4}$$

for the radial stresses.

In order to solve the boundary value problem (1)–(4), we select the total stress σ to be the key function and, by making use of Eqs. $(1)_1$ and (2) in view of the boundary conditions (4), express the normal stresses as

$$\sigma_{rr}(r) = -p_i\left(\frac{r_i}{r}\right)^2 + \frac{1}{r^2}\int_{r_i}^{r}\eta\sigma(\eta)\mathrm{d}\eta, \qquad \sigma_{\theta\theta}(r) = \sigma(r) + p_i\left(\frac{r_i}{r}\right)^2 - \frac{1}{r^2}\int_{r_i}^{r}\eta\sigma(\eta)\mathrm{d}\eta. \tag{5}$$

The first formula of (5) along with the second condition (4) yield

$$\int_{r_i}^{1} r\sigma(r)\mathrm{d}r = p_i r_i^2 - p_1. \tag{6}$$

Putting Eqs. (2) and (3) into Eq. $(1)_2$ yields the compatibility equation

$$\frac{d}{dr}\left(\frac{1-\nu(r)}{G(r)}(\sigma(r) + 2\alpha(r)E(r)T(r))\right) = \sigma_{rr}\frac{d}{dr}\left(\frac{1}{G(\rho)}\right), \tag{7}$$

where $2G(r) = E(r)/(1+\nu(r))$. In view of formula $(5)_1$, Eq. (7) can be given as

$$\sigma(r) = f(r) + \int_{r_i}^{r} \sigma(\xi)\mathcal{K}(r,\xi)d\xi, \tag{8}$$

where

$$f(r) = \frac{2AG(r)-\alpha(r)E(r)T(r)}{1-\nu(r)} - p_i r_i^2 \frac{G(r)}{1-\nu(r)} \int_{r_i}^{r} \frac{1}{\eta^2}\frac{d}{d\eta}\left(\frac{1}{G(\eta)}\right) d\eta,$$

$$\mathcal{K}(r,\xi) = \frac{\xi G(r)}{1-\nu(r)} \int_{\xi}^{r} \frac{1}{\eta^2}\frac{d}{d\eta}\left(\frac{1}{G(\eta)}\right) d\eta,$$

and A is a constant of integration, which can be determined from condition (6).

Equation (8) can be solved by making use of an appropriate analytical, semi-analytical or numerical method [4] in order to determine the total stress σ. After the key function is found, the remaining stress-tensor components can be computed by means of the formulae (5).

2.3 A Two-Dimensional Problem

Assume the external loadings of the cylinder $\mathcal{C}$ to be irrespective of the axial coordinate z. In this case, we arrive at the state of plane strain, which implies the elastic equilibrium to be governed by the following set of equations

$$\begin{gathered} \frac{1}{r}\frac{\partial}{\partial r}(r^2\sigma_{rr}) + \frac{\partial \sigma_{r\theta}}{\partial \theta} = \sigma, \quad \frac{1}{r}\frac{\partial}{\partial r}(r^2\sigma_{r\theta}) + \frac{\partial \sigma_{\theta\theta}}{\partial \theta} = 0, \\ \frac{\partial^2(r\varepsilon_{r\theta})}{\partial r \partial \theta} = \frac{\partial^2 \varepsilon_{rr}}{\partial \theta^2} + \frac{\partial}{\partial r}\left(r^2 \frac{\partial \varepsilon_{\theta\theta}}{\partial r}\right) - r\frac{\partial \varepsilon_r}{\partial r} \end{gathered} \tag{9}$$

under the constitutive equations

$$\begin{gathered} E\varepsilon_{rr} = \sigma_{rr} - \nu(\sigma_{\theta\theta} + \sigma_{zz}) + \alpha ET, \quad E\varepsilon_{\theta\theta} = \sigma_{\theta\theta} - \nu(\sigma_{rr} + \sigma_{zz}) + \alpha ET, \\ E\varepsilon_{zz} = \sigma_{zz} - \nu(\sigma_{\theta\theta} + \sigma_{rr}) + \alpha ET, \quad G\varepsilon_{r\theta} = \sigma_{r\theta}. \end{gathered} \tag{10}$$

Here, the total stress is given by formula (2). The boundary conditions are imposed in the form

$$\sigma_{rr}(r_i,\theta) = -p_i, \quad \sigma_{rr}(1,\theta) = -p_1, \quad \sigma_{r\theta}(r_i,\theta) = q_i, \quad \sigma_{r\theta}(1,\theta) = q_1. \tag{11}$$

In order to solve the problem (9)–(11), we represent Eqs. $(9)_1$ and $(9)_2$, in view of formula (2), in the form

$$\mathcal{D}\sigma_{rr} = \frac{\partial(r^2\sigma)}{\partial r} + r\frac{\partial^2\sigma}{\partial\theta^2}, \quad \mathcal{D}\sigma_{\theta\theta} = r\frac{\partial^2(r^2\sigma)}{\partial r^2}, \quad \mathcal{D}\sigma_{r\theta} = -r\frac{\partial^2(r\sigma)}{\partial r\partial\theta}, \tag{12}$$

and Eq. $(9)_3$ as

$$\Delta\left(\frac{1-\nu(r)}{G(r)}\sigma+2\alpha(r)(1+\nu(r))T\right)=\sigma_{rr}\frac{d^2}{dr^2}\left(\frac{1}{G(r)}\right)+\frac{\sigma_{\theta\theta}}{r}\frac{d}{dr}\left(\frac{1}{G(r)}\right), \tag{13}$$

where $\mathcal{D}=\frac{\partial}{\partial r}\left(r\frac{\partial}{\partial r}r^2\right)+r\frac{\partial^2}{\partial\theta^2}$ and $\Delta=\frac{1}{r}\frac{\partial}{\partial r}\left(r\frac{\partial}{\partial r}\right)+\frac{1}{r^2}\frac{\partial^2}{\partial\theta^2}$.

By determining the normal stresses from Eqs. $(12)_{1,2}$ and putting the results into Eq. (13), we can arrive at an integral equation similar to Eq. (8), which can be solved in the same manner [1].

2.4 An Axisymmetric Problem

Assume the external loadings to be irrespective of the angular coordinate θ that refers to the case of axial symmetry. The problem is governed by the equilibrium and compatibility equations

$$\begin{aligned}&\frac{\partial(r\sigma_{rr})}{\partial r}+r\frac{\partial\sigma_{rz}}{\partial z}=\sigma_{\theta\theta}, \quad \frac{\partial}{\partial r}(r\sigma_{rz})+r\frac{\partial\sigma_{zz}}{\partial z}=0,\\&\frac{r\partial\varepsilon_{\theta\theta}}{\partial r}=\varepsilon_{rr}-\varepsilon_{\theta\theta}, \quad \frac{r\partial^2\varepsilon_{\theta\theta}}{\partial z^2}-\frac{\partial\varepsilon_{rz}}{\partial z}+\frac{\partial\varepsilon_{zz}}{\partial r}=0\end{aligned} \tag{14}$$

under the constitutive ones given by Eq. (10). For the simplicity sake impose $r_i=0$ and the thermal loading to be absent. The force loadings are represented by the conditions

$$\sigma_{rr}(1,z)=-p(z), \quad \sigma_{rz}(1,z)=q(z). \tag{15}$$

The total stress in this case is to given by the formula

$$\sigma=\sigma_{rr}+\sigma_{\theta\theta}+\sigma_{zz}. \tag{16}$$

Making use of Eqs. (10) and (16) allows for representing Eqs. $(14)_{3,4}$ in terms of stresses as

$$\begin{aligned}&r\frac{\partial}{\partial r}\left(\frac{\sigma_{\theta\theta}-\nu(r)(\sigma_{rr}+\sigma_{zz})}{E(r)}\right)+\frac{\sigma_{\theta\theta}-\sigma_{rrr}}{2G(r)}=0,\\&r\frac{\partial^2}{\partial z^2}\left(\frac{\sigma_{\theta\theta}-\nu(r)(\sigma_{rr}+\sigma_{zz})}{E(r)}\right)-\frac{1}{G(r)}\frac{\partial\sigma_{rz}}{\partial z}+\frac{\partial}{\partial r}\left(\frac{\sigma_{zz}-\nu(r)(\sigma_{rr}+\sigma_{\theta\theta})}{E(r)}\right)=0.\end{aligned} \tag{17}$$

Having applied the integral Fourier transform with respect to coordinate z to Eqs. (14)–(17), expressing the shearing stress from Eqs. $(14)_{1,2}$ and putting them into Eqs. $(17)_{1,2}$, the latter ones can be given, after some algebra, as

$$\begin{aligned}&\bar{\sigma}_{rz}(r)=AI_1(sr)+\int_0^r(\mathcal{K}_{11}(r,\xi)\bar{\sigma}_{rz}(\xi)+\mathcal{K}_{12}(r,\xi)\bar{\sigma}(\xi))d\xi,\\&\bar{\sigma}(r)=B\frac{E(r)I_0(sr)}{1-\nu(r)}+\int_0^r(\mathcal{K}_{21}(r,\xi)\bar{\sigma}_{rz}(\xi)+\mathcal{K}_{22}(r,\xi)\bar{\sigma}(\xi))d\xi,\end{aligned} \tag{18}$$

where the overline denotes a function in the Fourier mapping domain, A and B are constants of integration, which can be determined from the conditions (15), and

$$\mathcal{K}_{11}(r,\xi) = \frac{d}{d\xi}\left(\xi\chi_{11}(r,\xi)\frac{d}{d\xi}\left(\ln\left(\frac{1}{G(\xi)}\right)\right)\right) + s^2\xi^2\int_{\xi}^{r}\frac{\chi_{11}(r,\eta)}{\eta}\frac{d}{d\eta}\left(\ln\left(\frac{1}{G(\eta)}\right)\right)d\eta,$$

$$\mathcal{K}_{12}(r,\xi) = \frac{2is}{E(\xi)}\frac{d}{d\xi}(\xi G(\xi)\chi_{11}(r,\xi)) + is\xi\int_{\xi}^{r}\frac{\chi_{11}(r,\eta)}{\eta}\frac{d}{d\eta}\left(\ln\left(\frac{1}{G(\eta)}\right)\right)d\eta,$$

$$\mathcal{K}_{22}(r,\xi) = \frac{s\xi E(r)}{1-\nu(r)}\int_{\xi}^{\rho}\frac{\chi_{01}(r,\eta)}{\eta}\frac{d}{d\eta}\left(\frac{1}{2G(\eta)}\right)d\eta,$$

$$\mathcal{K}_{21}(r,\xi) = \frac{iE(r)}{1-\nu(r)}\left((s\xi\chi_{00}(r,\xi) - \chi_{01}(r,\xi))\frac{d}{d\xi}\left(\frac{1}{2G(\xi)}\right)\right.$$

$$\left.-\frac{s^2\xi^2}{2}\int_{\xi}^{r}\frac{\chi_{01}(r,\eta)}{\eta}\frac{d}{d\eta}\left(\frac{1}{G(\eta)}\right)d\eta\right), \quad \chi_{nm}(r,\xi) = I_n(sr)K_m(s\xi) - (-1)^{n+m}K_n(sr)I_m(s\xi).$$

Having found the key stresses $\bar{\sigma}_{rz}$ and $\bar{\sigma}(r)$ from Eq. (18), the remaining stresses can be restored by making use of the relations derived from Eqs. $(14)_{1,2}$ [5].

Acknowledgement. This research was supported by the budget program of Ukraine "Support for the Development of Priority Research Areas" (CPCEC 6451230)

References

1. Tokovyy, Y.V.: Direct integration method. In: Hetnarski, R.B. (ed.) Encyclopedia of Thermal Stresses, vol. 2, pp. 951–960. Springer, Dordrecht (2014)
2. Tokovyy, Y.V., Kalynyak, B.M., Ma, C.C.: Nonhomogeneous solids: integral equations approach. In: Hetnarski, R.B. (ed.) Encyclopedia of Thermal Stresses, vol. 7, pp. 3350–3356. Springer, Dordrecht (2014)
3. Kushnir, R.M., Yasinskyy, A.V.: Identification of the temperature field and stresses in a thermosensitive cylinder according to the surface strains. Mater. Sci. **43**(6), 814–822 (2007)
4. Tokovyy, Y.V., Chyzh, A.I., Ma, C.C.: An analytical solution to the axisymmetric thermoelasticity problem for a cylinder with arbitrarily varying thermomechanical properties. Acta Mech. **230**(4), 1469–1485 (2019)
5. Tokovyy, Y.V., Ma, C.C.: Axisymmetric stresses in an elastic radially inhomogeneous cylinder under length-varying loadings. J. Appl. Mech. **83**, 111007-1–7 (2016)

The Stress State Near the Corner Point of the Interface of Piecewise Homogeneous Plane in the Presence of Interfacial Shear Cracks

V. M. Nazarenko and A. L. Kipnis(✉)

S.P. Timoshenko Institute of Mechanics of the National Academy of Science of Ukraine, Kiev, Ukraine
a.l.kipnis@gmail.com

Abstract. In the framework of plane deformation, a piecewise homogeneous isotropic elastic body with interfacial shear cracks near the corner point of the interface is considered. The exact solution of the corresponding problem of the theory of elasticity for a wedge-shaped body is constructed by the Wiener-Hopf method. Based on the constructed solution, the behavior of stresses near the corner point is investigated.

Keywords: Interface of media · Corner point · Interfacial shear crack · Wiener – Hopf method

Modern structures made from composite materials often contain elements in which the surface between the layers of the composite is not smooth. Such elements are modeled by a piecewise-homogeneous body with a media interface in the form of the sides of the angle, the corner point of which can be a stress concentrator and can be dangerous from the point of view of the origin of interfacial cracks in it. Therefore, studies of the stress state of piecewise homogeneous bodies weakened by interfacial cracks. born at a corner point of the interface is a topical problem in the mechanics of a deformable solid.

Below, using the Wiener – Hopf method, an exact solution of the plane static problem of the theory of elasticity for a piecewise-homogeneous isotropic elastic body with small-scale interfacial shear cracks at the corner point of the interface is constructed. Based on the constructed solution, the stress state of the body at the corner point is investigated.

1 Formulation of the Problem

Under the conditions of plane deformation, within the framework of the static symmetric problem, we consider a piecewise homogeneous body with a media interface in the form of sides of the angle, which is composed of isotropic elastic parts with Young's moduli E_1, E_2 ($E_1 > E_2$) and Poisson's ratios ν_1, ν_2 (see Fig. 1).

E. Gdoutos and M. Konsta-Gdoutos (Eds.): ICTAEM 2020, STIN 16, pp. 216–221, 2020.
https://doi.org/10.1007/978-3-030-47883-4_40

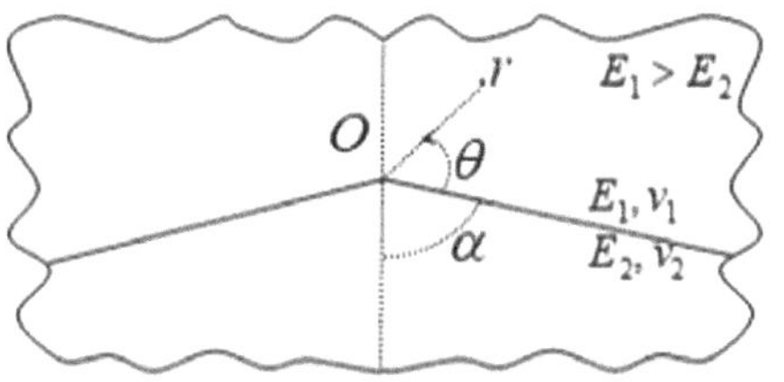

Fig. 1. Piece-homogeneous isotropic body with the interface of media in the form the sides of angle

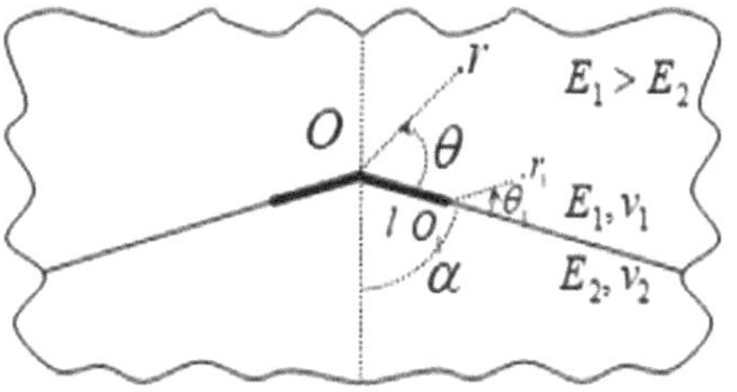

Fig. 2. Piece-homogeneous isotropic body with small-scale interior interfacial shear cracks

In accordance with the general provisions on the behavior of stresses at the corner points of elastic bodies, the corner point of the interface is a stress concentrator with a power-law singularity. In this case, the following representations take place for stresses in the vicinity of the corner point:

$$\begin{aligned}\tau_{r\theta}(r,0) &= Cg_1(\alpha, e_0, v_1, v_2)r^{\lambda_0} + o\left(r^{\lambda_0}\right),\\ \sigma_\theta(r,0) &= Cg_2(\alpha, e_0, v_1, v_2)r^{\lambda_0} + o\left(r^{\lambda_0}\right), \quad e_0 = \frac{E_1}{E_2}\\ &(r \to 0),\end{aligned}$$

where

$$g_1 = \lambda_0 g_{11} \sin \lambda_0\alpha - g_{12} \sin(\lambda_0 + 2)\alpha,$$

$$\begin{aligned}g_{11} = {} & (1-e)\lambda_0^2 \sin^2 2\alpha \cos(\lambda_0+2)\alpha - (1-æ_1-2e)\lambda_0 \sin 2\alpha \cos(\lambda_0+2)\alpha \cos\lambda_0(\pi-\alpha)\\ & \times \sin[\lambda_0(\pi-\alpha)-2\alpha] + [2-(1-æ_2)e]\lambda_0 \sin 2\alpha \cos\lambda_0\alpha \cos(\lambda_0+2)\alpha \cos\lambda_0(\pi-\alpha)\\ & \times \sin[\lambda_0(\pi-\alpha)-2\alpha] + (1+æ_1)\lambda_0 \sin 2\alpha \sin(\lambda_0+2)\alpha \cos\lambda_0(\pi-\alpha) \cos[\lambda_0(\pi-\alpha)-2\alpha]\\ & + (1+æ_1)(1-æ_2) \cos\lambda_0\alpha \sin^2(\lambda_0+2)\alpha \cos\lambda_0(\pi-\alpha) \cos[\lambda_0(\pi-\alpha)-2\alpha] - (1+æ_2)\lambda_0\\ & \times \sin 2\alpha \cos\lambda_0\alpha \sin(\lambda_0+2)\alpha \cos(\lambda_0+2)\alpha(1+æ_1)(1-æ_2) \times \cos\lambda_0\alpha \sin(\lambda_0+2)\alpha\\ & \times \cos(\lambda_0+2)\alpha \cos\lambda_0(\pi-\alpha) \sin[\lambda_0(\pi-\alpha)-2\alpha],\end{aligned}$$

$$\begin{aligned}g_{12} = {} & (1-e)(1-æ_2+\lambda_0)\lambda_0^2 \sin^2 2\alpha \cos\lambda_0\alpha - (1-æ_1-2e)\lambda_0(1-æ_2+\lambda_0) \sin 2\alpha \cos\lambda_0\alpha\\ & \times \cos\lambda_0(\pi-\alpha) \sin[\lambda_0(\pi-\alpha)-2\alpha] + [2-(1-æ_2)e]\lambda_0(1-æ_2+\lambda_0) \sin 2\alpha \cos^2\lambda_0\alpha\\ & \times \sin(\lambda_0+2)\alpha - 2[1-æ_1-(1-æ_2)e](1-æ_2+\lambda_0)\cos^2\lambda_0\alpha \sin(\lambda_0+2)\alpha \cos\lambda_0(\pi-\alpha)\\ & \times \sin[\lambda_0(\pi-\alpha)-2\alpha] + (1+æ_1)\lambda_0^2 \sin 2\alpha \sin\lambda_0\alpha \cos\lambda_0(\pi-\alpha)\cos[\lambda_0(\pi-\alpha)-2\alpha]\\ & + (1+æ_1)(1-æ_2)\lambda_0 \sin\lambda_0\alpha \cos\lambda_0\alpha \sin(\lambda_0+2)\alpha \cos\lambda_0(\pi-\alpha)\cos[\lambda_0(\pi-\alpha)-2\alpha]\\ & - (1+æ_2)\lambda_0^2 \sin 2\alpha \sin\lambda_0\alpha \cos\lambda_0\alpha \cos(\lambda_0+2)\alpha + (1-æ_1)(1+æ_2)\lambda_0 \sin\lambda_0\alpha \cos\lambda_0\alpha\\ & \times \cos(\lambda_0+2)\alpha \cos\lambda_0(\pi-\alpha) \sin[\lambda_0(\pi-\alpha)-2\alpha],\end{aligned}$$

$$e = \frac{1+\nu_2}{1+\nu_1} e_0,\ æ_{1,2} = 3 - 4\nu_{1,2}$$

The degree of stress singularity is the unique root of the characteristic equation in the interval (−1; 0):

$$\Delta(-\lambda - 1) = 0,\ \Delta(z) = \delta_0(z) + \delta_1(z)e + \delta_2(z)e^2,$$

$$\delta_0(z) = (\sin 2z\alpha + z\sin 2\alpha)[æ_1 \sin 2z(\pi - \alpha) + z\ \sin 2\alpha],$$

$$\delta_1(z) = (1+æ_1)(1+æ_2)\sin^2 z\pi - (\sin\ 2z\alpha + z\ \sin 2\alpha)[æ_1 \sin\ 2z(\pi - \alpha) + z\ \sin 2\alpha] - [\sin\ 2z(\pi - \alpha) - z\ \sin 2\alpha](æ_2 \sin\ 2z\alpha - z\ \sin 2\alpha),$$

$$\delta_2(z) = [\sin 2z(\pi - \alpha) - z\ \sin 2\alpha](æ_2 \sin 2z\alpha - z\ \sin 2\alpha)$$

Constant C, which is included in these representations, is determined by the solution of each specific problem of the theory of elasticity, which is depicted in Fig. 1. The constant C, which depends on the external load, can be regarded as the stress intensity factor at the corner point of the interface.

If $C < 0$ (it is shown that this condition must be fulfilled) and $\alpha \in (\alpha_1; \pi/2) \cup (\alpha_2; \pi)$ then, according to the information about the function and the formula given above, $\sigma_\theta(r,0) \to -\infty$ when $r \to 0$, therefore, at the interface between the media at the corner point, normal stresses are compressive. In this case, due to the high concentration of stresses at the corner point, it is possible to initiate interfacial shear cracks with completely contacting faces extending from this point, the length of which is much less than the sizes of the body (see Fig. 2). The larger the ratio of Young's moduli $e_0 = E_1/E_2 > 1$, the wider the range of angle α values at which such cracks should be expected to form.

Taking into account the smallness of the cracks, we arrive at the plane static symmetric problem of the theory of elasticity for a piecewise homogeneous isotropic plane with a media interface in the form of sides of an angle containing cuts of finite length starting from the corner point and located on this boundary (see Fig. 3).

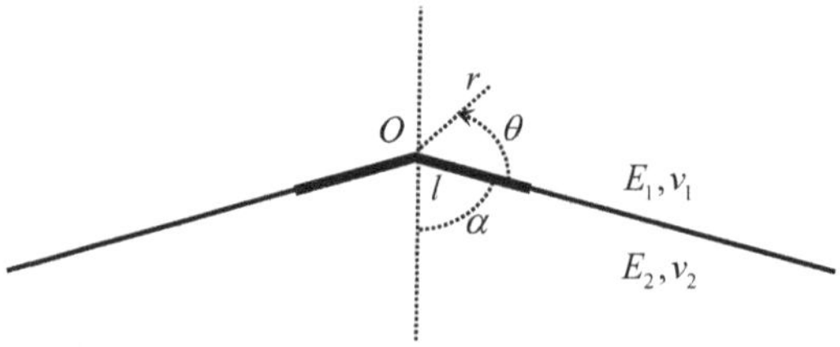

Fig. 3. Piece-homogeneous isotropic plane with interior interfacial shear cracks

At infinity, the asymptotic behavior of the stress field is given, which is a solution to a similar problem without cuts (problem K), a homogeneous problem of the theory of elasticity for a piecewise homogeneous plane with a media interface in the form of the sides of the angle, which is generated by the unique root λ_0 of its characteristic equation on the interval $(-1;0)$.

An arbitrary constant C included in the indicated solution is considered given. It characterizes the intensity of the external field and should be determined from the solution of the external problem.

Taking into account the symmetry, we write the boundary conditions of the problem (Fig. 3) as follows:

$$\theta = \pi - \alpha,\ \tau_{r\theta} = 0,\ u_\theta = 0;\ \theta = 0,\ \langle\sigma_\theta\rangle = \langle\tau_{r\theta}\rangle = 0,\ \langle u_r\rangle = 0; \tag{1}$$

$$\theta = 0,\ r < l,\ \tau_{r\theta} = 0;\ \theta = 0,\ r > l,\ \langle u_r\rangle = 0; \tag{2}$$

$$\theta = 0,\ r \to \infty,\ \tau_{r\theta} = Cg_1 r^{\lambda_0} + o(1/r)\,(-\alpha \le \theta \le \pi - \alpha). \tag{3}$$

The solution of the formulated problem of the theory of elasticity (Fig. 3) is the sum of the solutions of the following two problems. The first (problem 1) differs from it in that instead of the first of conditions (2) we have

$$\theta = 0,\ r < l,\ \tau_{r\theta} = -Cg_1 r^{\lambda_0}, \tag{4}$$

and at infinity, the stresses fade as $o(1/r)$ (in (3) the first term is absent). The second problem is problem K. Since the solution to the second problem is known, it is enough to construct a solution to the first. To construct an exact solution to problem 1, we will use the Wiener – Hopf method [1] in combination with the Mellin integral transform apparatus [2].

2 Solution of the Wiener – Hopf Equation

Applying the Mellin transforms with the complex parameter p to the equilibrium equations, the compatibility conditions for the strains, Hooke's law, conditions (1) and taking into account the second of conditions (2) and condition (4), we obtain the Wiener – Hopf functional equation

$$\Phi^+(p) + \frac{\tau}{p + \lambda_0 + 1} = A\mathrm{ctg}p\pi G(p)\Phi^-(p) \tag{5}$$

$$A = \frac{(1 + æ_1)[1 + æ_1 + (1 + æ_2)e]}{2[æ_1 + (1 + æ_1æ_2)e + æ_2e^2]},\ G(p) = \frac{G_1(p)}{G_2(p)}$$

$$G_1(p) = \left[æ_1 + (1 + æ_1æ_2)e + æ_2e^2\right]\left[a_0(p) + a_1(p)e\right]\sin p\pi,$$

$$G_2(p) = [1 + æ_1 + (1 + æ_2)e]\left[b_0(p) + b_1(p)e + b_2(p)e^2\right] \cos p\pi,$$

$$a_0(p) = (1 + æ_1)[\cos 2p(\pi - \alpha) - \cos 2\alpha](\sin 2p\alpha + p \sin 2\alpha),$$

$$a_1(p) = (1 + æ_2)(\cos 2p\alpha - \cos 2\alpha)[\sin 2p(\pi - \alpha) - p \sin 2\alpha],$$

$$b_0(p) = (\sin 2p\alpha + p \sin 2\alpha)[æ_1 \sin 2p(\pi - \alpha) + p \sin 2\alpha],$$

$$b_1(p) = (1 + æ_1)(1 + æ_2) \sin^2 p\pi - (\sin 2p\alpha + p \sin 2\alpha)[æ_1 \sin 2p(\pi - \alpha) + p \sin 2\alpha] - [\sin 2p(\pi - \alpha) - p \sin 2\alpha](æ_2 \sin 2p\alpha - p \sin 2\alpha),$$

$$b_2(p) = [\sin 2p(\pi - \alpha) - p \sin 2\alpha](æ_2 \sin 2p\alpha - p \sin 2\alpha), \ \tau = -Cg_1 l^{\lambda_0},$$

$$\Phi^+(p) = \int_1^\infty \tau_{r\theta}(\rho l, 0)\rho^p d\rho, \Phi^-(p) = \frac{E_1}{4\left(1 - \nu_1^2\right)} \int_0^1 \left\langle \frac{\partial u_r}{\partial r} \right\rangle \Big|_{\substack{r = \rho l \\ \theta = 0}} \rho^p d\rho$$

$$(-\varepsilon_1 < \mathrm{Re} p < \varepsilon_2).$$

The factorization of the equation coefficient on the imagine axis carried out by its splitting into two functions: the function which factorizes using gamma-functions [3] and the function which factorizes according Gakhov's formula [4]. The solution of Eq. (5) is the follows.

$$\Phi^+(p) = \frac{\tau K^+(p)G^+(p)}{p + \lambda_0 + 1}\left[\frac{1}{K^+(-\lambda_0 - 1)G^+(-\lambda_0 - 1)} - \frac{1}{K^+(p)G_1^+(p)}\right] (\mathrm{Re} p < 0),$$

$$\Phi^-(p) = \frac{\tau p G^-(p)}{AK^+(-\lambda_0 - 1)G^+(-\lambda_0 - 1)(p + \lambda_0 + 1)K^-(p)} \ (\mathrm{Re} p > 0) \qquad (6)$$

$$\exp\left[\frac{1}{2\pi i}\int_{-i\infty}^{+i\infty} \frac{\ln G(z)}{z - p} dz\right] = \begin{cases} G^+(p), & \mathrm{Re} p < 0 \\ G^-(p), & \mathrm{Re}\, p > 0 \end{cases}, \quad K^\pm(p) = \frac{\Gamma(1 \pm p)}{\Gamma(1/2 \pm p)}$$

3 Investigation of the Behavior of Stresses in the Vicinity of a Corner Point

Based on the solution (6) of the Wiener – Hopf Eq. (5), expressions for Mellin stress transformants are obtained. As a result of applying Mellin's inversion formulas to these expressions, the stress in problem 1 is found. The following formula holds:

$$\sigma_\theta(r, \theta) = r^{\lambda_1} S(\theta) C_1 + o\left(r^{\lambda_1}\right), \qquad (7)$$

$$C_1 = \frac{g_1 \Gamma(\lambda_0 + 3/2)\Gamma(\lambda_1 + 2)G_1^+(-\lambda_1 - 1)}{(\lambda_1 - \lambda_0)s\Gamma(\lambda_0 + 2)\Gamma(\lambda_1 + 3/2)G_1^+(-\lambda_0 - 1)} Cl^{\lambda_0 - \lambda_1}, \tag{8}$$

$$\begin{aligned} s = (1 + æ_1)\{[\cos 2(\lambda_1 + 1)(\pi - \alpha) - \cos 2\alpha][2\alpha\ \cos 2(\lambda_1 + 1)\alpha + \sin 2\alpha] \\ - 2(\pi - \alpha)[\sin 2(\lambda_1 + 1)\alpha + (\lambda_1 + 1)\sin 2\alpha] \sin 2(\lambda_1 + 1)(\pi - \alpha)\} \\ + (1 + æ_2)\{[2(\pi - \alpha)\cos 2(\lambda_1 + 1)(\pi - \alpha) - \sin 2\alpha][\cos 2(\lambda_1 + 1)\alpha - \cos 2\alpha] \\ - 2\alpha[\sin 2(\lambda_1 + 1)(\pi - \alpha) - (\lambda_1 + 1)\sin\ 2\alpha]\sin 2(\lambda_1 + 1)\alpha\}e, \end{aligned}$$

where $S(\theta)$ is known function.

Formula (7) gives the main term of the expansion of the normal stress $\sigma_\theta(r, \theta)$ in the asymptotic series for $r \to 0$ for material 1 ($0 < \theta < \pi - \alpha$). Similar representations take place for $\sigma_\theta(r, \theta)$ when $-\alpha < \theta < 0$, as well as for $\tau_{r\theta}(r, \theta)$ and $\sigma_r(r, \theta)$.

In expression (7) λ_1 is the unique root of the following equation in the interval $(-1; 0)$

$$\begin{aligned} (1 + æ_1)[\cos\ 2(\lambda + 1)(\pi - \alpha) - \cos\ 2\alpha][\sin\ 2(\lambda + 1)\alpha + (\lambda + 1)\sin\ 2\alpha] \\ + (1 + æ_2)[\cos\ 2(\lambda + 1)\alpha - \cos\ 2\alpha][\sin\ 2(\lambda + 1)(\pi - \alpha) - (\lambda + 1)\sin\ 2\alpha]e = 0. \end{aligned} \tag{8}$$

It follows from (7) that the corner point O is a special point of the boundary-value problem of the elasticity theory under consideration (Fig. 3). It is a stress concentrator. As the point of the region approaches the point O, stress goes to infinity. The degree of singularity of stresses depends on the angle, the ratio of the Young's modulus, and the Poisson's coefficients. This degree r is the unique one in the interval $(-1; 0)$ of roots of a certain transcendental equation (Eq. (8)).

As the angle α increases from zero to $\pi/2$ the stress concentration near the corner point decreases, and with increasing it from $\pi/2$ up to π – increases.

If the angle α tends to zero or to π, the degree of the singularity of stresses tends to -1. If the angle α tends to $\pi/2$, then the degree of the singularity of stresses tends to zero. If the angle α is acute, then as the ratio of Young's modulus $e_0 = E_1/E_2 > 1$ increases, the stress concentration near the angular point decreases, and if the angle obtuse – increases.

References

1. Nobl, B.: Applying the Wiener – Hopf Method for Solving Partial Derivatives Differential Equations. Inostr. Lit., Moscow (1962). (in Russian)
2. Uflyand, Ya.S.: Integral Transformations in Elasticity Theory Problems. Nauka, Leningrad (1967). (in Russian)
3. Lavrentiev, M.A., Shabat, B.V.: The Theory of Complex Variable Methods. Nauka, Moscow (1973). (in Russian)
4. Gakhov, F.D.: Boundary-value problems. Nauka, Moscow (1977). (in Russian)

Coupled Non-stationary Longitudinal Vibrations of an Infinite Electromagnetoelastic Rod

Thong D. Pham[1], Dmitry V. Tarlakovskii[1,2(✉)], and Vladimir A. Vestyak[1]

[1] Moscow Aviation Institute, National Research University, Volokolamskoe Shosse. 4, Moscow 125993, Russia
dtstudio.pro@gmail.com, tdvhome@mail.ru
[2] Institute of Mechanics, Lomonosov Moscow State University, Michurinskii Prospekt. 1, Moscow 119192, Russia

Abstract. We study the associated non-stationary longitudinal vibrations of an infinite electromagnetoelastic rod. It is assumed that the material of the rod is a homogeneous isotropic conductor. In this case, the initial electromagnetic field, the Lorentz force, Maxwell's equations and the generalized Ohm's law are taken into account. Assuming that the desired functions depend only on the longitudinal coordinate, using the corresponding relations for the shells, a closed system of equations of motion is obtained. The desired functions are assumed to be bounded, and the initial conditions to be zero. The Fourier transform in the coordinate and the Laplace transform in time, as well as the small parameter method characterizing the relationship of mechanical and electromagnetic fields, are used to solve this problem. The solution is presented in integral form with the cores in the form of Green functions for which images and originals are constructed. An example of calculation is given.

Keywords: Non-stationary electromagnetoelasticity · Infinite rod · Longitudinal vibrations · Green's function · Laplace and Fourier transforms

1 Problem Statement

In a rectangular Cartesian coordinate system *Oxyz* we consider non-stationary longitudinal movements of a straight-line homogeneous isotropic infinite electromagnetoelastic rod with an *Ox* axis under zero initial conditions and the absence of an initial electric field. The equations of the corresponding process are written in the space of Laplace transforms [2] in time (additional superscript «*L*» indicates image; *s* - corresponding parameter) as a special case of the general equations for a thin shell constructed in [1]:

E. Gdoutos and M. Konsta-Gdoutos (Eds.): ICTAEM 2020, STIN 16, pp. 222–227, 2020.
https://doi.org/10.1007/978-3-030-47883-4_41

$$s^2u^L = \frac{\partial^2 u^L}{\partial x^2} + \zeta\frac{\partial \psi_3^L}{\partial x} - \alpha\gamma H_0\left(H_0 u^L + e_3^L\right) + q_{*1}^L,$$
$$s^2\psi_3^L = \eta^{-2}\frac{\partial^2\psi_3^L}{\partial x^2} - r^{-2}\left(\zeta\frac{\partial u^L}{\partial x} + \psi_3^L\right) + \alpha\frac{H_0^2\gamma s}{s+\gamma}\left(\frac{\partial\psi_3^L}{\partial x} - s\psi_3^L\right) + m_*^L; \quad (1)$$

$$\frac{\partial h_2^L}{\partial x} = \eta_e^2\left[(s+\gamma)e_3^L + \gamma s H_0 u^L\right], \ \frac{\partial e_3^L}{\partial x} = sh_2^L + \frac{\gamma s}{s+\gamma}H_0\psi_3^L. \quad (2)$$

where u - longitudinal movement; ψ_3 - compression (deformation in the Oz direction); e_3 и h_2 - changes in electric and magnetic fields on the axis of the rod, respectively, in the directions Oz and Oy; H_0 - surface charge, the intensity of the initial magnetic field, respectively, in the direction; q_{*1} и m_* - external linear load and moment in the Oxy plane.

Assuming that the required functions are limited.

Hereinafter, the following dimensionless quantities are used (for the same style, the dimension parameters are indicated by a prime symbol):

$$x = \frac{x'}{L}, \tau = \frac{c_1 t}{L}, u = \frac{u'}{L}, q_{*1} = \frac{q'_{*1}L}{(\lambda+2\mu)h},$$
$$m_* = \frac{m'_*L^2}{(\lambda+2\mu)I}, I = \frac{h^3}{12}, r^2 = \frac{I}{L^2 h},$$
$$\eta = \frac{c_1}{c_2}, \zeta = \frac{\lambda}{\lambda+2\mu} = 1 - \frac{2}{\eta^2}, c_1^2 = \frac{\lambda+2\mu}{\rho}, c_2^2 = \frac{\mu}{\rho},$$
$$e_3 = \frac{e'_3}{E_*}, h_2 = \frac{h'_2\mu_e c_1}{cE_*}, H_0 = \frac{H'_0\mu_e c_1}{cE_*}, \eta_e = \frac{c_1}{c_e},$$
$$\alpha = \frac{\varepsilon_e E_*^2}{4\pi(\lambda+2\mu)}, \gamma = \frac{4\pi\sigma L}{\varepsilon_e c_1}, c_e^2 = \frac{c^2}{\mu_e\epsilon_e},$$

here t - time; L и E_* - some characteristic linear dimension and the electric field intensity; λ, μ - Lame elastic constants; h - cross section height; ρ - material density; σ, ε_e и μ_e - conductivity, dielectric and magnetic permeability coefficients.

2 The Solution in the Image Space

To construct a solution to the system of Eq. (1), (2), we additionally apply the Fourier transform [2] along the x coordinate (additional superscript «F» indicates image; q - corresponding parameter):

$$\left(k_1^2 + \alpha\gamma H_0^2 s\right)u^{FL} + iq\zeta\psi_3^{FL} + \alpha\gamma H_0 e_3^{FL} = q_{*1}^{FL},$$
$$\left[\left(k_2^2 + \eta^2 r^{-2}\right)(s+\gamma) + \alpha H_0^2\eta^2\gamma s(iq+s)\right]\psi_3^{FL} - iq\zeta\eta^2 r^{-2}(s+\gamma)u^{FL} = \eta^2(s+\gamma)m_*^L;$$

$$\eta_e^2\gamma s H_0 u^{FL} + \eta_e^2(s+\gamma)e_3^{FL} + iqh^{FL} = 0, \ \gamma s H_0\psi_3^{FL} + iq(s+\gamma)e_3^{FL} + s_e^2 h^{FL} = 0,$$

Here

$$k_1 = \sqrt{q^2+s^2},\ k_2 = \sqrt{q^2+\eta^2 s^2},\ s_e = \sqrt{s(s+\gamma)},\ \mathrm{Re}\sqrt{\cdot} > 0$$

Eliminating functions e_3^{FL} and h^{FL} from this system of equations, we obtain a system of linear algebraic equations

$$a_{11}(q,s)u^{FL} + a_{12}(q,s)\psi_3^{FL} = (s+\gamma)k_e^2 q_{*1}^{FL},$$
$$a_{21}(q,s)u^{FL} + a_{22}(q,s)\psi_3^{FL} = \eta^2(s+\gamma)m_*^{FL},$$

where

$$a_{11}(q,s) = a_{11}(q,s) = (s+\gamma)\left[k_1^2k_e^2 + \alpha\gamma s H_0^2\left(q^2+\eta_e^2 s_e^2 - \gamma\eta_e^2 s\right)\right],$$
$$a_{12}(q,s) = iq\left[\zeta(s+\gamma)k_e^2 + \alpha\gamma^2 H_0^2 s\right],\ a_{21}(q,s) = -ir^{-2}\zeta\eta^2(s+\gamma)q,$$
$$a_{22}(q,s) = (s+\gamma)\left(k_2^2+\eta^2 r^{-2}\right) + \alpha\eta^2 H_0^2\gamma s(iq+s).$$

Its solution is written as follows:

$$\begin{aligned} u^{FL} &= G_{uq}^{FL}(q,s)q_{*1}^{FL} + G_{um}^{FL}(q,s)m_*^{FL}, \\ \psi_3^{FL} &= G_{\psi_3 q}^{FL}(q,s)q_{*1}^{FL} + G_{\psi_3 m}^{FL}(q,s)m_*^{FL}. \end{aligned} \tag{3}$$

where

$$G_{uq}^{FL}(q,s) = \frac{(s+\gamma)k_e^2 a_{22}(q,s)}{D(q,s)},\ G_{um}^{FL}(q,s) = -\frac{\eta^2(s+\gamma)a_{12}(q,s)}{D(q,s)},$$
$$G_{\psi_3 q}^{FL}(q,s) = -\frac{(s+\gamma)k_e^2 a_{21}(q,s)}{D(q,s)},\ G_{\psi_3 m}^{FL}(q,s) = \frac{\eta^2(s+\gamma)a_{11}(q,s)}{D(q,s)},$$
$$D(q,s) = a_{11}(q,s)a_{22}(q,s) - a_{12}(q,s)a_{21}(q,s).$$

Here $G_{uq}^{FL}(q,s)$, $G_{um}^{FL}(q,s)$, $G_{\psi_3 q}^{FL}(q,s)$ and $G_{\psi_3 m}^{FL}(q,s)$ - images of influence functions.

Similarly, we can represent functions e_3^{FL} and h_2^{FL}.

3 Originals of the Influence Functions and Solutions

The originals of the Laplace transform are quite simple, since the images of the influence functions are rational functions of the parameter s. However, the analytic inversion of the Fourier transform is impossible. An attempt at numerical inversion does not lead to a satisfactory result due to the presence of rapidly oscillating functions.

Therefore, a simplified model is used in which the compression of the rod is neglected. For this, it suffices to set $\psi_3^{FL} = 0$ in (3) and exclude m_*^{FL}. We obtain the following result

$$u^{FL} = \Gamma_{uq}^{FL}(q,s)q_{*1}^{FL}, \Gamma_{uq}^{FL}(q,s) = \frac{k_e^2}{k_1^2k_e^2 + \alpha\gamma s H_0^2\left(q^2 + \eta_e^2 s^2\right)}, \quad (4)$$

where $\Gamma_{uq}^{FL}(q,s)$ - image of the influence function for a simplified model.

In this case, the above-mentioned difficulties remain. Therefore, we use the method of a small parameter, for which we take the coupling coefficient of the mechanical and electromagnetic fields α.

Limiting the linear approximation, we obtain

$$\Gamma_{uq}^{FL}(q,s) = \Gamma_{uq0}^{FL}(q,s) + \Gamma_{uq1}^{FL}(q,s)\alpha,$$

where

$$\Gamma_{uq0}^{FL}(q,s) = \frac{1}{k_1^2}, \ \Gamma_{uq1}^{FL}(q,s) = -\gamma H_0^2\frac{\left(q^2 + \eta_e^2 s^2\right)s}{k_1^4 k_e^2}.$$

Originals are found using transformation properties and tables [2–4]:

$$\begin{gathered}\Gamma_{uq}(x,\tau) = \Gamma_{uq0}(x,\tau) + \alpha\Gamma_{uq1}(x,\tau),\\ \Gamma_{uq0}(x,\tau) = \frac{1}{2}H(\tau - |x|), \ \Gamma_{uq1}(x,\tau) = -\gamma H_0^2\left[f_{uq}(x,\tau) + f_{uq3}(x,\tau)\right],\end{gathered} \quad (5)$$

where

$$f_{uq}(x,\tau) = \frac{1}{4\left(1-\eta_e^2\right)}\left\{2(\tau - |x|)e^{\gamma_e(\tau-|x|)} - \frac{1+\eta_e^2}{\gamma_e}\left[e^{\gamma_e(\tau-|x|)} - 1\right]\right\}H(\tau - |x|),$$

$$f_{uq3}(x,\tau) = \frac{\gamma_e\tau +}{2\eta_e\left(1-\eta_e^2\right)}e^{\gamma_e\tau} * e^{-\gamma\tau/2}I_0\left(\frac{\gamma}{2}\sqrt{\tau^2 - \eta_e^2x^2}\right)H(\tau - \eta_e|x|), \ \gamma_e = \frac{\eta_e^2\gamma}{1-\eta_e^2}.$$

Here $I_v(x)$ - modified Bessel function of the first kind of order v[5], $H(\tau)$ - Heaviside function.

Note that the function $\Gamma_{uq0}(x,\tau)$ corresponds to a purely elastic solution.

The original of movement according to (4) is written as follows (asterisks symbol indicate convolutions in time and coordinate):

$$u(x,\tau) = \Gamma_{uq}(x,\tau) * *q_{*1}(x,\tau)$$

4 Examples for the Calculation

As an example, consider an aluminum rod, the material of which is characterized by the following values [6]:

$$\eta = 2,04;\ \zeta = 0,519,\ \eta_e = 0,111 \cdot 10^{-4};\ \gamma = 5,06;\ \alpha = 0,0806.$$

It is accepted here $E_* = 100\,\mathrm{B/M}$ and $L = 1\ \ M$. We also assume that $H_0 = 1$ and the rod has a square cross section with the following geometric parameters:

$$h = 0,05\ \ \mathrm{M};\ r^2 = \frac{h^2}{12L^2} = 0,208 \cdot 10^{-3}$$

In Fig. 1 and 2 show the dependences of the $\Gamma_{uq1}(x, \tau)$ function obtained using (5) on the coordinate and time, respectively, for different values of τ and x.

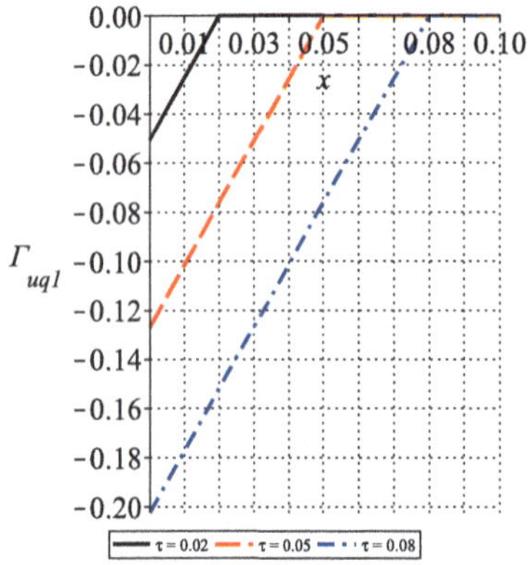

Fig. 1. The dependences of Γ_{uq1} on x

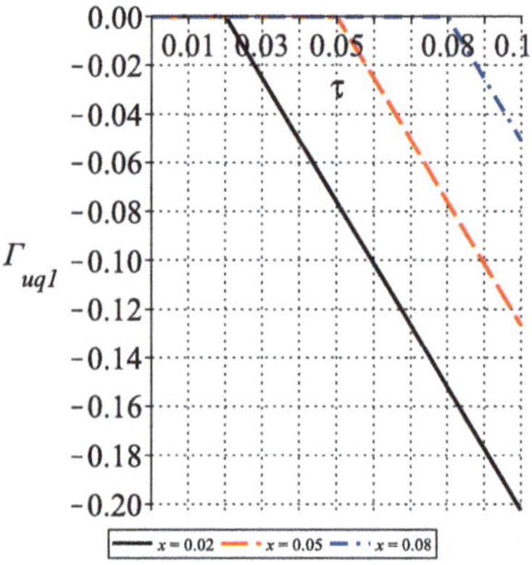

Fig. 2. The dependences of Γ_{uq1} on τ

In Fig. 3 and 4 show similar dependences for the influence function $\Gamma_{uq}(x, \tau)$.

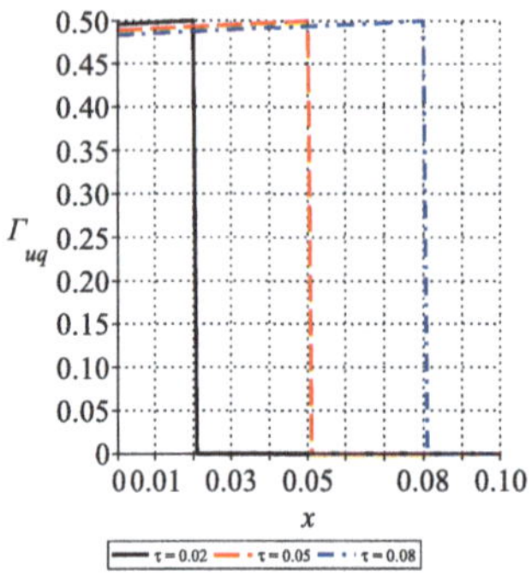

Fig. 3. The dependences of Γ_{uq} on x

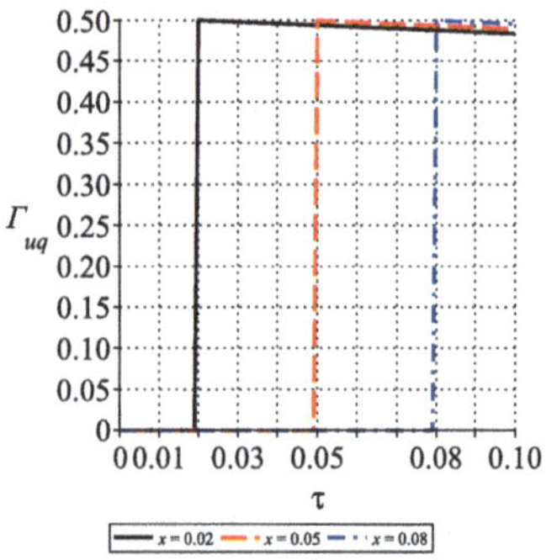

Fig. 4. The dependences of Γ_{uq} on τ

Note that, as follows from (5), for the material under consideration the $f_{uq3}(x,\tau)$ function has the order of a small coefficient η_e, then for $\tau > \eta_e|x|$ the contribution of this term to the Γ_{uq1} function is insignificant. Taking into account the connectivity of the process gives a difference from the elastic solution of the order of α, which in these calculations corresponds to approximately eight percent.

5 Conclusion

An approach to the study of coupled non-stationary longitudinal vibrations of an electromagnetoelastic rod is proposed. A small correction introduced by taking into account the connectivity is explained by the parameters of the material under consideration.

Similarly, the problems of longitudinal vibrations of a semi-infinite and final rod, as well as transverse vibrations, can be solved.

References

1. Vestyak, V.A., Tarlakovskii, D.V.: The model of thin electromagnetoelastic shells dynamics. In: Proceedings of the Second International Conference on Theoretical, Applied and Experimental Mechanics. Structural Integrity, pp. 254–258. Springer, Switzerland AG (2019)
2. Gorshkov, A.G., Medvedskii, A.L., Rabinskii, L.N., Tarlakovskii, D.V.: Waves in Continuous Media. Physmatlit, Moscow (2004)
3. Prudnikov, A.P., Brychkov, Yu.A., Marichev, O.I.: Integrals and Series. Gordon and Breach Science Publishers, New York (1986)
4. Ditkin, V.A., Prudnikov, A.P.: Handbook of operational calculus. "Higher School", Moscow (1965)
5. Abramowitz, M., Stegun, I.A.: Handbook of Mathematical Functions with Formulas, Graphs and Mathematical Tables. U.S. Department of Commerce, National Bureau of Standards (1964)
6. Vestyak, V.A., Gachkevich, A.R., Musiy, R.S., Tarlakovskii, D.V., Fedotenkov, G.V.: Two-dimensional non-stationary waves in electromagnetoelastic bodies. Physmatlit, Moscow (2019)

The Effect of Cracks Interaction in Materials Under Loading Along Cracks

Viacheslav Bogdanov(✉) and Vladimir Nazarenko

S. P. Timoshenko Institute of Mechanics of the National Academy of Sciences of Ukraine, Kyiv 02053, Ukraine
bogdanov@nas.gov.ua

Abstract. The non-classical problems of the fracture of materials loaded along the cracks they contain are analyzed. Two fracture mechanisms are considered, namely, the fracture of materials with initial (residual) stresses acting in parallel to the surfaces of cracks location and the fracture of the bodies compressed along cracks. In the study of such problems, a new combined approach developed in the framework of the three-dimensional linearized mechanics of deformable bodies is used. The results of the investigation of two- and three-dimensional problems for typical configurations of parallel coaxial Mode I and Mode II cracks are presented. The effect of residual stresses on stress intensity factors is analyzed for highly elastic materials and composites. The compressive strength for bodies with interacting cracks located in parallel planes under compression is calculated using the approach mentioned. The crack length (diameter), mechanical properties of materials and the mutual position of cracks were found to influence the critical fracture parameters.

Keywords: Non-classical problems of fracture mechanics · Residual stresses · Compression along cracks · Parallel coaxial cracks · Highly elastic materials · Composites

1 Introduction

Initial (or residual, technological) stresses are nearly always present in actual structural materials and elements as a result of technological processes of their production, joining, mechanical and thermal surface treatment, exploitation, producing a considerable effect on the fracture processes in bodies with cracks. The non-classical nature of the problems on fracture of pre-stressed materials, when initial (residual) stresses are directed in parallel to the surfaces of cracks location, is accounted for by the fact that the solutions of respective problems of linear elasticity theory show that these initial stresses are not involved in the expressions for stress intensity factors, J-integral and crack opening displacements. As a result, the stress components directed along cracks cannot be taken into consideration in Griffith–Irwin and Cherepanov–Rice classical models and fracture criteria, critical crack opening displacements or their generalizations [1]. At the same time, this is contrary to the data of numerous experimental studies, which reveal the effect of the load component directed along cracks on fracture parameters under both static and cyclic loads. In [2] an approach based on the three-

E. Gdoutos and M. Konsta-Gdoutos (Eds.): ICTAEM 2020, STIN 16, pp. 228–234, 2020.
https://doi.org/10.1007/978-3-030-47883-4_42

dimensional linearized mechanics of deformable bodies to solving the problems of the fracture of bodies with residual stresses was proposed and fracture criteria for the abovementioned problems were formulated.

The non-classical nature of fracture mechanics problems on body compression along cracks is related to the fact that under uniform compression strictly along the planes of cracks location in an isotropic and orthotropic material, a uniform stress-strain state emerges, which leads to the absence of singularities at crack tips in corresponding solutions. Hence, it is also impossible to directly apply the classical criteria of fracture, since in linear fracture mechanics both the stress intensity factors and crack opening displacements are equal to zero in this situation. In the problem at issue the most likely mechanism of the onset (start) of the fracture process, by analogy with the problem on the compression of structural elements along the symmetry axes, is the local loss of equilibrium stability of the material that surrounds cracks. In [3] it was proposed to determine the critical compressive loads that correspond to local buckling of the material near cracks by solving eigenvalue problems in the framework of the three-dimensional linearized theory of the stability of deformable bodies.

As follows from the above, the common point in investigating these two fracture mechanics problems on loading along cracks is the employment of related mathematical apparatuses within the three-dimensional linearized mechanics of deformable bodies. For both classes of problems the linearized equilibrium equations and constitutive stress-strain relations are identical, as are, within corresponding design diagrams, the considered configurations of bodies and defect location patterns. But a significant difference in the formulation and solution of problems on the fracture of the bodies compressed along the arrays of cracks located in parallel planes and problems on the fracture of materials with residual stresses acting along cracks, is that in the former case we end up investigating eigenvalue problems with zero boundary conditions on crack faces, while in the latter we examine boundary problems with non-zero boundary conditions on crack faces. Taking into account those considerations, a combined approach to the investigation of the problems on the fracture mechanics of cracked materials compressed along cracks and the problems on the fracture of bodies with residual stresses was proposed in [4, 5]. In this approach, critical loading parameters in the problems on the bodies compressed along the cracks they contain can be determined *via* the solution of corresponding non-homogenous problems of the fracture mechanics of materials with residual stresses, as the values of compressive residual loads which, when achieved, lead to a dramatic "resonance-like" change in stresses and displacements near crack contours. This situation can be interpreted relying on the following physical considerations. Consider a body with initial stresses acting along a single inner crack. When the initial compressive stresses achieve the values that correspond to the surface instability of the half-space, a neutral equilibrium state emerges in the vicinity of the crack tip. In this situation, a minor increase in the external load is sufficient to break the neutral equilibrium and launch the fracture process characterized with a local loss of material's stability in the vicinity of the crack.

Below we demonstrate the employment of this combined approach by investigating 2D and 3D problems on the fracture of bodies containing interacting cracks (two parallel coaxial cracks or a periodic array of coaxial parallel cracks). In the investigation of composite materials it is assumed that crack sizes are significantly larger than those of the structural elements of composites (i.e. macrocracks are considered); besides, we only analyze fracture processes in which the composite does not demonstrate properties of the piecewise-uniform medium (like fracture on the media interface etc.). Under such assumptions, we use the continuum composite model with the abovementioned characteristics of an orthotropic body (in the case of 2D problems) or a transversely isotropic body (in the case of 3D problems).

2 Problems Formulations

Let us consider 3D problems (the procedure of solving 2D problems is similar). We use the Cartesian coordinates y_j, $j = \overline{1,3}$ of the initial state induced by the initial stresses S_{ij}^0 that are the components of a symmetric stress tensor per unit area of the non-deformed configuration. Let the initial stresses (or compressive forces) act strictly in the crack planes $y_3 = const$ resulting in the following homogeneous initial stress-strain state:

$$S_{33}^0 = 0, S_{11}^0 = S_{22}^0 = const \neq 0, u_j^0 = \lambda_j^{-1}\left(\lambda_j - 1\right) y_j,\ \lambda_1 = \lambda_2 \neq \lambda_3, j = \overline{1,3}, \quad (1)$$

where u_j^o are the components of the displacement vector in the initial configuration and $\lambda_j = const$ are coefficients of elongations (or contractions) along the coordinate axes caused by the initial tensile (or compressive) stresses S_{ij}^0. We also denote by Q'_{ij} the components of the asymmetric stress tensor per unit area in the initial configuration, and by u_j – the associated components of the displacement vector.

General solutions of the linearized equilibrium equations expressed in terms of harmonic potential functions have the following form in circular cylindrical coordinates (r, θ, y_3) derived from the Cartesian coordinates [6]:

$$\begin{gathered} u_r = \frac{\partial(\varphi_1 + \varphi_2)}{\partial r},\ u_3 = \frac{m_1}{\sqrt{n_1}}\frac{\partial \varphi_1}{\partial z_1} + \frac{m_2}{\sqrt{n_2}}\frac{\partial \varphi_2}{\partial z_2}, \\ Q'_{3r} = C_{44}\left(\frac{d_1}{\sqrt{n_1}}\frac{\partial^2 \varphi_1}{\partial r \partial z_1} + \frac{d_2}{\sqrt{n_2}}\frac{\partial^2 \varphi_2}{\partial r \partial z_2}\right),\ Q'_{33} = C_{44}\left(d_1 l_1 \frac{\partial^2 \varphi_1}{\partial z_1^2} + d_2 l_2 \frac{\partial^2 \varphi_2}{\partial z_2^2}\right), \end{gathered} \quad (2)$$

where values C_{44}, n_i, m_i, d_i and l_i depend on material's properties and residual stresses. Below we formulate boundary conditions on the faces of cracks and on the body boundary for different types of problems. E.g., in the case of an unbounded body containing a periodic array of parallel coaxial Mode I penny-shaped cracks of the same radii a located in parallel planes, $y_3 = const$: $\{r < a,\ 0 \le \theta < 2\pi,\ y_3 = 2hn;\ n = 0, \pm 1, \pm 2, \ldots\}$ the boundary conditions take the forms

$$\begin{gathered} u_3 = 0\ \ (y_3 = 0, r > a),\ Q'_{33} = -\sigma(r)\ \ (y_3 = 0, r < a), \\ Q'_{3r} = 0, (y_3 = 0,\ \ 0 \le r < \infty), u_3 = 0, Q'_{3r} = 0, (y_3 = h,\ \ 0 \le r < \infty). \end{gathered} \quad (3)$$

(Note that additional loads on the crack faces must be zero in problems on the fracture of cracked materials compressed along cracks).

3 Fredholm Integral Equations. Stress Intensity Factors

By applying Bessel integral transformation (Fourier transformation for plane problems) to the harmonic potential functions in (2), we reduce the above problems first to the systems of dual integral equations and then to Fredholm integral equations of the second kind, which can be effectively analyzed numerically. E.g., by solving the problem for the body containing a periodic array of parallel coaxial Mode I cracks we obtain the resolving system of Fredholm integral equation in dimensionless form

$$f(\xi) - \frac{1}{\pi}\int_0^1 f(\eta)K(\xi,\eta)d\eta = \frac{2}{\pi}\frac{k_1}{kC_{44}d_1l_1}\xi\int_0^{\pi/2}\sigma(\xi\,\sin\theta)\sin\theta d\theta,\ 0\le\xi,\eta\le1,$$

$$K(\xi,\eta) = R(\xi-\eta) - R(\xi+\eta),\ R(z) = \frac{1}{k}\left[\frac{k_1}{\beta_1}\mathrm{Re}\psi\left(1+\frac{iz}{2\beta_1}\right) - \frac{k_2}{\beta_2}\mathrm{Re}\psi\left(1+\frac{iz}{2\beta_2}\right)\right]. \tag{4}$$

In (4) the following notation has been introduced:
$\xi\equiv a^{-1}x$, $\eta\equiv a^{-1}t$, $k_1 = l_1(n_1)^{-1/2}$, $k_2 = l_2(n_1)^{-1/2}$, $k = k_1 - k_2$ $\beta_k = n_k^{-1/2}ha^{-1}$, $k = 1,2$.

As is customary in LEFM [1], we define stress intensity factors as the coefficients of the singularities in the distributions of the stress components in the neighbourhood of the crack tips. By solving Fredholm integral Eq. (4), we can obtain the representations of potential functions φ_1 and φ_2, and then from (2) we can determine the distribution of the components of the stress tensor $Q'_{ij}(r,y_3)$. Analyzing the values of these components in the crack plane beyond the crack, we can derive the following expressions for the stress intensity factors (SIFs)

$$K_I = -C_{44}d_1l_1\frac{k}{k_1}\sqrt{\pi a}f(1), K_{II} = 0,\ K_{III} = 0, \tag{5}$$

where function f should be found by solving (4). It can be seen from (5) that the value of K_I depends on the initial stresses, cracks radii a and distance h between the cracks because f, as it follows from (4), also depends on these parameters.

In the similar way, we can obtain the expressions of SIFs for the periodic array of parallel co-axial Mode II and Mode III cracks in an unbounded body as well as for two parallel coaxial Mode I, Mode II and Mode III cracks in an infinite body [4–6].

4 Numerical Results

We present the results obtained for the cracks subjected to uniform stresses on their faces. Consider a highly elastic material with Bartenev–Khazanovich potential [6]. For 3D problems the dependences of the ratio of SIFs K_I/K_I^{∞} (where K_I^{∞} is the SIF for a single isolated Mode I crack in an unbounded body) on dimensionless half-distance between cracks $\beta = h/a$ are shown in Fig. 1 for the values of initial stress parameters $\lambda_1 = 1.2$ (tensile initial stresses), $\lambda_1 = 0.8$ (compressive initial stresses), $\lambda_1 = 1.0$ (no initial stresses). These data show that the interaction of cracks in an unbounded body with a periodic array of coaxial Mode I cracks (the solid lines), as in the case of two parallel coaxial Mode I cracks (the dashed lines), leads to a decrease of K_I values as compared to K_I^{∞}. On the other hand, when the separation distance between cracks increases, their mutual influence quickly weakens and the respective values of SIFs tend to the values obtained for a single isolated crack in an unbounded material. For the distances between cracks which exceed 8 crack radii, the interaction between them can be neglected in practical calculations, since in that case the difference between the values of K_I and K_I^{∞} is less than 5% for all considered values of parameter λ_1.

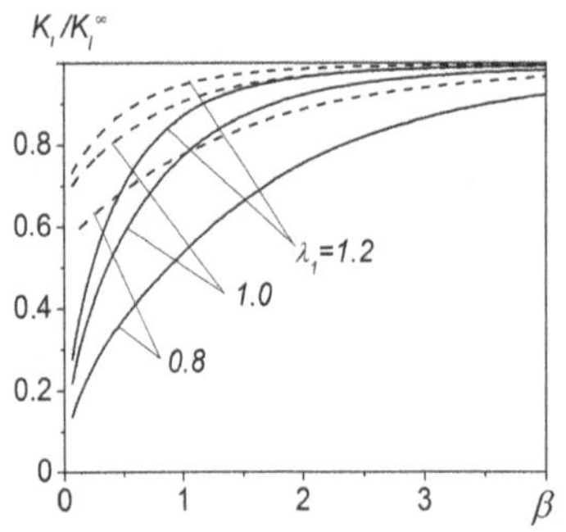

Fig. 1. Dependence of SIF ratio on dimensionless distance between cracks for elastic material with Bartenev–Khazanovich potential

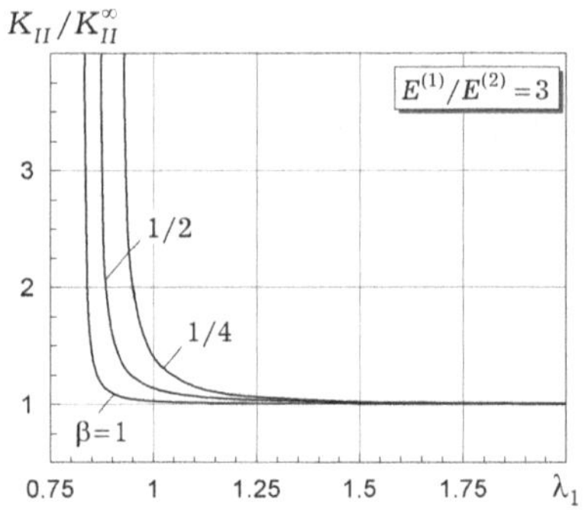

Fig. 2. Dependence of SIF ratio on the parameter of initial stresses for a laminated (layered) composite material

Figure 2 presents results for a laminated two-component composite material with isotropic layers for the values of the ratio of elastic moduli of the layers $E^{(1)}/E^{(2)} = 3$, Poisson's ratios of the layers $\nu^{(1)} = \nu^{(2)} = 0.3$, and the volume concentration of the first layer $c_1 = 0.3$. This figure shows the ratio K_{II}/K_{II}^{∞} (where K_{II}^{∞} is the SIF for a Mode II crack in an unbounded material) *versus* parameter of initial stress λ_1 for the periodic array of parallel coaxial Mode II cracks for different values of β. As can be seen, the curves in Fig. 2 have vertical asymptotes representing the resonant behavior of the SIFs when the initial compressive stresses are so high as to cause local buckling of the material near the cracks. According to the combined approach described in the Introduction, this phenomenon allows us to determine the critical loads for a material compressed along a periodic array of parallel coaxial cracks. Figure 3 shows the critical dimensionless compressive stress $\bar{\sigma} = S^0_{11}/E$ (the stress divided by the effective elastic modulus of the laminated composite) obtained this way *versus* the ratio of the elastic moduli of the layers $E^{(1)}/E^{(2)}$ for $\beta = 0.5$. The results are presented for the periodic array of cracks (solid line), as well as for two parallel cracks (dashed line).

For a highly elastic material with Treloar potential [6], Fig. 4 illustrates the effect of the cracks interaction on the values of relative critical contraction parameter $\varepsilon_1 = 1 - \lambda_1$ in the 2D problem (the condition of plane strain state is assumed) for an unbounded body containing a periodic array of parallel coaxial cracks under compression along cracks. It can be seen that at small distances between cracks β their mutual influence results in a significant reduction of the critical compression parameter ε_1. At the same time, with the increased distance between cracks the values of relative critical contraction parameter ε_1 tend to $\varepsilon_1^* = 0.456$, which for Treloar potential corresponds to the critical contraction parameter for a single crack in an infinite body.

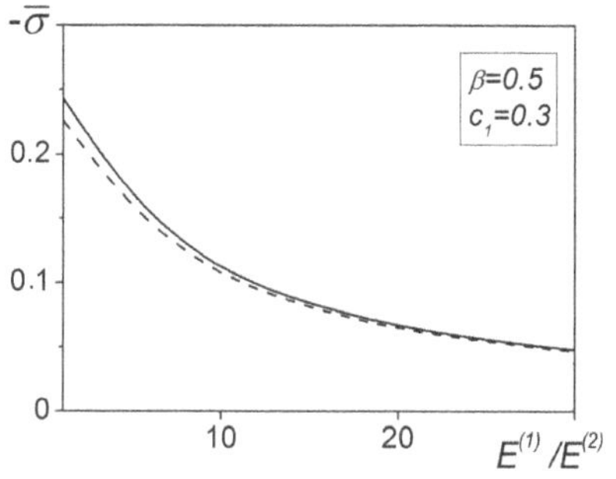

Fig. 3. Dependence of the critical dimensionless compressive stress on the ratio of elastic moduli of layers for the laminated composite

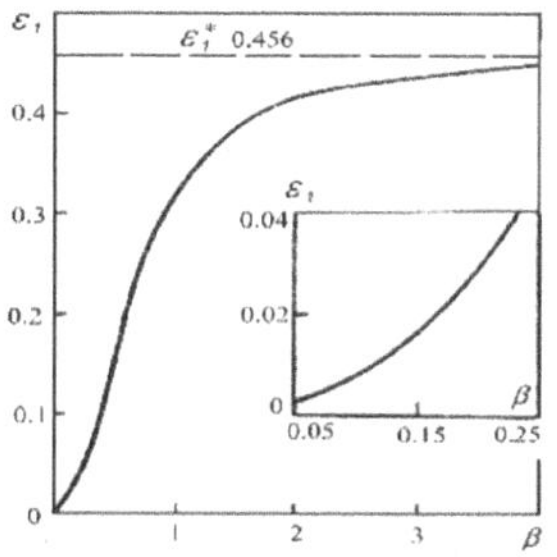

Fig. 4. Dependence of the relative critical contraction parameter on dimensionless distance between cracks for the material with Treloar potential

5 Conclusions

The results obtained allow the following conclusions: (1) for all the problems considered the stress intensity factors are substantially dependent on the initial (residual) stresses; (2) the values of the stress intensity factors change abruptly (the "resonance-like" effect) when the initial contraction ratio λ_1 tends to the value with which there is a local loss of stability in the crack vicinity. According to the combined approach used here this effect enables one to determine critical loading parameters in problems on bodies compressed along cracks; (3) the geometrical parameters of the problems and materials' properties produce a significant effect both on the values of SIFs and on the critical compression parameters.

References

1. Gdoutos, E.E.: Fracture Mechanics: An Introduction. Springer, Dordrecht (2005)
2. Guz, A.N.: Brittle Fracture of Materials with Initial Stresses. Naukova Dumka, Kyiv (1991). (in Russian)
3. Guz, A.N., Dyshel, M.S., Nazarenko, V.M.: Fracture and Stability of Materials with Cracks. Naukova Dumka, Kyiv (1992). (in Russian)
4. Guz, A.N., Nazarenko, V.M., Bogdanov, V.L.: Combined analysis of fracture under stresses acting along cracks. Arch. Appl. Mech. **83**(9), 1273–1293 (2013)
5. Bogdanov, V.L., Guz, A.N., Nazarenko, V.M.: Spatial problems of the fracture of materials loaded along cracks (Review). Int. Appl. Mech. **51**(5), 489–560 (2015)
6. Bogdanov, V.L., Guz, A.N., Nazarenko, V.M.: Unified Approach in Non-Classical Problems of Fracture Mechanics. LAP LAMBERT Acad. Publ, Saarbrücken (2017). (in Russian)

Safe Loading of the Orthotropic Plate with Periodic System of Collinear Cracks

Olga Bogdanova(✉)

S.P.Timoshenko Institute of Mechanics of National Academy of Sciences of Ukraine, Nesterov Str, 3, 02053 Kyiiv, Ukraine
o.bogdanova@i.ua

Abstract. On the basis of a modified δ_c-model of crack, the safe loading of an orthotropic plate with periodic system of collinear cracks is studied. The plate material is satisfying the strength condition in general form. The relations for the determination of major parameters of the crack model (the size of process zones, stresses in these zones, and the crack-tip opening displacements) are deduced. The mechanism of fracture of the plate containing a periodic system of collinear cracks is investigated. The influence of the degree of anisotropy and geometric parameters of the problem on the formation of the process zones and limiting state of the elastic orthotropic plate is revealed. The region of safe loading of an orthotropic viscoelastic plate with cracks is determined. The influence of the rheological parameters of the material on the region of safe loading is analyzed.

Keywords: Periodic system of collinear cracks · Critical loading · Safe loading · Orthotropic materials

1 Limiting State of an Orthotropic Plate with Periodic System of Collinear Cracks

Consider a thin orthotropic plate with periodic system of collinear cracks of length $2l$ located along the axis of orthotropy which coincides with the Ox-axis. The centers of the cracks are located at the points $x_n = 2nD(x_n = 0, \pm 1, \pm 2, \ldots, y = 0)$. The plate is stretched by a homogeneous load applied at infinity

$$\sigma_y = p > 0, \quad \sigma_x = 0, \ \tau_{xy} = 0, \ \ z \to \infty, \ (z = x + iy),$$

Replace the process zones formed under the action of the load near the crack tips by additional cuts of length d on the continuations of the cracks whose lips are subjected to the action of stresses σ_x^0 , σ_y^0. Assume that the limiting state of the material in the process zones is described by a strength criterion

$$F(\sigma_1, \sigma_2, C_i) = 0, \tag{1}$$

where σ_1, σ_2 are the principal stresses and C_i are constants of the material. For the numerical analysis and conclusions the Mises–Hill condition of strength is used [1].

E. Gdoutos and M. Konsta-Gdoutos (Eds.): ICTAEM 2020, STIN 16, pp. 235–241, 2020.
https://doi.org/10.1007/978-3-030-47883-4_43

In view of the symmetry of the problem, the directions x and y are principal. Therefore, the stresses σ_x^0 , σ_y^0 satisfy the condition of strength (1) in the process zone. These stresses are found from the solution of the system of two Eqs. [1]:

$$\sigma_x^0 = \beta\left(\sigma_y^0 - p\right), \quad F\left(\sigma_x^0, \sigma_y^0, C_i\right) = 0 \tag{2}$$

$\beta = \sqrt{E_1/E_2}$; E_1, E_2 - are the elasticity moduli of the material in directions 1 and 2.

The crack opening displacement $\delta(x, l, L)$ at a point x from the segments $|x - x_n| \leq L,\ y = 0$ is given by the formula [2]:

$$\delta(x) = -\frac{2T_0\sigma_y^0 l}{\pi}\left\{\frac{\operatorname{arc} sin(\tilde{x})}{\operatorname{arc} sin(\tilde{l})} F\left(\frac{\tilde{x}}{\tilde{l}}, \rho\right) + 2\frac{\sin\rho}{\operatorname{arc} sin(\tilde{l})}\int\limits_{\tilde{x}/\tilde{l}}^{\sec\rho}\frac{\operatorname{arc} sin\left(t\tilde{l}\right)}{(t^2-1)\sqrt{1-t^2\cos^2\rho}}dt\right\}, \tag{3}$$

where $F(x, r) = \ln\dfrac{\left(1 + x\cos^2 r + \sqrt{1 - x^2\cos^2 r}\sin r\right)(x-1)}{\left(1 - x\cos^2 r + \sqrt{1 - x^2\cos^2 r}\sin r\right)(x+1)}$,

$$T_0 = \frac{1}{\sqrt{E_1 E_2}}\sqrt{2\left(\sqrt{\frac{E_1}{E_2}} - \nu_{12}\right) + \frac{E_1}{G_{12}}},\ \tilde{x} = \sin\frac{\pi x}{2D},\ \tilde{l} = \sin\frac{\pi l}{2D},\ \rho = \frac{\pi p}{2\sigma_y^0},$$

It is clear that the integral in relation (3) is computed in the finite form as $D \to \infty$, which corresponds to the case of a single crack [1].

The size of process zone is determined by the ratio [2]

$$\frac{d}{l} = \frac{2}{\pi}\frac{D}{l}\operatorname{arc}\sin\left\{\frac{\sin\tilde{l}}{\cos\rho}\right\} \tag{4}$$

In accordance with the δ_c - criterion, a crack starts when $\delta(l) = \delta_c$. Then, basing on (3), the field of ultimate loads p_* can be defined by [2]

$$\sigma_y^0(p_*)\frac{\tilde{l}}{\operatorname{arc}\sin\left(\tilde{l}\right)}\int\limits_1^{\tilde{L}/\tilde{l}} F(t, \rho_*)\frac{dt}{\sqrt{1 - t^2\left(\tilde{l}\right)^2}} = 2\sigma_y^0\left(p_*^{(1)}\right)\ln\sec\frac{\pi p_*^{(1)}}{2\sigma_y^0\left(p_*^{(1)}\right)}, \tag{5}$$

where $\rho_* = \frac{\pi p_*}{2\sigma_y^0(p_*)}$, $p_*^{(1)}$ - the limiting load in the case of a plate with a single crack. The change $p_*^{(1)}$ from 0 to σ_{02} corresponds to a change in the crack length from infinity to zero.

It should be noted that the boundary condition of the plane, weakened by the system of cracks, is not always determined by the fracture criteria of type (5). If the external load is such that the condition $d = D - l$ is fulfilled, then the fusion of the areas for the elimination of neighboring cracks occurs, which can also be considered as a condition of the boundary state. Taking into account equality (4), this condition will be written as:

$$\frac{D}{l} = \frac{\pi}{2\arcsin(\cos\rho^{(d)})}, \quad \rho^{(d)} = \frac{\pi p_*^{(d)}}{2\sigma_y^0\left(p_*^{(d)}\right)}, \tag{6}$$

The relation (6) determines the load $p_*^{(d)}$ at which the areas of process zones of neighboring cracks occur. The dependence of this load on the dimensionless distance between cracks D/l is constructed (Fig. 1). Increasing the degree of anisotropy of the material reduces the level of the maximum allowable load at which the merge of process zones occurs, but in general, the effect of the degree of anisotropy on this process is small.

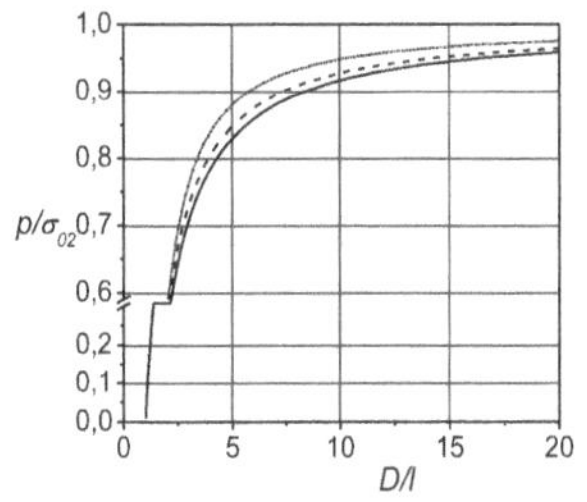

Fig. 1. Dependence of the load $p_*^{(d)}/\sigma_{02}$ on the dimensionless distance between the cracks D/l: solid line - $\beta = 0,3$; dashed line - $\beta = 0,5$; dash-dotted line - $\beta = 0,9$

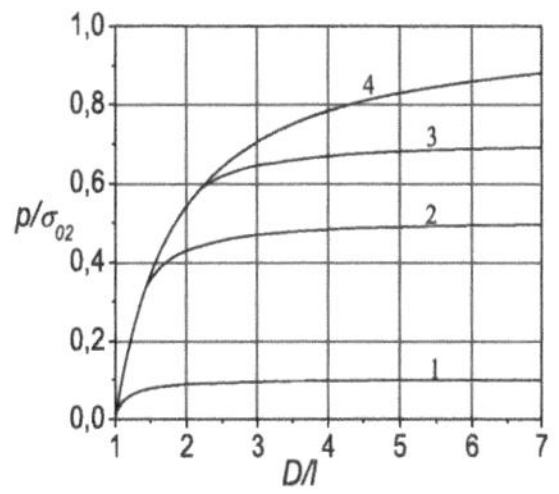

Fig. 2. Dependence of the load p_*/σ_{02} on the dimensionless distance between the cracks D/l: $p_*^{(1)}/\sigma_{02}$ = 0,1 (1); 0,5 (2); 0,7 (3); 4 – load when the process zones are joined

Thus, the boundary condition of the plate with the periodic system of collinear cracks is determined by the set of conditions (5) and (6):

$$p_* = \min\left(p_*^{(d)}, p_*\right), \tag{7}$$

The dependence of the critical load on the dimensionless distance between the cracks D/l obtained on the basis of the relations (5) and (6) for $\beta = 0.3$ was constructed (Fig. 2). It is obvious that for arbitrary crack lengths and the distance between them, there is such a "critical" level of load when there is a partial destruction of the entire intersection along the line of location of the cracks. This "critical" load is higher, the smaller the length of the cracks.

2 Safe Loading of a Viscoelastic Orthotropic Plate, Weakened by a Periodic System of Collinear Cracks

Consider a thin orthotropic plate of viscoelastic material, weakened by a periodic system of collinear cracks, located along one of the orthotropy axes. Assume that the deformation properties of a material are described by linear relationships up to destruction. During the destruction of such materials areas of destruction formed at the tops of cracks, inside which there is a destruction of a less durable binder and a part of high-strength fittings.

We will replace these zones with additional cuts of length $d(t)$ on the continuation of cracks, on the banks of which there is a uniformly distributed self-equilibrium stress. Since these stresses are determined on the basis of the condition (1), which is constant in time, they also do not depend on time and can be determined at the time $t = 0$ from the solution of the corresponding problem of the theory of elasticity, that is, the stresses in the process zones are determined by a system of Eqs. (2).

Thus, it is necessary to investigate the stress-strain state of an orthotropic viscoelastic plate with a periodic system of collinear incisions $|x - x_n| \leq L(t),\ y = 0$ along the axis of orthotropy Ox, under the action of a uniformly distributed load independent of time $\sigma_y = p > 0,\ \sigma_x = 0,\ \tau_{xy} = 0,\ |x + iy| \to \infty$.

As is known [3, 4], the Volterra principle can be applied to the problem, according to which the viscoelastic opening of the shore of the crack can be written as:

$$\delta(x,t) = T^* \delta_0(x,t), \tag{8}$$

where $\delta_0(x,t)$ - function of power and geometric parameters, T^*- the function of integral operators whose form is determined from the solution of the corresponding problem of the theory of elasticity. For this problem we have [3]:

$$T^* = \frac{1}{\sqrt{E_1^* E_2^*}} \sqrt{2\left(\sqrt{\frac{E_1^*}{E_2^*}} - v_{12}^*\right) + \frac{E_1^*}{G_{12}^*}}, \tag{9}$$

$$\delta_0(x,t) = \frac{2\sigma_y^0 l(t)}{\pi} \frac{\tilde{l}(t)}{\arcsin\left(\tilde{l}(t)\right)} \int\limits_{\tilde{x}/\tilde{l}(t)}^{\tilde{L}/\tilde{l}(t)} F(\tau,\rho) \frac{d\tau}{\sqrt{1 - \tau^2 \left(\tilde{l}(t)\right)^2}} \tag{10}$$

It has been experimentally shown that for some orthotropic viscoelastic materials it is possible to disregard the creep along the orthotropic axes and take into account only creep at displacement, that is, to put [3]:

$$E_1^* \approx E_1^0 = const,\ E_2^* \approx E_2^0 = const,\ v_{12}^* \approx v_{12}^0 = const,\ \frac{1}{G_{12}^*} = \frac{1}{G_{12}^0}\left[1 + \lambda R^*(\gamma)\right], \tag{11}$$

where $E_1^0, E_2^0, v_{12}^0, G_{12}^0$ - instantaneous values of relevant characteristics, R^*- resovent integral operator of the form $R^*f = \int_0^t R(t-\tau)f(\tau)d\tau$.

Using the method of integral continued fractions, the operator (9) can be represented as a linear combination of resolvent integral operators [3]:

$$T^* = T_0\left[1 + \sum_{i=1}^{N} A_i R^*(\gamma_i)\right] \tag{12}$$

with the kernel $T(t-\tau) = \sum_{i=1}^{N} A_i R(t-\tau, \gamma_i)$, $R(t-\tau, \gamma_i)$ - kernels of operators $R^*(\gamma_i)$.

If the criterion of fracture is to choose a criterion for the critical crack opening, we obtain the equation for determining the moment of the beginning of the slow crack development t_* (or the duration of the incubation period of the crack development in the viscoelastic body) if the external load is constant over time in the form [3]:

$$\int_0^{t_*} T(\theta)d\theta = \frac{\delta_c}{\delta(l_0)} - 1, \tag{13}$$

where $\delta(x) = T_0\delta_0(x)$ - elastic opening of the crack at the point x.

For bodies with limited creep, there are "*safe*" loads, for which the value $\delta(l_0, t)$ can't reach the critical value δ_c, and the development of the crack does not happen. In the general case, the safe load field is determined from the ratio [3]:

$$\frac{\delta_c}{\delta(l, p_S, q_S)} = \frac{T_\infty}{T_0}, \; T_\infty = T_* \cdot 1|_{t=\infty}, \tag{14}$$

If, as the kernel of an integral operator, we choose the fractional-exponential function of Yu.M. Rabotnov [3], then from (4) we obtain:

$$\begin{aligned} T_\infty &= \frac{1}{\sqrt{E_1^0 E_2^0}}\sqrt{2\left(\sqrt{\frac{E_1^0}{E_2^0}} - v_{12}^0\right) + \frac{E_1^0}{G_{12}^\infty}} \\ &= \frac{1}{\sqrt{E_1^0 E_2^0}}\sqrt{2\left(\sqrt{\frac{E_1^0}{E_2^0}} - v_{12}^0\right) + \frac{E_1^0}{G_{12}^\infty}\left(1 + \frac{\lambda}{|\gamma|}\right)}. \end{aligned} \tag{15}$$

where, γ – the rheological parameters of the material.

Then, on the basis of (5), (10) and (14), we obtain an equation for determining a safe load p_S:

$$\sigma_y^0(p_S)\frac{\tilde{l}}{\arcsin(\tilde{l})}\int_1^{\tilde{L}/\tilde{l}} F(\tau,p_S)\frac{d\tau}{\sqrt{1-\tau^2(\tilde{l})^2}} = 2\frac{T_0}{T_\infty}\sigma_y^0\left(p_*^{(1)}\right)\ln\sec\frac{\pi p_*^{(1)}}{2\sigma_y^0\left(p_*^{(1)}\right)}. \quad (16)$$

The value $p_*^{(1)}$ uniquely determines the length of the crack l_0 and, consequently, Eq. (16) determines the safe load p_S for a given initial crack length. Instantaneous fracture of the plate with the system of cracks occurs at a critical load p_*, which is determined by the relation (5). In addition, there is some load $p_*^{(d)}$ on which there is an instantaneous partial destruction of the material on the line of location of the cracks. This load is determined from the equation:

$$\cos\frac{\pi p_*^{(d)}}{2\sigma_y^0\left(p_*^{(d)}\right)} = \sin\frac{\pi l_0}{2D}. \quad (17)$$

Thus, the development of a crack in a viscoelastic body will occur only in the region of load change, which is limited by curves, which are determined by the relations (5), (16) and (17).

Numerical results are obtained for a material with the following characteristics [3] $E_1^0 = 1,553\cdot 10^9 Pa$, $E_2^0 = 2,315\cdot 10^9 Pa$, $G_{12}^0 = 0,3086\cdot 10^9 Pa$, $\nu_{12}^0 = 0,08$, $\alpha = 0,85$, $\lambda = 0,275\ \sec^{1-\alpha}$, $\gamma = -0,181\ \sec^{1-\alpha}$. Figure 3 show the dependence of the "safe" load (16) on the dimensionless distance between the cracks D/l for $p_*^{(1)}/\sigma_{02}$ = 0,1 (curves 1); $p_*^{(1)}/\sigma_{02}$ = 0,5 (curves 2); 3–$p_*^{(1)}/\sigma_{02}$ = 0,7 (curves 3); curve 4 show loading when the process zones are joined; dashed line – critical loading p_*/σ_{02}. The dependence of the safe load on the rheological parameters of the material are shown on Fig. 4 and Fig. 5 for $D/l = 3,0$.

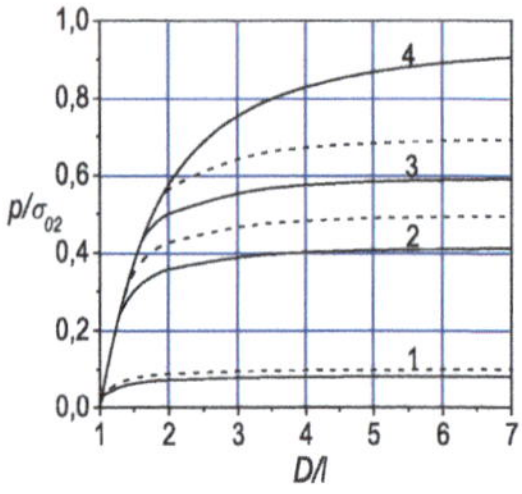

Fig. 3. Dependence of the safe load p_S/σ_{02} on the dimensionless distance between the cracks D/l

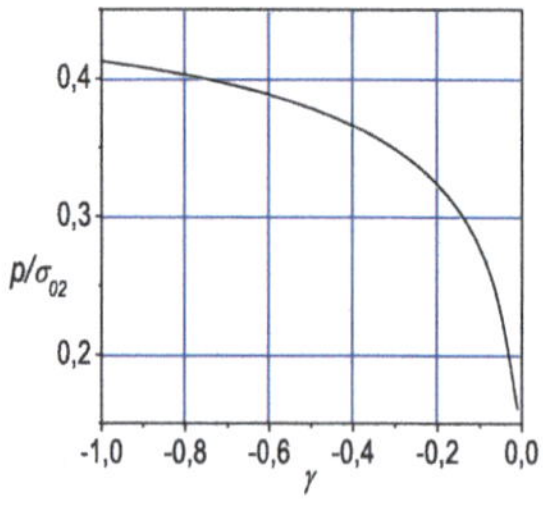

Fig. 4. Dependence of the safe load p_S/σ_{02} on the rheological parameter γ

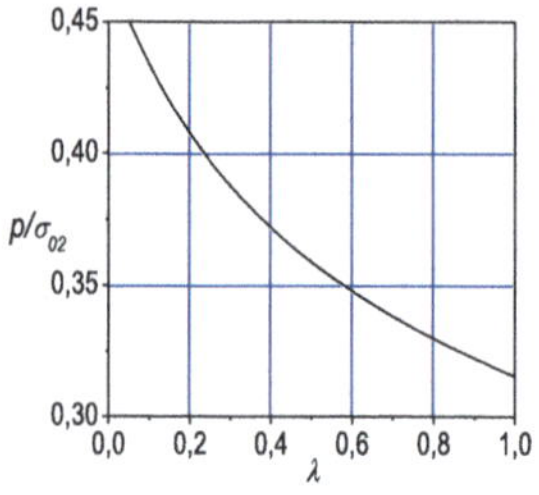

Fig. 5. Dependence of the safe load p_S/σ_{02} on the rheological parameter λ

It can be noted that the change in the rheological parameters of the material λ and γ significantly affects the area of safe load: the decrease λ or increase $|\gamma|$ in other permanent characteristics leads to an increase in the level of safe loading. The parameter α does not affect the level of safe load, since it is not part of the ratio (15). Thus, by changing the rheological parameters of the material, it is possible to achieve a decrease in the area of loading, in which the development of cracks occurs.

3 Conclusions

The proposed modification of the Leonov-Panasyuk-Dagdale crack model to the case of orthotropic materials allows to effectively solve problems of the destruction of orthotropic bodies with cracks, the material of which satisfies the condition of the strength of the general form. With this model, the boundary state of the orthotropic plate, relaxed by the periodic system of collinear cracks, the material of which is elastic or viscoelastic, is investigated. The mechanism of fracture of the plate with cracks and the influence on the boundary and safe loading of the properties of the material have been investigated.

References

1. Kaminsky, A.A., Bogdanova, O.S., Bastun, V.N.: On modeling cracks in orthotropic plates under biaxial loading: synthesis and summary. Fatigue Fract. Eng. Mater. Struct. **5**(34), 345–355 (2011)
2. Bogdanova, O.S.: Limiting state of an elastoplastic orthotropic plate with a periodic system of collinear cracks. Int. Appl. Mech **5**(43), 539–546 (2008)
3. Kaminsky, A.A.: Fracture mechanics of viscoelastic bodies with cracks. Nauk. Dumka, Kyiv (1990)

Boundary Element Analysis of Partially Debonded Shell-Like Rigid Inclusions in Anisotropic Medium

Heorhiy Sulym[1], Nataliia Ilchuk[2], and Iaroslav Pasternak[2(✉)]

[1] Pidstryhach Institute for Applied Problems of Mechanics and Mathematics of NASU, Naukova Street 3b, 79060 Lviv, Ukraine
[2] Lutsk National Technical University, Lvivska Street 75, 43018 Lutsk, Ukraine
iaroslav.m.pasternak@gmail.com

Abstract. The paper presents boundary element models of anisotropic thermoelastic medium containing partially debonded shell-like rigid inclusions, which possess high rates of heat conduction. Boundary integral equations are derived, which account for partial debonding of a shell-like inclusion at one of its faces. Special attention is paid to full debonding of a rigid shell at two faces simultaneously (bilateral debonding). Obtained equations are implemented in a fast and accurate boundary element approach.

Keywords: Boundary element method · Anisotropic · Thermoelastic · Stress intensity factor

1 Introduction

Structural materials, including composite and anisotropic ones, generally contain different natural or artificial inhomogeneities, which influence field distribution and can even initiate fracture processes. In particular, composite materials consisting of fibre, ribbon-like or plate-like filament are widely used in modern mechanical and airspace engineering due to their high performance. However, the non-homogeneous structure of such materials produces redistribution of mechanical and thermal field. In this case partial debonding of a filament can cause high stress concentration, which influence safety and structural integrity. Another example of stress concentrator are anchors in rock mechanics, which are in general, partially bonded, which significantly influences stress distribution. Therefore, these raises interest to the problems for solids with partially debonded inhomogeneities.

The problems on debonding of spatial thin rigid inclusions dates back to Selvadurai et al. [1], who considered rigid disc inclusion with circular debonded region. Since then the interest to problems of partially debonded inclusions has only raised. In particular, recently Shahmohamadi et al. [2] considered a frictionless contact of a rigid disk with crack faces, Nategh et al. [3] considered bonded contact of a rigid disk with a penny-shaped crack. Nevertheless, the studies of debonded inclusions are limited to analytic or semi-analytic methods, which constrains the shape of inclusion and a debonded region to simple geometries, i.e. planar circular or elliptic domains.

E. Gdoutos and M. Konsta-Gdoutos (Eds.): ICTAEM 2020, STIN 16, pp. 242–247, 2020.
https://doi.org/10.1007/978-3-030-47883-4_44

Therefore, this study is concerned with the development of numerical boundary element approach, which can hold problems of partially debonded 3D rigid shell-like inclusions of arbitrary smooth shape. It is natural continuation of recent studies on cracks [4] and perfectly bonded inclusions [5] in anisotropic thermoelastic medium.

2 Modelling of Partially Debonded Rigid Shell-Like Inclusions

2.1 Boundary Integral Equations of Anisotropic Thermoelasticity

Consider linear thermoelastic anisotropic solid $\mathfrak{B}$ in the reference coordinate system $Ox_1x_2x_3$, which contains internal smooth opened surface C of field discontinuity. According to [4, 5] the boundary integral equations of steady-state heat conduction and thermoelasticity of anisotropic solids are as follows

$$\begin{aligned}\frac{1}{2}\sum\theta(\mathbf{x}_0) = &\iint_S \Theta^*(\mathbf{x},\mathbf{x}_0)\sum h_n(\mathbf{x})dS(\mathbf{x}) - \mathrm{CPV}\iint_S H^*(\mathbf{x},\mathbf{x}_0)\Delta\theta(\mathbf{x})dS(\mathbf{x})\\ &-\iiint_{\mathfrak{B}} \Theta^*(\mathbf{x},\mathbf{x}_0)f_h(\mathbf{x})dV(\mathbf{x}),\end{aligned}\tag{1}$$

$$\begin{aligned}\frac{1}{2}\Delta h_n(\mathbf{x}_0) = n_j(\mathbf{x}_0)\Bigg[&\mathrm{CPV}\iint_{\partial\mathfrak{B}} \Theta_i^{**}(\mathbf{x},\mathbf{x}_0)\sum h_n(\mathbf{x})dS(\mathbf{x})\\ &-\mathrm{HFP}\iint_{\partial\mathfrak{B}} H_i^{**}(\mathbf{x},\mathbf{x}_0)\Delta\theta(\mathbf{x})dS(\mathbf{x}) - \iiint_{\mathfrak{B}} \Theta_i^{**}(\mathbf{x},\mathbf{x}_0)f_h(\mathbf{x})dV(\mathbf{x})\Bigg],\end{aligned}\tag{2}$$

$$\begin{aligned}\frac{1}{2}\sum u_i(\mathbf{x}_0) = &\iint_S U_{ij}(\mathbf{x},\mathbf{x}_0)\sum t_j(\mathbf{x})dS(\mathbf{x}) - \mathrm{CPV}\iint_S T_{ij}(\mathbf{x},\mathbf{x}_0)\Delta u_j(\mathbf{x})dS(\mathbf{x})\\ &+\iint_S \left[R_i(\mathbf{x},\mathbf{x}_0)\Delta\theta(\mathbf{x}) + V_i(\mathbf{x},\mathbf{x}_0)\sum h_n(\mathbf{x})\right]dS(\mathbf{x})\\ &+\iiint_{\mathfrak{B}} U_{ij}(\mathbf{x},\mathbf{x}_0)f_j(\mathbf{x})dV(\mathbf{x}) - \iiint_{\mathfrak{B}} V_i(\mathbf{x},\mathbf{x}_0)f_h(\mathbf{x})dV(\mathbf{x}),\end{aligned}\tag{3}$$

$$\begin{aligned}\frac{1}{2}\Delta t_i(\mathbf{x}_0) = n_j(\mathbf{x}_0)\Bigg[&\mathrm{CPV}\iint_{\partial\mathfrak{B}} D_{ijk}(\mathbf{x},\mathbf{x}_0)\sum t_k(\mathbf{x})dS(\mathbf{x}) - \mathrm{HFP}\iint_{\partial\mathfrak{B}} S_{ijk}(\mathbf{x},\mathbf{x}_0)\Delta u_k(\mathbf{x})dS(\mathbf{x})\\ &+\mathrm{CPV}\iint_{\partial\mathfrak{B}} Q_{ij}(\mathbf{x},\mathbf{x}_0)\Delta\theta(\mathbf{x})dS(\mathbf{x}) + \iint_{\partial\mathfrak{B}} W_{ij}(\mathbf{x},\mathbf{x}_0)\sum h_n(\mathbf{x})dS(\mathbf{x})\\ &+\iiint_{\mathfrak{B}} D_{ijk}(\mathbf{x},\mathbf{x}_0)f_k(\mathbf{x})dV(\mathbf{x}) - \iiint_{\mathfrak{B}} W_{ij}(\mathbf{x},\mathbf{x}_0)f_h(\mathbf{x})dV(\mathbf{x})\Bigg],\end{aligned}\tag{4}$$

where $\mathbf{x}_0 \in S$; $S = C \cup \partial\mathfrak{B}$; and $\sum\theta = \Delta\theta = \theta$, $\sum h_n = \Delta h_n = h_n$, $\sum u_i = \Delta u_i = u_i$, $\sum t_i = \Delta t_i = t_i$ on $\partial\mathfrak{B}$; $\sum f = f^+ + f^-$; $\Delta f = f^+ - f^-$; $h_n = h_i n_i$; $\sum h_n = h_i^+ n_i^+ + h_i^- n_i^- = \left(h_i^+ - h_i^-\right) n_i^+$; $\mathbf{n}^+(\mathbf{x}) = -\mathbf{n}^-(\mathbf{x})$ are unit outwards normal vectors to the faces C^+ and C^- of the mathematical cut C; $t_i = \sigma_{ij} n_j$ is a stress vector; σ_{ij} is a stress tensor; f_i is a body force vector; h_i is a heat flux; $h_n = h_i n_i$; f_h is a distributed heat source density; u_i is a displacement vector; θ is a temperature change with respect to the reference temperature. CPV stands for the Cauchy Principal Value of the integral and HFP stands for the Hadamard Finite Part of the integral. The indices vary from 1 to 3. Here and further the Einstein summation convention is used. A comma at subscript denotes differentiation with respect to a coordinate indexed after the comma, i.e. $u_{i,j} = \partial u_i / \partial x_j$. The kernels of Eqs. (1)–(4) are listed in [4, 5].

2.2 Inclusion Model and Boundary Conditions

Integral Eqs. (1)–(4) are derived from the extended Somigliana identity for thermoelastic material [4] and relate given and unknown boundary conditions at the boundary $\partial\mathfrak{B}$ of the solid $\mathfrak{B}$; e.g. when the traction vector is given on $\partial\mathfrak{B}$ the sought function is displacement vector and vice versa. Consequent usage of extended Somigliana identity allow determination of the stress field in the entire solid $\mathfrak{B}$. When cracks or shell-like inclusions are considered, the sought functions are field discontinuities $\Delta\theta$, $\sum h_n$, Δu_i, $\sum t_i$ at the surface C. The same functions are convenient to use as sought functions in the problems of partially debonded inclusions. Therefore, boundary conditions should be rewritten through Δf and $\sum f$.

Prior to consideration of the boundary conditions at the discontinuity surface C consider the model of a rigid perfectly conducting non-deformable inclusion, which is derived in [5]:

$$\theta^{\mathrm{incl}}(\mathbf{x}) = \theta^0, \quad -\iint\limits_S h_n^{\mathrm{incl}}(\mathbf{x}) dS(\mathbf{x}) - H^0 = 0 \tag{5}$$

$$\begin{gathered} u_i^{\mathrm{incl}}(\mathbf{x}) = u_i^0 + e_{ijm}\omega_j^0 x_m, \\ -\iint\limits_S t_i^{\mathrm{incl}}(\mathbf{x}) dS(\mathbf{x}) - P_i^0 = 0, \quad -\iint\limits_S \varepsilon_{ijm} x_j t_m^{\mathrm{incl}}(\mathbf{x}) dS(\mathbf{x}) - M_i^0 = 0, \end{gathered} \tag{6}$$

where θ^0 is a temperature of the inclusion; H^0 is the constant heat (actually heat generation rate) applied to the inclusion; u_i^0 is a rigid displacement of the inclusion and ω_j^0 is its rigid rotation about the origin; e_{ijm} is the permutation tensor; S is a surface of the inclusion; P_i^0 and M_i^0 are resultants of forces and couples applied to the inclusion, respectively.

Now consider the following types of the boundary conditions at the discontinuity surface:

a. perfect thermal and mechanical contact at $S_a \subset C$;
b. partial debonding at $S_b \subset C^+$;
c. partial debonding at $S_c \subset C^-$;
d. bilateral debonding at $S_d \subset C$.

Also assume that $S_a \cup S_b \cup S_c \cup S_d = C$ and $S_i \cap S_j = 0$ for i $\neq$ j, i.e. the domains of different type of boundary conditions do not intersect.

Accounting for inclusion models (5), (6), these boundary conditions can be written in terms of temperature, heat flux, displacement and traction on the faces C^+ and C^- of the cut as follows,

$$\theta^+(\mathbf{x}) = \theta^-(\mathbf{x}) = \theta^0, \quad u_i^+(\mathbf{x}) = u_i^-(\mathbf{x}) = u_i^0 + \varepsilon_{ijm}\omega_j^0 x_m, \quad \forall \mathbf{x} \in S_a, \tag{7}$$

$$\theta^-(\mathbf{x}) = \theta^0, \quad h_n^+(\mathbf{x}) = 0, \quad u_i^-(\mathbf{x}) = u_i^0 + \varepsilon_{ijm}\omega_j^0 x_m, \quad t_i^+(\mathbf{x}) = 0, \quad \forall \mathbf{x} \in S_b, \tag{8}$$

$$\theta^+(\mathbf{x}) = \theta^0, \quad h_n^-(\mathbf{x}) = 0, \quad u_i^+(\mathbf{x}) = u_i^0 + \varepsilon_{ijm}\omega_j^0 x_m, \quad t_i^-(\mathbf{x}) = 0, \quad \forall \mathbf{x} \in S_c, \tag{9}$$

$$h_n^+(\mathbf{x}) = h_n^-(\mathbf{x}) = 0, \quad t_i^+(\mathbf{x}) = t_i^-(\mathbf{x}) = 0, \quad \forall \mathbf{x} \in S_d, \tag{10}$$

In this case inclusion equilibrium equations do not depend on the type of boundary conditions and can be written through the sought functions on the cut as

$$\begin{gathered}\iint\limits_S \sum h_n(\mathbf{x})dS(\mathbf{x}) - H^0 = 0,\\ \iint\limits_S \sum t_i(\mathbf{x})dS(\mathbf{x}) - P_i^0 = 0, \quad \iint\limits_S \varepsilon_{ijm}x_j \sum t_m(\mathbf{x})dS(\mathbf{x}) - M_i^0 = 0.\end{gathered} \tag{11}$$

Since

$$f^+ = \frac{1}{2}\left(\sum f + \Delta f\right), \quad f^- = \frac{1}{2}\left(\sum f - \Delta f\right) \tag{12}$$

boundary conditions (7)–(10) on the cut C can be rewritten through the sought discontinuity functions as

$$\frac{1}{2}\sum \theta(\mathbf{x}) = \theta^0, \quad \Delta\theta(\mathbf{x}) = 0, \quad \frac{1}{2}\sum u_i(\mathbf{x}) = u_i^0 + \varepsilon_{ijm}\omega_j^0 x_m, \quad \Delta u_i(\mathbf{x}) = 0, \quad \forall \mathbf{x} \in S_a; \tag{13}$$

$$\frac{1}{2}\sum\theta(\mathbf{x}) = \theta^0 + \frac{1}{2}\Delta\theta(\mathbf{x}), \quad \Delta h_n(\mathbf{x}) = -\sum h_n(\mathbf{x}),$$
$$\frac{1}{2}\sum u_i(\mathbf{x}) = u_i^0 + \varepsilon_{ijm}\omega_j^0 x_m + \frac{1}{2}\Delta u_i(\mathbf{x}), \quad \Delta t_i(\mathbf{x}) = -\sum t_i(\mathbf{x}) \quad \forall\mathbf{x}\in S_\mathrm{b}; \tag{14}$$

$$\frac{1}{2}\sum\theta(\mathbf{x}) = \theta^0 - \frac{1}{2}\Delta\theta(\mathbf{x}), \quad \Delta h_n(\mathbf{x}) = \sum h_n(\mathbf{x}),$$
$$\frac{1}{2}\sum u_i(\mathbf{x}) = u_i^0 + \varepsilon_{ijm}\omega_j^0 x_m - \frac{1}{2}\Delta u_i(\mathbf{x}), \quad \Delta t_i(\mathbf{x}) = \sum t_i(\mathbf{x}) \quad \forall\mathbf{x}\in S_\mathrm{c}; \tag{15}$$

$$\Delta h_n(\mathbf{x}) = \sum h_n(\mathbf{x}) = 0, \quad \Delta t_i(\mathbf{x}) = \sum t_i(\mathbf{x}) = 0, \quad \forall\mathbf{x}\in S_\mathrm{d} \tag{16}$$

Boundary conditions (13)–(16) are readily substituted into Eqs. (1)–(4) to obtain the system of boundary integral equations for the sought discontinuity functions $\Delta\theta$, $\sum h_n$, Δu_i, $\sum t_i$ at the surface C, which model debonded shell-like inclusion. To reduce the number of equations one can use only Eqs. (1) and (3) with boundary conditions (13) (Type a) and Eqs. (2) and (4) with boundary conditions (16) (Type d).

3 Boundary Element Approach

At the first step (preprocessing) of the boundary element solution of derived boundary integral equations for a particular problem the surface $\partial\mathfrak{B}$ of the solid along with the discontinuity surface C (which model partially debonded inclusion) is meshed with quadrilateral quadratic discontinuous boundary elements. The local curvilinear coordinate system $O\xi\eta$ is associated with each boundary element, moreover, $-1\leq\xi\leq 1$, $-1\leq\eta\leq 1$. The collocation points are placed at nodes $\xi = (-2/3; 0; 2/3)$; $\eta = (-2/3; 0; 2/3)$. Therefore, there are 9 collocation points associated with each boundary element.

Boundary conditions along with unknown boundary and discontinuity functions are interpolated within the collocation points at each boundary element Γ_N as

$$\mathbf{b}_N(\xi,\eta) = \sum_{i=1}^{3}\sum_{j=1}^{3}\mathbf{b}_N^{i,j}\phi_i(\xi)\phi_j(\eta), \tag{17}$$

where $\mathbf{b} = \left(\theta, \Delta\theta, h_n, \sum h_n, u_i, \Delta u_i, t_i, \sum t_i\right)^\mathrm{T}$, and the discontinuous shape functions are given as [4]

$$\phi_1(\xi) = \xi\left(\frac{9}{8}\xi - \frac{3}{4}\right), \quad \phi_2(\xi) = \left(1 - \frac{3}{2}\xi\right)\left(1 + \frac{3}{2}\xi\right), \quad \phi_3(\xi) = \xi\left(\frac{9}{8}\xi + \frac{3}{4}\right) \tag{18}$$

If the side of the boundary element models the inclusion front, which is perfectly bonded, special shape functions are used for heat flux and traction to capture the square root singularity arising at its front [5]

$$\phi_i^{\Sigma}(\xi) = \frac{1}{\sqrt{1 \pm \xi}} \left(\Phi_{i1}^{\Sigma} + \sum_{j=2}^{3} \Phi_{ij}^{\Sigma} (1 \pm \xi)^{j-1} \right). \tag{19}$$

Constants Φ_{ij}^{Σ} are determined from the system of equations $\phi_i(\xi_j) = \delta_{ij}$, where $\xi_j = (-2/3; 0; 2/3)$.

Substituting (17) along with appropriate boundary conditions (13)–(16) into the boundary integral Eqs. (1)–(4) one obtains the system of linear algebraic equations for unknown nodal values of the boundary and discontinuity functions.

Evaluation of kernels, regular, singular and hypersingular integrals is the same as in the case of cracks and perfectly bonded inclusions and is discussed in details in Refs [4, 5].

4 Discussion and Further Studies

Obtained boundary element approach was programmed to object-oriented C ++ code, which was used in calculation of solids with debonded non-deformable shell-like inclusions. Verification was made through comparison of the numerical results with analytic solution of Selvdurai et al. [1]. Good agreement was observed.

Further studies are assumed to be concerned with smooth or frictional contact conditions on the surfaces of the inclusion, since it is essential to account for these conditions in the cases of non-planar shapes of inclusions.

References

1. Selvadurai, A.P.S., Singh, B.M., Au, M.C.: Axial loading of a rigid disc inclusion with a debonded region. Int. J. Solids Struct. **25**(7), 783–795 (1989)
2. Shahmohamadi, M., Khojasteh, A., Rahimian, M.: Frictionless contact of a rigid disk with the face of a penny-shaped crack in a transversely isotropic solid. Int. J. Solids Struct. **106**, 274–283 (2017)
3. Nategh, S., Khojasteh, A., Rahimian, M.: Bonded contact of a rigid disk inclusion with a penny-shaped crack in a transversely isotropic solid. J. Eng. Math. **110**, 123–146 (2018)
4. Pasternak, I.A., Pasternak, R., Pasternak, V., Sulym, H.: Boundary element analysis of 3D cracks in anisotropic thermomagnetoelectroelastic solids. Eng. Anal. Bound Elem **74**, 70–78 (2017)
5. Pasternak, I.A., Sulym, H., Ilchuk, N.: Boundary element analysis of 3D shell-like rigid electrically conducting inclusions in anisotropic thermomagnetoelectroelastic solids. Z Angew Math Mech, e201800319 (2019)

The Transmission of an Acoustic Wave Through a Composite Plate with Energy Dissipation in a Material Taken into Account Based on Three-Dimensional Equations of Elasticity Theory

V. N. Paimushin[1,2] and R. K. Gazizullin[1(✉)]

[1] Kazan National Research Technical University named after A. N. Tupolev - KAI, Kazan, Russia
vpajmushin@mail.ru, gazizullin.rk@yandex.ru
[2] Kazan Federal University, Kazan, Russia

Abstract. The problem of a plane monoharmonic sound wave passage through a thin composite rectangular plate hinged in the opening of an absolutely stiff dividing wall has been considered on the basis of two-dimensional in spatial coordinates motion equations. These equations have been constructed by the reduction of three-dimensional equations on the basis of the discrete layered model of a multilayer plate deformation at small displacements and deformations with account of the internal damping of layers according to the Kelvin–Voight model. Behavior of acoustic media has been described by the classical wave equations based on the model of an ideal compressible fluid. The exact analytical solutions of the formulated problem have been constructed. The dependence of sound insulation properties and parameters of the stress-strain state of a composite plate reinforced with carbon fiber textile on the incident sound wave frequency was studied. It is shown that under high-frequency acoustic impact, the deformation mechanics of the structures made of fiber reinforced composite must be described by refined equations of motion as they have a high degree of accuracy and intensionality since the formation of the stress-strain state in them is almost three-dimensional with components of the same order.

Keywords: Multilayer plate · Discrete layered model · Internal friction of the material · Kelvin–Voigt model · Acoustic medium · Sound wave · Analytical solution

1 Introduction

The sound waves passage through thin plates, including composite ones, has been studied in numerous scientific works [1–5 et al.]. However, in these works the simplest relations of plate theory are used to describe the plate deformation mechanics. As a rule, these models are based on the introduction of a number of well-known simplifying hypotheses and the assumption that only the tangential components of the stress tensor

E. Gdoutos and M. Konsta-Gdoutos (Eds.): ICTAEM 2020, STIN 16, pp. 248–253, 2020.
https://doi.org/10.1007/978-3-030-47883-4_45

are formed in the plate during the sound wave passage. In order to further develop the results obtained earlier [6] and to study the influence of the incident sound wave frequency on the parameters of a dynamic stress-strain state forming in a plate, we use the developed earlier [6] discrete layered model to formulate the considered problem of sound waves passage through multilayer thin plates. This model makes it possible to obtain solutions that are practically approaching a solution based on the equations of the three-dimensional elasticity theory.

2 Statement of the Problem and the Construction of its Solution

Let us consider a multilayer rectangular plate having dimensions a, b and consisting of $M = N - 1$ layers of thickness $h_{[k]}$ (here $k = \overline{1, M}$ is the layer number). For displacement components $U_i^{[k]}$ $(i = 1, 2)$ and $U_3^{[k]}$ approximations

$$U_i^{[k]} = \frac{u_i^{(k+1)}}{2} + \frac{z_{[k]}}{h_{[k]}}\left(u_i^{(k+1)} - u_i^{(k)}\right), \quad U_3^{[k]} = \frac{w^{(k+1)}}{2} + \frac{z_{[k]}}{h_{[k]}}\left(w^{(k+1)} - w^{(k)}\right).$$

are accepted within the thickness of each layer $(-h_{[k]}/2 \leq h_{[k]}/2)$.

We assume that the plate is placed in the opening of an absolutely stiff dividing wall separating two adjacent spaces V_1 and V_2. This plate is overrun by a flat harmonic wave of pressure p_* and frequency ω. As a result of it interaction with the plate, acoustic waves are excited in the half-spaces V_1 and V_2 around the plate, which are reflected in the first medium and are radiated in the second medium. To determine the pressures p_*, p_1 applied to the boundary plane of the 1st layer of the plate and the pressure p_2 applied to the boundary plane of the Nth layer of the plate, it is necessary to find solutions of the wave equations

$$\Phi_{*,zz} - c_1^{-2}\ddot{\Phi}_* = 0, \quad \Phi_{s,xx} + \Phi_{s,yy} + \Phi_{s,zz} - c_s^{-2}\ddot{\Phi}_s = 0, \ s = 1, 2 \tag{1}$$

written in the potentials Φ_*, Φ_1 and Φ_2 of velocities. Here, c_1 and c_2 are sound velocities. The pressure p_*, p_s and the velocity components v_z^*, v_x^s, v_y^s, v_z^s in the spaces V_1 and V_2 are determined in terms of the functions Φ_* and Φ_s:

$$p_* = -\rho_1\dot{\Phi}_*, \quad p_s = -\rho_s\dot{\Phi}_s, \quad v_z^* = \Phi_{*,z}, \quad v_x^s = \Phi_{s,x}, \quad v_y^s = \Phi_{s,y}, \quad v_z^s = \Phi_{s,z}$$

where ρ_s are the densities of acoustic media in the spaces.

Solutions to the equations of plate motion and Eqs. (1) have to satisfy the kinematic conjugation conditions

$$\dot{w}^{(1)} = \left(v_z^* + v_z^1\right)\big|_{z=0}, \quad \dot{w}^{(N)} = v_z^2\big|_{z=0}, \tag{2}$$

where $\dot{w}^{(1)}$ and $\dot{w}^{(N)}$ are the deflections of the boundary planes of the 1st and $[N-1]$th layers, respectively.

The process of dynamic deformation of a plate is described by the equations of the multilayer plate theory taking into account transverse shear and transverse compression in each layer obtained in [6]. The internal friction of the material in each layer of the plate is taken into account according to the Kelvin–Voigt hysteresis model. For such a model, the stresses $\sigma_{\alpha\alpha}^{[k]}$, $\sigma_{\alpha\beta}^{[k]}$; $\alpha \neq \beta$; $\alpha = \overline{1,3}$; $\beta = \overline{1,3}$ and strains $\varepsilon_{\alpha}^{[k]}$, $\gamma_{\alpha\beta}^{[k]}$ arising in the plate can be connected by the relationships (δ_α and $\delta_{\alpha\beta}$ are logarithmic decrements of vibrations under axial and shear deformations, respectively; τ is time):

$$\sigma_{\alpha\alpha}^{[k]} = \tilde{g}_{\alpha 1}^{[k]}\varepsilon_1^{[k]} + \tilde{g}_{\alpha 2}^{[k]}\varepsilon_2^{[k]} + \tilde{g}_{\alpha 3}^{[k]}\varepsilon_3^{[k]}, \quad \sigma_{\alpha\beta}^{[k]} = \tilde{G}_{\alpha\beta}^{[k]}\gamma_{\alpha\beta}^{[k]},$$

where

$$\tilde{E}_{\alpha}^{[k]} = E_{\alpha}^{[k]}\left(1 + \frac{\delta_{\alpha}^{[k]}}{\pi\omega}\frac{\partial}{\partial\tau}\right), \; \tilde{G}_{\alpha\beta}^{[k]} = G_{\alpha\beta}^{[k]}\left(1 + \frac{\delta_{\alpha\beta}^{[k]}}{\pi\omega}\frac{\partial}{\partial\tau}\right), \; \tilde{g}_{11}^{[k]} = \frac{\tilde{E}_1^{[k]}\left(1 - \nu_{23}^{[k]}\nu_{32}^{[k]}\right)}{\Delta^{[k]}},$$

$$\tilde{g}_{12}^{[k]} = \frac{\tilde{E}_1^{[k]}\left(\nu_{21}^{[k]} + \nu_{23}^{[k]}\nu_{31}^{[k]}\right)}{\Delta^{[k]}}; \quad \underleftrightarrow{1,2,3}, \; \Delta^{[k]} = 1 - \nu_{12}^{[k]}\nu_{21}^{[k]} - \nu_{23}^{[k]}\nu_{32}^{[k]} - \nu_{31}^{[k]}\nu_{13}^{[k]} - 2\nu_{12}^{[k]}\nu_{23}^{[k]}\nu_{31}^{[k]}.$$

The components of the displacements $u_1^{(k)}$, $u_2^{(k)}$ and $w^{(k)}$ of the points of first and last layers' front planes as well as the points of the conjugation planes of inner layers are accepted as unknowns. For the case of hinged support of the plate in the opening of an absolutely stiff dividing wall under the action of a monoharmonic sound pressure on it, for these functions we take the representations

$$u_1^{(k)} = e^{i\omega\tau}\sum_{m,n=1,3,\ldots}\tilde{u}_{1mn}^{(k)}\cos\lambda_m x \cdot \sin\lambda_n y, \quad u_2^{(k)} = e^{i\omega\tau}\sum_{m,n=1,3,\ldots}\tilde{u}_{2mn}^{(k)}\sin\lambda_m x \cdot \cos\lambda_n y,$$
$$w^{(k)} = e^{i\omega\tau}\sum_{m,n=1,3,\ldots}\tilde{w}_{mn}^{(k)}\sin\lambda_m x \cdot \sin\lambda_n y, \quad \lambda_m = m\pi/a, \; \lambda_n = n\pi/b; \tag{3}$$

In view of the conditions of (2) and representations (3), for solutions of two latter Eqs. (1) representations have to be in the form

$$\Phi_s = e^{i\omega\tau}\sum_{m,n=1,3,\ldots}\tilde{\Phi}_s^{mn}(z)\sin\lambda_m x \; \sin\lambda_n y. \tag{4}$$

Based on the accepted representations (3)–(4), the analytical solution to the formulated problem is found. This solution should be considered as exect for $N \to \infty$, $M \to \infty$. The solution found allows us to calculate the sound transmission loss $R_p^0 = -20\lg\left|\tilde{p}_2|_{z=0}/(\tilde{p}_* + \tilde{p}_1)|_{z=0}\right|$ of the plate and determine the amplitude values of all stress-strain state parameters. Let us note that at $\tilde{p}_* = 0$ the formulated problem becomes homogeneous. Its solution determines the eigenfrequencies and the corresponding eigenmodes of vibration for the hingedly supported composite plate, taking into account the internal energy dissipation in the plate material and its aerodynamic interaction with surrounding acoustic media.

3 Calculation Results and Their Analysis

On the basis of the solutions found, studies of a rectangular composite plate of thickness $t = 3$ mm, length $a = 480$ mm and width $b = 560$ mm, made of carbon fiber textile Porcher 3692 and a binder EDT-69NM were carried out. Such material under dynamic deformation processes has the following mechanical characteristics and logarithmic decrements of vibrations [7]: $E_{1,d} = E_{2,d} = 40$ GPa, $E_{3,d} = 5.9$ GPa, $G_{12,d} = G_{13,d} = G_{23,d} = 4$ GPa, $\nu_{12} = 0.163$, $\nu_{13} = \nu_{23} = 0.3$, $\rho = 1500$ kg/m^3, $\delta_1 = \delta_2 = \delta_3 = 0.012$, $\delta_{12} = \delta_{13} = \delta_{23} = 0.03$. The plate is surrounded on both sides by air characterized by the parameters $\rho_1 = \rho_2 = \rho = 1.225$ kg/m^3, $c_1 = c_2 = c = 340$ m/s.

Table 1 represents the results of the calculation of the real parts of eigenfrequencies $f_{mn} = \omega^r_{mn}/(2\pi)$ (here ω^r_{mn} is the real part of the circular frequency) for the hingedly supported composite plate taking into account the internal energy dissipation. Furthermore, the parameters ξ_{mn} characterizing the imaginary parts of the complex vibration frequencies determined by the formula $\xi_{mn} = \omega^i_{mn}/\omega^r_{mn}$ (ω^i_{mn} is the imaginary part of the circular frequency) are also presented in the table. Let us note that the results are given for a plate in vacuum, i.e. excluding interaction with acoustic media. Such results, as was shown in [8], practically do not differ from the results obtained by taking into account the influence of acoustic media, if the behavior of the media is described on the basis of three-dimensional wave equations.

Table 1. Real parts of eigenfrequencies f_{mn} and the parameters ξ_{mn} of the composite plate.

No of tone	f_{mn}, Hz	ξ_{mn}	M	n
1	44,4	0,0023	1	1
2	105,81	0,00219	1	2
3	133,314	0,00209	2	1
4	177,465	0,00231	2	2
5	217,008	0,00207	1	3
6	273,609	0,00229	2	3
7	286,483	0,002	3	1
8	321,336	0,00219	3	2
9	374,493	0,00201	1	4
10	398,778	0,00232	3	3

To establish the effect of the incident sound wave frequency on the sound transmission loss and the characteristics of stress-strain state of the plate, computational experiments were also carried out. The data in Fig. 1*a* illustrates dependence of the sound transmission loss R_p^0 values in the center of the plate at frequency $f = \omega/(2\pi)$. Moreover, Fig. 1*b* and 1*c* on a logarithmic scale along the axis $\tilde{\sigma}$ show the dependence of the maximum amplitude values of the normal stresses $\tilde{\sigma}_{11}$ (line 1 in Fig. 1*b* and 1*c*), $\tilde{\sigma}_{33}$ (line 2 in Fig. 1*b*) and tangential stresses $\tilde{\sigma}_{13}$ (line 3 in Fig. 1*c*) on the frequency f. As is seen, there are frequencies f^R of sound wave at which significant drop in sound transmission loss R_p^0 as well as significant splashes of stresses $\tilde{\sigma}_{11}$ and $\tilde{\sigma}_{13}$ arise. By comparing such frequencies f^R obtained through analysis of Fig. 1 with the data represented in the Table 1 we can see that in case of a plane sound wave action, the plate exhibits forced vibrations only with resonant frequencies corresponding to the eigenmodes with odd numbers of harmonics m and n in each direction. Along this, the harmonics with even number at least in one direction, do not objectify.

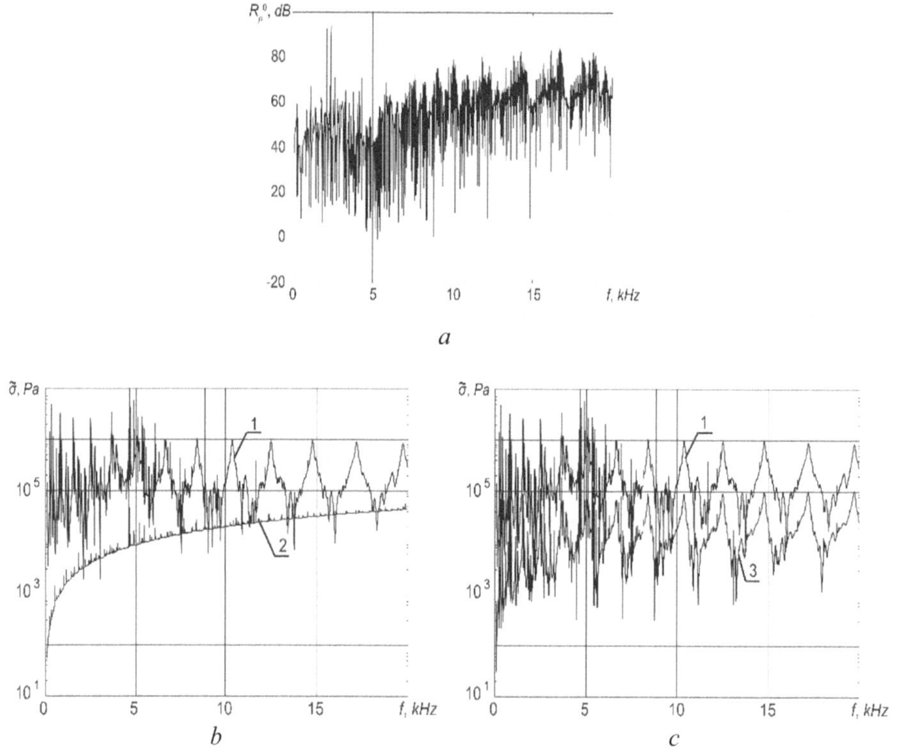

Fig. 1. Dependences of sound transmission loss R_p^0 (*a*) and stresses $\tilde{\sigma}_{11}$, $\tilde{\sigma}_{33}$ and $\tilde{\sigma}_{13}$ (*b*, *c*) on frequency.

It is important to note that at low frequencies of the sound wave, the main in the plate are the tangential components of the stress tensor ($\tilde{\sigma}_{33}$ values are by three orders of magnitude less than $\tilde{\sigma}_{11}$ values) and the description of the dynamic stress-strain state of the plate by classical two-dimensional equations of plate theory is quite acceptable. However, at high frequencies, stresses $\tilde{\sigma}_{33}$ formed in the plate are of the same magnitude as stresses $\tilde{\sigma}_{11}$ ($\tilde{\sigma}_{33}$ values is $25 \div 30\%$ of $\tilde{\sigma}_{11}$ values). Furthermore, stresses $\tilde{\sigma}_{33}$ can even exceed the $\tilde{\sigma}_{11}$ values in narrow frequency ranges in the spectrum of medium and high frequencies. Consequently, a correct description of the plate deformation mechanics at high frequencies setting the problem of the class in question is possible only on the basis of refined equations with the necessary allowance for transverse compression. This is especially important studying the acoustic effects on thin-walled structural elements made of fibrous reinforced composite materials. Such materials, as a rule, have elastic and strength characteristics in the transverse direction on several orders of magnitude less than similar characteristics in the tangential directions. It also should be noted that the obtained dependences $\tilde{\sigma}_{13} = \tilde{\sigma}_{13}(f)$ and $\tilde{\sigma}_{11} = \tilde{\sigma}_{11}(f)$ have the same character of change (Fig. 1*c*). The analysis of the calculation results showed that the maximum $\tilde{\sigma}_{13}$ values of shear stresses are an order of magnitude less than the maximum $\tilde{\sigma}_{11}$ values of stresses.

Acknowledgment. This research was supported by a grant from the Russian Science Foundation (project № 19-19-00058).

References

1. Pretlove, A.J.: Free vibrations of a rectangular panel backed by a closed rectangular cavity. J. Sound Vib. **2**(3), 197–209 (1965)
2. Pretlove, A.J.: Forced vibrations of a rectangular panel backed by a closed rectangular cavity. J. Sound Vib. **3**(3), 252–261 (1966)
3. Jeyaraj, P., Ganesan, N., Padmanabhan, C.: Vibration and acoustic response of a composite plate with inherent material damping in a thermal environment. J. Sound Vib. **320**(1), 322–338 (2009)
4. Li, X., Yu, K., Han, J., Song, H., Zhao, R.: Buckling and vibro-acoustic response of the clamped composite laminated plate in thermal environment. Int. J. Mech. Sci. **119**, 370–382 (2016)
5. D'Alessandro, V., Petrone, G., Franco, F., De Rosa, S.: A review of the vibroacoustics of sandwich panels: models and experiments. J. Sandwich Struct. Mater. **15**(5), 541–582 (2013)
6. Paimushin, V.N., Gazizullin, R.K.: Static and monoharmonic acoustic impact on a laminated plate. Mech. Compos. Mater. **53**(3), 283–304 (2017)
7. Paimushin, V.N., Firsov, V.A., Gyunal, I., Shishkin, V.M.: Identification of the elastic and damping characteristics of carbon fiber-reinforced plastic based on a study of damping flexural vibrations of test specimens. J. Appl. Mech. Tech. Phys. **57**(4), 720–730 (2016)
8. Paimushin, V.N., Tarlakovskii, D.V., Firsov V.A., Gazizullin, R.K.: Free and forced bending vibrations of a thin plate in a perfect compressible fluid with energy dissipation taken into account. Zeitschrift für Angewandte Mathematik und Mechanik, e201900102 (2020)

Vibration of the Plate with Integral Layer Damping: Experimental and Theoretical Studies

V. N. Paimushin[1,2], V. A. Firsov[1], V. M. Shishkin[3], and R. K. Gazizullin[1](✉)

[1] Kazan National Research Technical University named after A. N. Tupolev - KAI, Kazan, Russia
vpajmushin@mail.ru, gazizullin.rk@yandex.ru
[2] Kazan Federal University, Kazan, Russia
[3] Vyatka State University, Kirov, Russia

Abstract. The paper discusses the classical methods of surface damping of bending vibrations in thin-walled structures. Furthermore, the perspective integral version of a damping coating has been proposed. Such coating consists of two layers of material with pronounced viscoelastic properties with a thin reinforcing layer of high modulus material between them. Dynamic tests of cantilevered duralumin specimens under the damped bending vibrations were carried out using the created experimental setup. The purpose of these tests is to compare the effectiveness of the known and proposed methods of surface vibration damping. The influence of aerodynamic drag forces on the vibration damping of specimens is noted. A refined finite element model of an elongated plate with an integral layer damping is constructed on the basis of the four-layer finite element. This model allows taking into account the effect of transverse compression of damping layers under high-frequency deformation. The analysis of the stress-strain state of the damping layers of a simply supported elongated plate under resonance vibrations in several lower eigenmodes has been carried out. The analysis showed a significant increase in the transverse compression stresses of the damping layers with frequency increasing.

Keywords: Plate · Damping coating · Experiment · Logarithmic decrement of vibrations · Finite element model

1 Introduction

Traditional structural materials (metals and metal alloys) are characterized by large values of elastic and strength parameters, but, as a rule, they have low damping properties. In this connection, various types of coatings made of viscoelastic materials [1–3] are used to reduce the vibration activity and dynamic stress of thin-walled structures. The relevance of the problem and the constant attention of researchers to it are reflected in the fundamental monograph [4]. In this monograph authors discuss the

E. Gdoutos and M. Konsta-Gdoutos (Eds.): ICTAEM 2020, STIN 16, pp. 254–260, 2020.
https://doi.org/10.1007/978-3-030-47883-4_46

effect on damping of various physical factors such as temperature, vibration frequencies, layer thicknesses, loading level, etc. The results of these studies formed the basis of the American standard for an experimental method of material damping properties measurement [5].

According to [4], surface damping devices are divided into two classical methods depending on the dominant deformations in the damping layer. The first method is the free-layer damping. In accordance with this method, the damping layer of the viscoelastic material is fixedly connected to the damped thin-walled structure. In this case the bending under transverse vibrations causes cyclic tensile-compression deformations in the damping layer. However, this damping method has a very low efficiency [4], as the damping forces are maximum only in the areas of greatest curvature of the damping layer. The second method is constraining damping layer. In accordance with this method the layer of damping material is additionally covered with a thin reinforcing layer which under cyclic bending of the structure mainly experiences transverse shear deformations. In this case, the maximum damping forces arise only in the areas of the maximum shift of the damping layer. Therefore this method does not allow you to effectively damp vibrations throughout the design area.

In view of this, the integral damping coating consisting of two viscoelastic layers with thin reinforcing layer of high modulus material between them (Fig. 1a) seems more effective. In case of bending vibrations of a thin-walled structure the lower layer of the damping coating undergo a shear and acts as a constraining layer damping, and the upper layer undergo tensile-compression deformations and acts as a free damping layer (Fig. 1b). This allows you to integrate the area of active vibration damping, practically getting rid of the ballast coating areas for any form of bending vibrations in a wide range of operating frequencies. So, for the simply supported beam under second bending vibrations mode the active damping regions (marked in dark on the figure) cover almost the entire surface of the damped structure (Fig. 1c). Moreover, the outer layer protects the thin reinforcing layer from buckling and delamination in the phase of cyclic compression of bending vibrations, as such destruction is possible in the reinforcing layer in case of the use of constraining layer damping method. In turn, the presence of a thin reinforcing layer having an insignificant mass and bending stiffness practically does not increase the weight characteristics of the damping coating and does not reduce the possibility of the proposed method using for damping of curved surfaces structures.

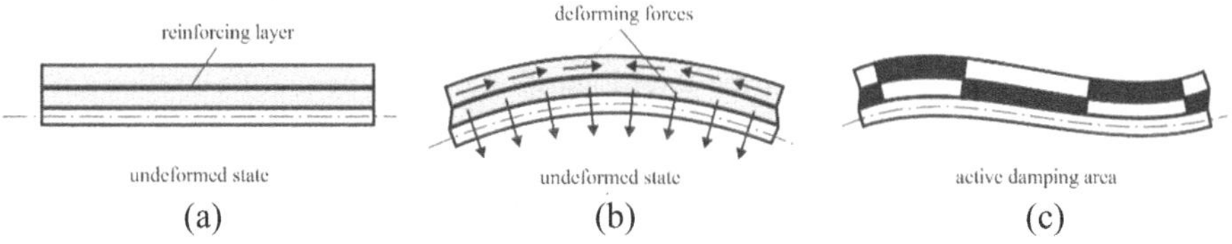

Fig. 1. The physical picture of surface vibration damping of the structure with an integral damping coating. (a) shows an undeformed state; (b) shows an deformed condition; (c) illustrates active damping areas.

2 Experimental Studies

Dynamic tests of cantilevered duralumin specimens under the damped bending vibrations were carried out using the created experimental setup [6]. The purpose of these tests is to compare the effectiveness of the known and proposed methods of surface vibration damping. Some calculation results for specimens with lengths of 500 (a), 300 (b) and 200 (c) mm are shown in Fig. 2 as relations between the logarithmic decrements of vibration $\delta(A)$ and the vibration amplitude A of specimens' free ends. Here, line 1 correspond to basic specimens without a damping coating; line 2 correspond to specimens with a free layer damping; line 3 correspond to specimens with constraining layer damping; line 4 correspond to specimens with integral layer damping. In order to more contrastly evaluate the effectiveness of the studied damping coatings, they were applied on both sides of the basic specimens. An oil-resistant rubber with a low static elastic modulus $E \approx 5.3$ MPa and manufactured in the form of plates 2H-1-MBS-S (GOST 7338-90) with a thickness of 1.8 mm and a density of $\rho = 1345$ kg/m^3 was used as a damping material. The damping properties of rubber are characterized by logarithmic decrements of vibration $\delta \approx 1.2$.

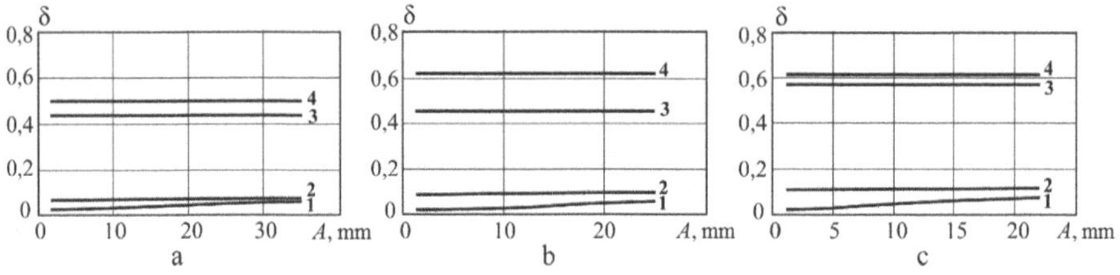

Fig. 2. Relation $\delta(A)$. Explanations in the text.

The above results indicate that in all the cases considered, the effectiveness of the proposed integral method of vibration damping is higher than the most effective of the previously known methods by an average of 20% even under conditions of the lowest modes of vibration. We should note that, in addition to the damping properties of the material, dependence δ(A) can be significantly affected by the air drag [6]. For duralumin basic specimens the aerodynamic component of damping is approximately 85–90% of the total (experimental) logarithmic decrements of vibration [7]. For specimens with a damping coating this effect becomes significantly less. The calculations showed that for specimens with a free layer damping the aerodynamic component of the damping does not exceed 12% of the total logarithmic decrements of vibration (at the maximum amplitude of oscillations). Moreover, for specimens with a constraining and integral layer damping this component becomes even significantly smaller.

3 Refined Finite Element Model of an Elongated Plate with Integral Layer Damping

Under static and low-frequency deformations of a plate with an integral damping coating, one can adopt the classical model in which the constraining layer is in a state of homogeneous transverse shear and the free layer is in a state of inhomogeneous tension-compression. However, in case of resonant vibrations at higher frequencies, the stress-strain state of the damping layers can be qualitatively different due to the influence of inertia forces on it. This leads to the necessity of taking into account all components of the stress state when determining the energy dissipation in damping layers. To analyze the stress-strain state of the marked layers under cyclic deformation in a wide frequency range, a refined finite element model of an elongated plate with an integral layer damping is constructed. This model is based on the developed four-layer compound finite element (Fig. 3) and uses the following models to describe the mechanics of layer deformation: the plate undergo a tension-compression and bending; both damping layers are in a plane stress state; the reinforcing layer undergo a tension-compression only. Independent nodal parameters of the element are displacements u_i, w_i $(i = 1, 2, \ldots, 6)$ as well as rotation angles ϕ_1, ϕ_4 of the cross sections of layer 1. They can be represented by a vector $\mathbf{r}^{(e)} = \{u_1\ w_1\ \varphi_1\ u_2\ w_2\ u_3\ w_3\ u_4\ w_4\ \varphi_4\ u_5\ w_5\ u_6\ w_6\}$.

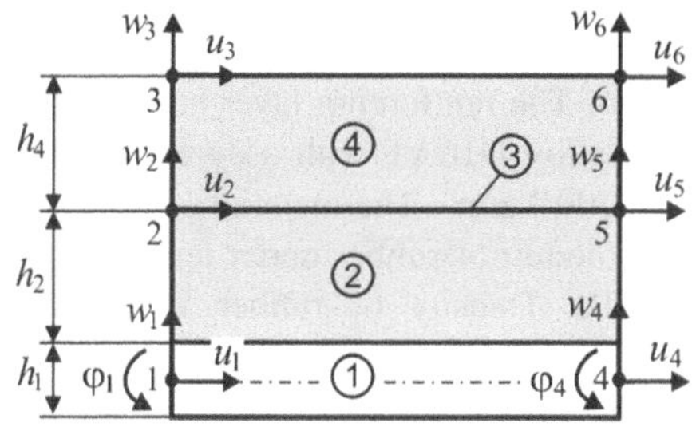

Fig. 3. The finite element of the elongated plate with integral layer damping.

The elastic and damping properties under cyclic deformation of the damping layer material are taken into account by the physical dependence

$$\boldsymbol{\sigma} = \mathbf{D}\boldsymbol{\varepsilon} + \mathbf{D}_g\dot{\boldsymbol{\varepsilon}} \tag{1}$$

representing a generalization of the well-known Kelvin-Voigt model under tension-compression of the material in the case of a complex stress state. Vectors $\boldsymbol{\sigma}$, $\boldsymbol{\varepsilon}$ contain components of the stress and strain states; $\mathbf{D}$, $\mathbf{D}_g$ are the stiffness matrix and the matrix of viscous damping of the material, respectively. For an isotropic viscoelastic material in a plane stress state, the matrices $\mathbf{D}$ and $\mathbf{D}_g$ will be as follows:

$$\mathbf{D} = \begin{bmatrix} E/(1-\nu^2) & E\nu/(1-\nu^2) & 0 \\ E\nu/(1-\nu^2) & E/(1-\nu^2) & 0 \\ 0 & 0 & G \end{bmatrix};\ \mathbf{D}_g = \frac{1}{\pi\omega}\begin{bmatrix} E\delta_\varepsilon/(1-\nu^2) & E\delta_\varepsilon\nu/(1-\nu^2) & 0 \\ E\delta_\varepsilon\nu/(1-\nu^2) & E\delta_\varepsilon/(1-\nu^2) & 0 \\ 0 & 0 & G\delta_\gamma \end{bmatrix}$$

Here E, G, δ_ε, δ_γ are the dynamic moduli of elasticity and the logarithmic decrements of vibrations of the material, respectively, under tension-compression and shear, ω is the deformation frequency, ν is the Poisson's ratio.

In the dynamic analysis of structures based on the FEM, usually, the task is to synthesize elastic, damping and inertial forces, which requires stiffness matrices $\mathbf{K}^{(e)}$, damping matrices $\mathbf{C}^{(e)}$ and mass matrices $\mathbf{M}^{(e)}$ constituting the finite element structure. To construct the matrices $\mathbf{K}^{(e)}$ и $\mathbf{C}^{(e)}$ we used dependence (1) and geometric relations which connect deformations $\boldsymbol{\varepsilon}$ with the vector of nodal displacements $\mathbf{r}^{(e)}$ of the finite element. To construct the matrix $\mathbf{M}^{(e)}$, we supposed that the kinetic energy of an element is determined only by the velocities $\dot{w}$.

For numerical experiments, we chose an elongated simply supported plate with an integral damping coating. Geometrical parameters of the plate are following: length $L =$ 200 mm; width $b = 20$ mm; thickness $h_1 = 1$ mm. The thickness of the damping layers are $h_2 = h_4 = 1.8$ mm. The reinforcing layer has a thickness $h_3 = 0.1$ mm. The plate material is aluminum alloy D16AT with a dynamic elastic modulus $E = 5.54 \cdot 10^4$ MPa and density $\rho = 2700$ kg/m^3. The material of the damping layers is technical rubber. Dynamic elasticity moduli of rubber under tension-compression and shear are $E = 20.1$ MPa, $G = 2.6$ MPa. Density of rubber is $\rho = 1345$ kg/m3, logarithmic decrement of vibrations under tensile-compression and shear are $\delta_\varepsilon = 1.2$, $\delta_\gamma = 1.1$, respectively. The reinforcing layer is made of unidirectional fiber composite based on carbon tape ELUR-P and binder HT-118 (dynamic elastic modulus $E = 10.5 \cdot 10^4$ MPa, density $\rho = 1200$ kg/m^3). The plate is divided into 100 finite elements of the same length.

We studied stress-strain state of damping layers under resonance vibrations of a plate by several lower eigenmodes with an amplitude of the normal coordinate $q_0 = 0.01$. Figure 4 and 5 illustrates the amplitudes of the normal and shear stresses σ_x, σ_z and τ_{xz} in the middle surface of the constraining and free damping layers, respectively, under resonance vibrations of the plate by the first, fifth, and seventh modes. The frequencies corresponding to these modes were as follows: f1 = 69.43 Hz; f5 = 856.43 Hz; f7 = 1588.66 Hz.

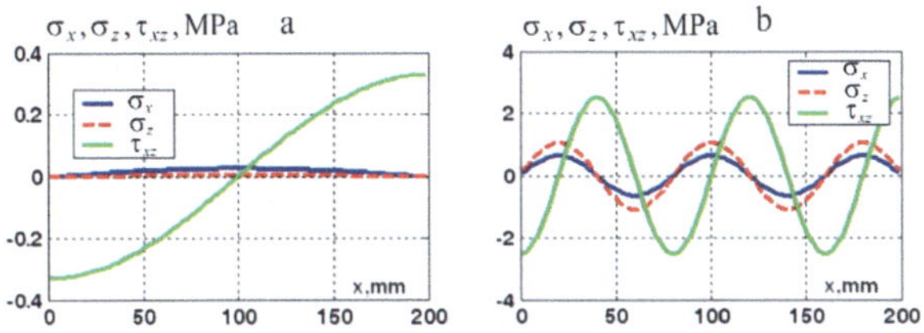

Fig. 4. Stresses in a constraining damping layer under resonance vibrations of the plate by the first (a), fifth (b) and seventh (c) modes.

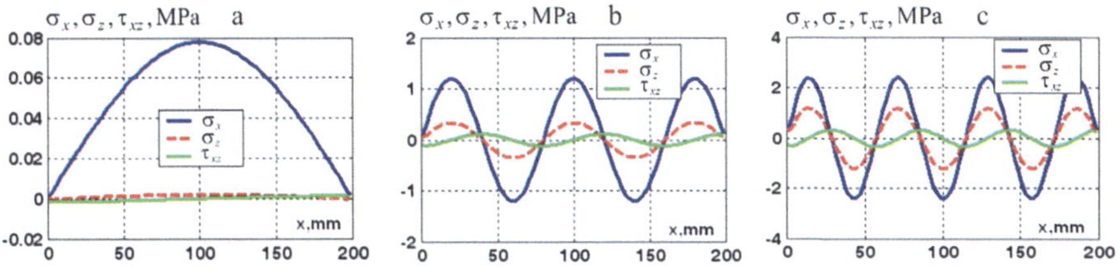

Fig. 5. Stresses in a free damping layer under resonance vibrations of the plate by the first (a), fifth (b) and seventh (c) modes.

An analysis of the obtained stresses shows that the transverse shear stresses τ_{xz} dominate (Fig. 4a) in the constraining damping layer when the plate vibrates in the first (main) mode as well as only normal stresses σ_x mainly act (Fig. 5a) in the free damping layer. As the shape number and, correspondingly, the frequency f increase, stresses σ_x, σ_z increasing appear against the background of stresses τ_{xz} in the constraining damping layer. Thus, on the seventh mode (Fig. 4c) stress σ_x is approximately 65% of the τ_{xz} value and stress σ_z even slightly exceeds the shear stress τ_{xz}. With increasing frequency f in the free damping layer (Fig. 5) in addition to stress σ_x two other stresses σ_z and τ_{xz} appear, of which the stress σ_z is the most significant.

Acknowledgment. This research was supported by a grant from the Russian Science Foundation (project № 19-19-00058).

References

1. Oberst, H., Becker, G.W., Frankenfeld, K.: Über die Dampfung der Biegeschwingunden dunner Bleche durch fest haftende Belage. Acustica **4**(2), 181–184 (1952)
2. Kerwin, E.M.: Damping of flexural waves by a constrained viscoelastic layer. J. Acoust. Soc. Am. **31**(7), 952–964 (1959)
3. Ross, D., Ungar, E.E., Kerwin, E.M.: Damping of plate flexural vibrations by means of viscoelastic laminate. In: Structural Damping. Section 3, pp. 49–88. ASME, New York (1959)

4. Nashif, A.D., Johnes, D.I.G., Henderson, J.P.: Vibration Damping. Wiley, New York (1984)
5. ASTM E756-05: Standard test method for measuring vibration-damping properties of materials. S. I.: ASTM Intern. (2017)
6. Paimushin, V.N., Firsov, V.A., Gyunal, I., Egorov, A.G.: Theoretical-experimental method for determining the parameters of damping based on the study of damped flexural vibrations of test specimens. 1. Experimental basis. Mech. Compos. Mater. **50**(2), 127–136 (2014)
7. Paimushin, V.N., Firsov, V.A., Gyunal, I., Shishkin, V.M.: Theoretical-experimental method for evaluating the elastic and damping characteristics of soft materials based on studying the resonance flexural vibrations of test specimens. Mech. Compos. Mater. **52**(5), 571–582 (2016)

Thermomagnetoelectroelasticity of Bimaterial Solids with High Temperature Conducting Interface and Thin Internal Inhomogeneities

Andrii Vasylyshyn[1(✉)], Iaroslav Pasternak[2], and Heorhiy Sulym[3]

[1] Ivan Franko National University of Lviv, Universytetska Str. 1, Lviv 79000, Ukraine
vasylyshyn.c.h@gmail.com
[2] Lutsk National Technical University, Lvivska Str. 75, Lutsk 43018, Ukraine
iaroslav.m.pasternak@gmail.com
[3] Pidstryhach Institute for Applied Problems of Mechanics and Mathematics of NASU, Naukova Str. 3b, Lviv 79060, Ukraine
gtsulym@gmail.com

Abstract. This work studies the problem of a thermomagnetoelectroelastic anisotropic bimaterial with imperfect high temperature-conducting coherent interface, which components contain thin inclusions. Using the extended Stroh formalism and complex variable calculus the Somigliana type integral formulae and corresponding boundary integral equations for the anisotropic thermo-magnetoelectroelastic bimaterial with high temperature-conducting coherent interface are obtained. These integral equations are introduced into the modified boundary element approach. The numerical analysis of new problems is held and results are presented.

Keywords: Bimaterial · Thermomagnetoelectroelasticity · Thin inclusion · Boundary element method · Coherent imperfect interface

1 Introduction

Modern advanced high-tech industries, especially precision mechanics, are increasingly using multifield (thermomagnetoelectoelastic) materials in the manufacture of various devices. The latter are smart composites based on the mechanical combination of pyroelectric (ferroelectric) and magnetostrictive (piezomagnetic) materials. When creating such materials, a thin intermediate layer is usually formed that affects the stress and temperature fields. The influence of this layer is reduced to the conditions of imperfect thermal and magneto-electro-mechanical contact of the bimaterial components. In general, in the scientific literature [1, 2] there are two types of imperfect thermal conditions and two types imperfect mechanical conditions of contact of a thin layer with the medium.

Thus, there arises the scientific problem of development integral mathematical models and methods of analysis of bimaterial thermomagnetoelectroelastic solids, which can contain both defects of structure and purposefully introduced thin layers that change the properties of these bodies.

E. Gdoutos and M. Konsta-Gdoutos (Eds.): ICTAEM 2020, STIN 16, pp. 261–267, 2020.
https://doi.org/10.1007/978-3-030-47883-4_47

At present, methods of analyzing the interaction of mechanical, electrical and magnetic fields in anisotropic smart materials are intensively being developed.

An effective method for solving such problems is an approach based on methods of the complex variable function theory and the Stroh formalism. Thus, in [3], the Somigliana type boundary integral equations along with the explicit closed form of expressions for the kernels were obtained and for the boundary element analysis of the anisotropic thermoelectroelastic bimaterial with holes, cracks and thin inclusions. This approach was also applied in [4] for a bimaterial solid with a Kapitsa type interface. The paper [5] obtained a two-dimensional Green's function for anisotropic bimaterials with imperfect thermal and spring-type mechanical contact.

In this work a mathematical model of a thermomagnetoelectroelastic bimaterial solid with high temperature conductive coherent interface, which parts can contain thin deformable inclusions, is developed.

The boundary integral equations of the formulated problem are constructed in a closed form. The boundary element method for their fast and accurate numerical solution is developed.

2 Governing Equations of Thermomagnetoelectroelasticity

In a fixed rectangular system of coordinates $Ox_1x_2x_3$ the balance equations for stress, electric displacement, magnetic induction and heat flux, and constitutive laws in absence of volumetric loading can be written in the following compact form

$$\tilde{\sigma}_{ij,j} + \tilde{f}_i \equiv \tilde{C}_{ijkm}\tilde{u}_{k,jm} + \tilde{f}_i = 0, \quad h_{i,i} - f_h = 0, \tag{1}$$

$$\tilde{\sigma}_{ij} = \tilde{C}_{ijkm}\tilde{u}_{k,m} - \tilde{\beta}_{ij}\theta, \quad h_i = -k_{ij}\theta_{,j}. \tag{2}$$

The nomenclature is the same as in Refs. [3, 4].

According to the extended Stroh formalism, the general homogeneous solution of Eqs. (1) and (2) can be expressed in terms of complex analytic functions as follows

$$\theta = 2\mathrm{Re}\{g'(z_t)\}, \vartheta = 2k_t \,\mathrm{Im}\,\{g'(z_t)\} \tag{3}$$

$$\tilde{\mathbf{u}} = 2\mathrm{Re}[\mathbf{Af}(z_*) + \mathbf{c}g(z_t)], \tilde{\boldsymbol{\varphi}} = 2\mathrm{Re}[\mathbf{Bf}(z_*) + \mathbf{d}g(z_t)], \tag{4}$$

$$\mathbf{f}(z_*) = [F_1(z_1), F_2(z_2), F_3(z_3), F_4(z_4), F_5(z_5)]^{\mathrm{T}}. \tag{5}$$

where ϑ is a heat flux function; θ is a change in temperature compared with actual; $\tilde{\mathbf{u}}$ is an extended displacement vector; $\tilde{\boldsymbol{\varphi}}$ is an extended stress function; $g(z_t)$ and $F_\alpha(z_\alpha)$ are complex analytic functions with respect to their arguments.

Based on Eqs. (3), (4) one can derive the following relation between the function $g(z_t)$ and temperature and heat flux function:

$$g'(z_t) = \frac{1}{2}\left(\theta + i\frac{\vartheta}{k_t}\right). \tag{6}$$

3 Problem Statement

Consider an anisotropic bimaterial medium consisting of two thermomagnetoelectroelastic anisotropic half-spaces $x_1 > 0$ and $x_2 < 0$ (which in terms of 2D problem are modeled as two half-planes S_1 $(x_1 > 0)$ and S_2 $(x_2 < 0)$ as depicted in Fig. 1), and which are connected along the boundary $x_2 = 0$. On the boundary, the conditions of imperfect thermal contact are satisfied, and can be written as follows:

$$\begin{aligned} \vartheta^{(1)}(x_1, x_2)\big|_{x_2=0} &= \vartheta(x_1) + \mu_0 \theta_{,1}(x_1), \forall x_2 = 0; \\ \vartheta^{(2)}(x_1, x_2)\big|_{x_2=0} &= \vartheta(x_1); \end{aligned} \tag{7}$$

$$\theta^{(1)}(x_1, x_2)\Big|_{x_2=0} = \theta^{(2)}(x_1, x_2)\Big|_{x_2=0} = \theta(x_1); \tag{8}$$

where μ_0 is constant thermal conductivity of the interface.

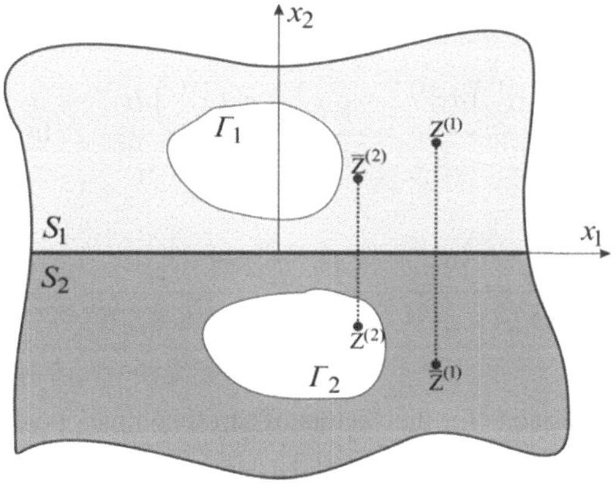

Fig. 1. The sketch of the problem

Also, the conditions of perfect mechanical contact of bimaterial components at the interface are as follows:

$$\tilde{\mathbf{u}}^{(1)}(x_1, x_2)\big|_{x_2=0} = \tilde{\mathbf{u}}^{(2)}(x_1, x_2)\big|_{x_2=0} = \tilde{\mathbf{u}}(x_1), \forall x_2 = 0 \tag{9}$$

$$\tilde{\boldsymbol{\varphi}}^{(1)}(x_1, x_2)\Big|_{x_2=0} = \tilde{\boldsymbol{\varphi}}^{(2)}(x_1, x_2)\big|_{x_2=0} = \tilde{\boldsymbol{\varphi}}(x_1), \forall x_2 = 0. \tag{10}$$

Here, indices (1) and (2) are used to denote the magnitudes of the fields acting in half-spaces S_1 and S_2 respectively. It is considered that each of the half-spaces contains a system of smooth closed contours $\Gamma_1 = \bigcup_i \Gamma_i^{(1)}$ and $\Gamma_2 = \bigcup_i \Gamma_i^{(2)}$ on which we can set different thermal or mechanical and electric conditions.

To derive integral formulas of Stroh complex potentials, the Cauchy integral formula is used [5]:

$$\frac{1}{2\pi i}\int\limits_{\partial S} \frac{\phi(\tau)d\tau}{\tau - z} = \begin{cases} \phi(z) & \forall z \in S, \\ 0 & \forall z \notin S. \end{cases} \tag{11}$$

It relates values of an arbitrary analytic function $\phi(\tau)$ at the boundary ∂S of the domain S with its value $\phi(z)$ inside this domain. Here $\tau, z \in \mathbb{C}$ are complex variables, which define the position of the source and field points, respectively. Herewith, if the domain S is infinite it is assumed that the function $\phi(\tau)$ vanishes at $z \to \infty$.

Thus, the expressions for complex temperature functions $g_i'\left(z_t^{(i)}\right)$ can be written as follows:

$$g_1'\left(z_t^{(1)}\right) = \frac{1}{2\pi i}\int\limits_{\Gamma} \frac{g_1'\left(\tau_t^{(1)}\right)d\tau_t^{(1)}}{\tau_t^{(1)} - z_t^{(1)}} + \frac{1}{2\pi i}\int\limits_{-\infty}^{+\infty} \frac{g_1'\left(x_t^{(1)}\right)dx_1^{(1)}}{x_1^{(1)} - z_t^{(1)}}, \forall \operatorname{Im}\left(z_t^{(1)}\right) > 0; \tag{12}$$

$$g_2'\left(z_t^{(2)}\right) = \frac{1}{2\pi i}\int\limits_{\Gamma} \frac{g_2'\left(\tau_t^{(2)}\right)d\tau_t^{(2)}}{\tau_t^{(2)} - z_t^{(2)}} + \frac{1}{2\pi i}\int\limits_{-\infty}^{+\infty} \frac{g_2'\left(x_t^{(2)}\right)dx_1^{(2)}}{x_1^{(2)} - z_t^{(2)}}, \forall \operatorname{Im}\left(z_t^{(2)}\right) < 0 \tag{13}$$

Similarly, the expressions for the vectors of Stroh complex potentials $\mathbf{f}^{(i)}\left(z_*^{(i)}\right)$ can be written. The procedure of obtaining boundary integral equations of the problem is in elimination of integral over the interface, which is done using complex variable calculus.

4 Numerical Examples

Using the obtained Somigliana type integral formulae incorporated with the boundary element approach [4] consider several problems for a thermoelectroelastic bimaterial with imperfect thermal and perfect electromechanical contact of its components. The considered material contains internal thin inclusion. Only 20 three-node boundary elements, including two special tip elements, were used to mesh the inclusion.

Consider a thermoelastic anisotropic bimaterial consisting of two half-planes. It contains a rectilinear isotropic thermally insulated deformable permeable inclusion of length $2a$. It is not subject to thermal deformation and it does not undergo thermal expansion. It is located in half-plane $x_2 < 0$ at a distance d parallel to the interface. The thickness of inclusion is $h = 0.01a$. The relative rigidity of inclusion is $k = G_i/C_{11}$, where G_i is inclusion shear modulus and C_{11} is material 2 elastic moduli. Half-plane $x_2 > 0$ contains two heat sources of equal intensity. They are located at a distance d_0 to the boundary.

Figure 2 and Fig. 3 show the relationship between the generalized stress intensity factors (SIF) and the parameter of thermal conductivity of the interface μ_0, $(\mu_0^* = ak_{11})$ when the relative rigidity is $k = 10^{-10}$. Figure 2 illustrates the case when bimaterial components are made of different materials. Namely, component $x_2 > 0$ is made of BaTiO3 and component is $x_2 < 0$ made of CdSe.

Figure 3 illustrates the case when bimaterial components are made of $BaTiO_3$, and have the same properties.

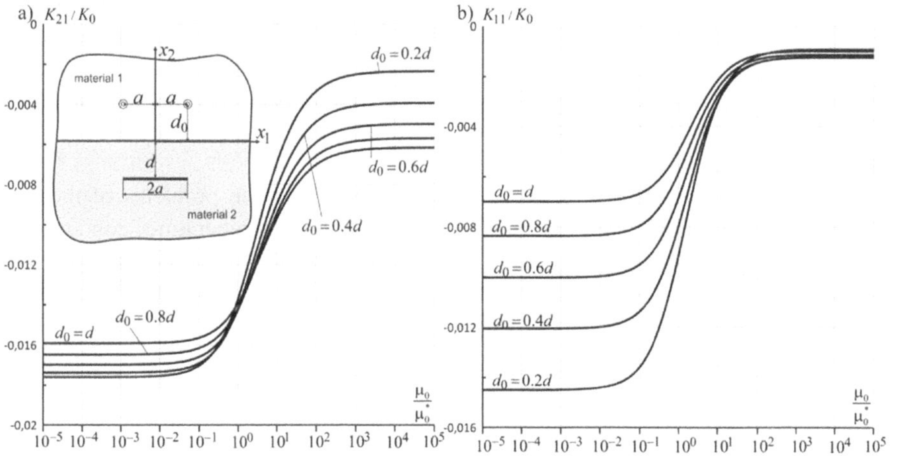

Fig. 2. The relationship between SIF K_{21}/K_0(a), K_{11}/K_0(b) and μ_0/μ_0^*

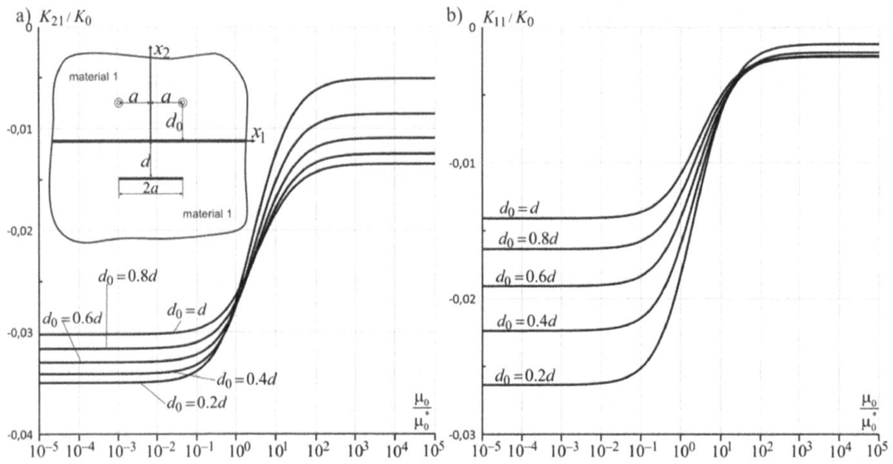

Fig. 3. The relationship between SIF K_{21}/K_0(a), K_{11}/K_0(b) and μ_0/μ_0^*

5 Conclusions

This paper extends the possibilities of an approach, methods of the theory of function of complex variable and Stroh formalism, to obtain integral formulas and equations for structurally heterogeneous bimaterial solids with imperfect thermal contact of its components.

A mathematical model of a thermomagnetoelectroelastic bimaterial solid with a high thermal conductivity interface and perfect electro-magneto-mechanical contact of components, which in turn can contain thin deformable inclusions, is developed. In closed form, the boundary integral equations of the formulated problem are constructed, as well as the boundary element method, which makes it possible to solve it effectively(accurate and fast).

All this makes it possible to solve with high precision the problems of thermomagnetoelectroelasticity for bimaterial solids with high-temperature conductivity interface containing thin ribbon-like deformable inclusions or cracks that have not yet been achieved using traditional numerical approaches, including boundary or finite element methods.

References

1. Kaessmair, S., Javili, A., Steinmann, P.: Thermomechanics of solids with general imperfect coherent interfaces. Arch. Appl. Mech. **84**(9–11), 1409–1426 (2014)
2. Benvensite, Y.: A general interface model for a three-dimensional curved thin anisotropic interface between two anisotropic media. J. Mech. Phys. Solids **54**, 708–734 (2006)
3. Pasternak, I., Pasternak, R., Sulym, H.: Boundary integral equations and Green's functions for 2D thermoelectroelastic bimaterial. Eng. Anal. Boundary Elem. **48**, 87–101 (2014)

4. Pasternak, I., Pasternak, R., Sulym, H.: 2D boundary element analysis of defective thermoelectroelastic bimaterial with thermally imperfect but mechanically and electrically perfect interface. Eng. Anal. Boundary Elem. **61**, 194–206 (2015)
5. Wang, X., Pan, E.: Thermal Green's functions in plane anisotropic bimaterials with spring-type and Kapitza-type imperfect interface. Acta Mechanica et Automatica **209**, 115–128 (2010)

Fracture of Composite Materials Under Compression Along Cracks

Mykhailo Dovzhyk(✉), Viacheslav Bogdanov, and Vladimir Nazarenko

S.P. Timoshenko Institute of Mechanics, National Academy of Sciences of Ukraine, Nesterov Str. 3, Kyiv 01057, Ukraine
Dovzhyk.M.V@ukr.net

Abstract. The axisymmetrical problems of fracture of bodies with near-surface penny-shaped cracks and two parallel penny-shaped cracks under compressive loads directed along cracks are considered. In the situation examined the start of the process of material fracture is determined by the local loss of stability of the equilibrium of the material surrounding cracks. There are two approaches that are used to investigate such problems, namely, so-called "beam approximation" based on applied theories of mechanics of thin-walled structures and the approach in the framework of the rigorous three-dimensional linearized theory of stability of deformable bodies. According to the second approach we reduce the problems to systems of integral Fredholm equations and then to system of algebraic linear equations with use the Bubnov-Galerkin method and numerically analytic technique. As an example we present the numerical calculation for a composite material. The values of critical loads corresponding to the start of fracture are obtained for small and large distance between the cracks (or between the crack and the body surface).

Keywords: Composite materials · Compression along cracks · Near-surface cracks · Parallel coaxial cracks

1 Compression Along Cracks

Fracture of material at compression along cracks is one of the nonclassical problems of fracture mechanics. In this case, the classical approaches of fracture mechanics such as Griffiths-Irwin approach do not work. In this situation, destruction process is identified as the moment of local stability loss of the equilibrium of a part of material in the vicinity of cracks. Currently, there are two general approaches used to investigate such problems [1]. The first of them is based on the use of approximate design diagrams and approximate theories of mechanics of thin-walled structures [2]. Within the framework of this approach, has the greatest application the "beam approximation", when the part between the crack and the free surface (between the cracks) is replaced by a thin-walled element: beam, plate or shell, which are investigated in the framework of the applied theory of stability of thin systems. However, this method has significant drawbacks: it is necessary to carry out separate investigations to determine the possibilities of its application depending on the distance between the cracks, but even having determined this distance,

E. Gdoutos and M. Konsta-Gdoutos (Eds.): ICTAEM 2020, STIN 16, pp. 268–272, 2020.
https://doi.org/10.1007/978-3-030-47883-4_48

there remains the question of choosing the conditions for fixing of thin-walled element. The second approach is based on the basic relationships and methods of rigorous three-dimensional linearized theory of stability of deformable bodies for finite and small sub-critical strains [3]. In [4] using the second approach the conditions of applicability of the "beam approximation" was found for isotropic elastic materials containing near-surface cracks. It is interesting to find such conditions for composite materials.

1.1 Problem Formulation

We considered two geometric schemes of cracks disposition in composites: a half-space with penny-shaped crack of radius a which is situated in the plane $x_3 = 0$ with center on Ox_3 (Fig. 1a) and unbounded body with two parallel penny-shaped cracks of radius a which are situated in the plane $x_3 = 0$ and $x_3 = -2h$ with center on Ox_3 (Fig. 1b). The initial stresses that operated along cracks correspond to biaxial uniform compression.

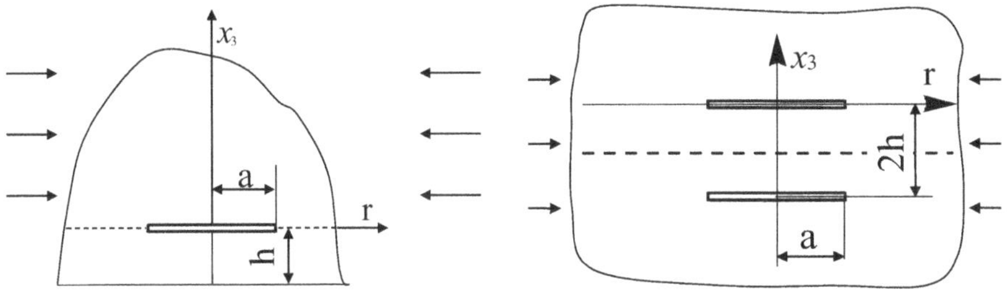

a) half-space with a near-surface crack b) material with two parallel cracks

Fig. 1. Compression of cracked bodies along cracks

Within the limits of the second approach the both problems were reduced to the solution of system of integral Fredholm equations with a side condition [5]

$$
\begin{aligned}
&f(\xi) + \frac{1}{\pi k}\int_0^1 M_1(\xi,\eta)f(\eta)d\eta + \frac{1}{\pi k}\int_0^1 N_1(\xi,\eta)g(\eta)d\eta = 0,\\
&g(\xi) + \frac{1}{\pi k}\int_0^1 M_2(\xi,\eta)g(\eta)d\eta + \frac{1}{\pi k}\int_0^1 N_2(\xi,\eta)f(\eta)d\eta + \tilde{C}_1 = 0,\\
&\int_0^1 g(\xi)d\xi = 0 \qquad (0 \le \xi \le 1, 0 \le \eta \le 1),\\
&f(\xi) \equiv \varphi(a\xi), \qquad g(\xi) \equiv \psi(a\xi).
\end{aligned}
\tag{1}
$$

1.2 Exploratory Procedure

We used the procedure based on the Bubnov-Galerkin method to solve integral Eqs. (1) and search of critical shortening and stress correspond to start of the fracture process. Power functions were used as a system of coordinate functions.

For further calculations we use the numerical analytical procedure proposed in [4]. This procedure allows to obtain results for elastic materials. The procedure allows to calculate analytically the integrals for the chosen system of coordinate functions using a package of symbolic computations. It is allowed to achieve at the further numerical calculations higher exactitude of evaluations at the expense of a numerical integration lapse exclusion. The recurrence relations were used to accelerate of integrals solutions.

Using the method offered in [4], Fredholm integral Eqs. (1) for both problems was transformed to systems of the equations with corresponding factors F_{1ji}, G_{1ji} and variables F_i, G_i, $\tilde{C}_1$, $i,j \in [0,N]$.

$$\sum_{i=0}^{N} F_i F_{1ji} + \sum_{i=0}^{N} G_i G_{1ji} = 0;$$

$$\sum_{i=0}^{N} F_i F_{2ji} + \sum_{i=0}^{N} G_i G_{2ji} + \tilde{C}_1 = 0; \qquad (2)$$

$$\sum_{i=0}^{N} \frac{1}{i+1} G_i = 0,\ 0 \le j \le N.$$

1.3 Results

Table 1. Critical compressive stress for small dimensionless distance between a crack and free surface in half-space with penny-shaped crack.

β	Critical compressive stress (σ_{11}^0/E)	A
$1 \cdot 10^{-2}$	$-1.268 \cdot 10^{-4}$	-1.268
$1 \cdot 10^{-3}$	$-1.290 \cdot 10^{-6}$	-1.290
$1 \cdot 10^{-4}$	$-1.287 \cdot 10^{-8}$	-1.287
$1 \cdot 10^{-5}$	$-1.285 \cdot 10^{-10}$	-1.285
$1 \cdot 10^{-6}$	$-1.1284 \cdot 10^{-12}$	-1.284
$1 \cdot 10^{-9}$	$-1.284 \cdot 10^{-18}$	-1.284

Table 2. Critical compressive stress for small dimensionless distance between cracks in material with two parallel cracks.

β	Critical compressive stress (σ_{11}^{0}/E)	A	A^{*}
$1 \cdot 10^{-2}$	$-5.01367 \cdot 10^{-4}$	−5.01367	−1,25342
$1 \cdot 10^{-3}$	$-5.16309 \cdot 10^{-6}$	−5.16309	−1,29077
$1 \cdot 10^{-4}$	$-5.15527 \cdot 10^{-8}$	−5.15527	−1,28882
$1 \cdot 10^{-5}$	$-5.14160 \cdot 10^{-10}$	−5.14160	−1,28540
$1 \cdot 10^{-6}$	$-5.13770 \cdot 10^{-12}$	−5.13770	−1,28443
$1 \cdot 10^{-9}$	$-5.13672 \cdot 10^{-18}$	−5.13672	−1,28418

As an example, we considered a laminate composite with isotropic layers. In macro-volumes such composite may be considered as transversely-isotropic medium. In the case considered, cracks are located in plane $x_3 = const$, parallel to interface boundary of layers. Dependence of critical dimensionless compressive stress σ_{11}^{0}/E on ratio of the dimensionless distance between a crack and free surface $\beta = h/a$ are given in Figs. 2, Table 1 and Table 2 (for $\nu = \nu_1 = \nu_2$, fiber concentration c_1 is 0.7 and fiber aspect ratio is 10).

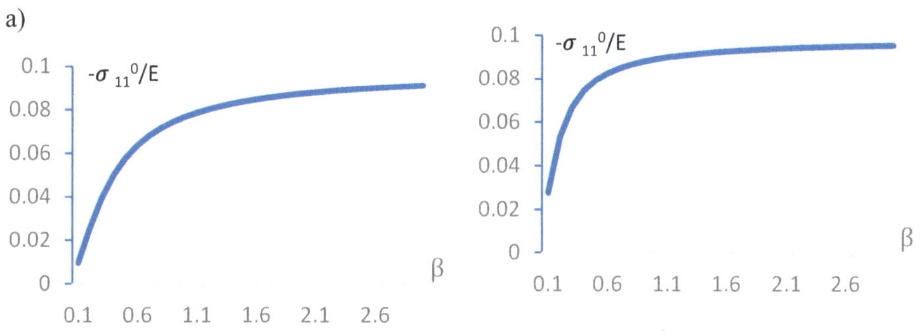

a) a half-space with near-surface crack **b)** a body with two parallel cracks

Fig. 2. Dependence of critical dimensionless compressive stress on ratio of the dimensionless distance between the crack and free surface of composite

For both problems the critical values of σ_{11}^{0}/E at $\beta \to \infty$ go asymptotically to values 0.0965, which are equal to respective critical values at surface instability of half-space.

For a thin plate critical stress can be found as $\sigma_{cr} = A_{cr}\beta^{2}$. For small dimensionless distance between the crack and the free surface the values of coefficient A are presented in Table 1 and Table 2. At both case $\beta = h/a$. For near-surface crack h is three distance between crack and free surface – plate thickness in the "beam approximation". For material with two cracks h is the half-distance between cracks – half of plate thickness in the "beam approximation". For a thin plate we can found critical stress as

$\sigma_{cr} = A^*_{cr} t^2/a^2$, where t is thickness of plate. At the case of near-surface crack $A^*_{cr} = A_{cr}$, for material with two parallel crack $A^*_{cr} = A_{cr}/4$ (A^* in Table 2). Values of A for half-space with near-surface crack are given in Table 1 and values of A^* for material with two parallel crack at $\beta \to 0$ go to same value -1.284.

Critical compressive stress was obtained for composite materials for large and small distances between the crack and free surface (between cracks). Analysis of the results allows to determine the conditions of applicability of the "beam approximation". Beam approximation good work for small distance between the crack and the free surface (when $\beta < 0.01$ computing error less than 1%) and bad work in else case (when $\beta > 0.1$ computing error more than 5%).

References

1. Guz, A.N.: Establishing the foundations of the mechanics of fracture of materials compressed along cracks (review). Int. Appl. Mech. **50**, 1–57 (2014)
2. Obreimoff, I.W.: The splitting strength of mica. Proc. R. Soc. London **127**, 290–297 (1930)
3. Guz, A.N.: On one criterion of fracture of solids in compression along the cracks. Plane problem. Docl. Akad. Nauk SSSR. **259**, 1315–1318 (1980). (in Russian)
4. Guz, A.N., Dovzhik, M.V., Nazarenko, V.M.: Fracture of a material compressed along a crack located at a short distance from the free surface. Int. Appl. Mech. **47**(6), 627–635 (2011)
5. Guz, A.N., Nazarenko, V.M.: Mechanics of fracture of materials in compression along the cracks (review) highly elastic materials. Sov. Appl. Mech. **25**(9), 3–32 (1989)

Development of Welded Elements of the Railway Freight Car Bogie with Increased Characteristics of Fatigue Resistance and Survivability

L. Lobanov, O. Makhnenko(✉), A. Pustovoy, and S. Solovey

E.O. Paton Electric Welding Institute, NAS of Ukraine, Kiev, Ukraine
makhnenko@paton.kiev.ua

Abstract. In last years in the railroad territory of 1520 mm accidents have often occurred due to the destruction (failure) of the casting elements of three-element freight car bogie. The quality and durability of load-bearing elements, which is traditionally produced by casting technology, is insufficient. The development of new welded designs of the load-bearing elements were carried out on the basis of wide use of mathematical modeling to determine the stress-strain state of the welded elements under the action of regulated loads and to make the assessment of strength according to Ukrainian standards and current world approaches. The accelerated fatigue tests on prototypes were carried out, which showed that the service life of the welded structure of the side frame is ten times longer than the service life of the cast structure, and its survivability is longer in several times.

Keywords: Railway freight car bogie · Low alloy steel · Side frame · Welded joints · Fatigue · Test · Service life

In recent years, on the area of 1520 mm gauge railways the accidents involving the destruction of the casting elements of three-element bogie of freight car has often had. The quality and durability of bogie elements such as the side frame and the beams traditionally made by casting technology are of insufficient level, so it is advisable to look for an alternative - the possibility of using welding technology in the manufacture of these parts to improve the characteristics of fatigue resistance [1].

In the E.O. Paton Electric Welding Institute of the NAS of Ukraine a work on the development of a new all-welded design of the side frame of a three-element bogie of a freight car was carried out on the basis of widespread use of mathematical modeling to determine the stress-strain state of the welded elements of the bogie under the influence of regular loads and the assessment of (fatigue) strength according to current standards [2] and modern approaches [3].

Calculation of the fatigue resistance of the developed welded structure of side frame of the bogie in accordance with the Standard [2] considering the coefficients of fatigue resistance for different evaluation zones (base metal and welds) taking into account the distribution of the coefficient of vertical dynamics over the operating speed ranges (range of operating speeds) taking into account the additional load spectrum

E. Gdoutos and M. Konsta-Gdoutos (Eds.): ICTAEM 2020, STIN 16, pp. 273–279, 2020.
https://doi.org/10.1007/978-3-030-47883-4_49

from the longitudinal compressive forces through the clutch, showed that the developed welded structure is workable with variable loading and rear frees the requirements of Standard [2] with the coefficient of fatigue resistance [n] = 2, both under the condition of not exceeding the calculated stresses of the magnitudes of the allowed stress amplitudes and under the condition of damage accumulation.

The fatigue resistance of the structure is estimated by the formula:

$$n = \frac{\sigma_{a,N}}{\sigma_{a,\ni}} \geq [n], \tag{1}$$

where $\sigma_{a,N}$ is the endurance limit (in amplitude) for a symmetric loading cycle on the base $N_0 = 10^7$ cycles; $\sigma_{a,e}$ - the calculated value of the amplitude of the dynamic stress of the conditional symmetric cycle, equivalent to the damaging effect of the real mode of operational stresses for the life of the part; *[n]* - the permissible minimum value of the fatigue resistance factor is taken in accordance with the Standard [2] for the new structure element of bogie - [n] = 2,0.

The endurance limit (in amplitude) for a symmetrical load cycle is determined by [2]:

$$\sigma_{a,N} = \frac{\overline{\sigma_{-1}}}{\overline{K}_\sigma}\left(1 - z_p v_\sigma\right), \tag{2}$$

where $\overline{\sigma_{-1}}$ is the average endurance limit of a smooth standard specimen; z_p is a quantile of the distribution $\sigma_{a,N}$ as a random variable; v_σ is the coefficient of variation of the endurance limit; $\overline{K_\sigma}$ - the average value of the reduction factor of the endurance limit of a part relative to the endurance limit of a smooth standard pattern.

The calculated value of the dynamic stress amplitude of the conditional symmetric cycle, equivalent to the damaging effect of the real mode of operational stresses over the life of the structure element, is calculated taking into account the distribution of the vertical dynamics coefficient over the 10 speed ranges [2]:

$$\sigma_{a,e} = \max(\sigma_a) \cdot \sqrt[m]{\frac{T_p f_e}{N_0} \sum_{i=1}^{10} P(v_i) k_i^m}, \tag{3}$$

where m is the slope of the fatigue strength curve in the amplitudes; T_p - the total time of dynamic load over the estimated life of the element; f_e - the effective frequency of the process of changing dynamic loads for the trimmed parts $f_e = \sqrt{4c/m_k}$, where c is the vertical rigidity of the spring kit under the gross weight of the car; m_k - weight of the loaded car body; N_0 is the base number of cycles of dynamical stresses; i - a speed range counter; $P(v_i)$ - the fraction of time in operation in the *i-th* speed range, v_i - the average value of the speed in the *i-th* range; k_i - the coefficient of vertical dynamics in the *i-th* speed range; $max(\sigma_a)$ - the maximum stress amplitude during a symmetrical load cycle.

From the formulas given above, the maximum permissible stress amplitude of the fatigue resistance criterion would be expressed as follows:

$$[\max(\sigma_a)] = \frac{\sigma_{a,N}}{[n]\sqrt[m]{\frac{T_p f_{\text{э}}'}{N_0}\sum_{i=1}^{10} P(v_i)k_i^m}}. \tag{4}$$

In the case of a welded structure of the side frame, the values of fatigue strength (stress amplitudes) are calculated for the several evaluation zones according to the fatigue resistance criterion. To assess the fatigue strength of the side frame the loads were applied corresponding to the mode of normal movement of the car in the train:

- vertical force reduced by the gross weight of the car body;
- transverse component of longitudinal quasi-static force.

The amplitudes of loads are determined by the coefficient of vertical dynamics in the range of speeds till the specified 120 km/h. The coefficient of vertical dynamics is assumed to be the same for the straight and curve sections of the rail path.

Additionally, the assessment of fatigue strength of the welded structure of the bogie side frame in accordance with Recommendations of the International Institute of Welding (IIW) [3] was carried out. The assessment is based on the fatigue fracture (macro crack) observation in different zones of the structure (zones of welded joints), taking into account the specified stress spectrum of load during long-term operation [2]. Results of assessment showed that the developed welded design of the side frame has a sufficient level of fatigue resistance with a safety factor γ_M = 1,1–1,4.

The Recommendations of IIW summarize a large amount of experimental research for typical welded joints, which allowed for each type of joint and load direction to formulate a methodology for determining the fatigue strength at regular loading in the form:

$$[\Delta\sigma] = \frac{FAT \cdot f_1(R) \cdot f_2(N) \cdot f_3(\partial) \cdot f_4(T)}{\gamma_M}, \tag{5}$$

where FAT – fatigue class or fatigue strength of classified welded joint at $2 \cdot 10^6$ cycles of regular loading (constant parameters of load cycle) at $f_1 = f_2 = f_3 = f_4 = \gamma_M = 1$, γ_M – safety factor. There is a table of FAT values for different classified welded joints [3].

Fatigue enhancement factor $f_1(R)$ depends on the stress ratio in a particular stress cycle and the level and direction of residual stresses. For stress relieved welded components with negligible residual stresses ($<0{,}2\sigma_T$), σ_T – Yield Stress (for low alloy steel 09Г2С $\sigma_T \approx 390$ MPa):

$f_1(R) = 1{,}6$ for $R < -1{,}0$;
$f_1(R) = -0{,}4R + 1{,}2$ for $-1{,}0 \leq R \leq 0{,}5$;
$f_1(R) = 1{,}0$ for $R > 0{,}5$.

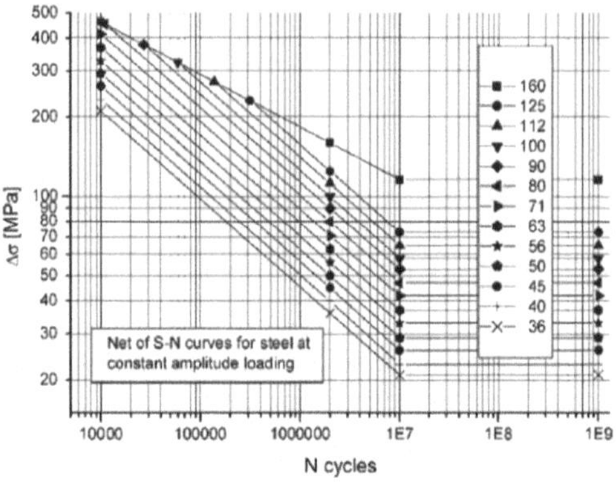

Fig. 1. Fatigue resistance S-N curves [3] for different classes FAT of welded joints (material – steel) for normal nominal stress at $N < 10^9$ cycles

For complex two- or three-dimensional components, components with global residual stresses, thick walled components like a welded side frame of railcar bogie $f_1(R) = 1{,}0$ and no enhancement is considered.

In the range of $10^4 < N < 10^8$ cycles the factor $f_2(N)$ (Fig. 1) is defined by equation:

$$f_2(N) = \left(\frac{C}{N}\right)^{\frac{1}{m}},$$

where N – the number of cycles to failure, $C = 2{\cdot}10^6$, $m = 3$ for $10^4 < N < 10^7$ cycles and $C = 5{,}8{\cdot}10^6$, $m = 5$ for $10^7 < N < 10^8$ cycles.

The influence of plate thickness on fatigue strength should be taken into account in cases where plates thicker than 25 mm. For weld joints of side frame with thickness of elements till 25 mm the correction factor can be taken equal $f_3(\delta) = 1{,}0$.

The reduction factor $f_4(T)$ is taken in consideration at elevated operation temperatures T. For $T < 100$ °C can be taken $f_4(T) = 1{,}0$.

The safety factor is recommended to take in the range $\gamma_M = 1\ldots1{,}4$ [3]. In most cases for normal fabrication quality and regular inspection in service, $\gamma_M = 1$ might be adequate. Safety factor value $\gamma_M = 1{,}4$ corresponds to the consequence of failure – loss of human life.

The load spectrum for the calculation of fatigue resistance is determined by the coefficient of vertical dynamics and the probability of movement in the range of speeds up to the maximum design (120 km/h).

Therefore, for the welded joints of the side frame the dependence (5) for determining the fatigue strength at regular loading can be represented as:

$$[\Delta\sigma] = \frac{FAT}{\gamma_M}\left(\frac{C}{N}\right)^{\frac{1}{m}}. \qquad (6)$$

Accordingly, the ultimate durability *[N]* at regular loading with a stress range $\Delta\sigma$ is expressed as follows

$$[N] = C \cdot \left(\frac{FAT}{\Delta\sigma \cdot \gamma_M}\right)^m. \tag{7}$$

At variable amplitude loading (stress spectrum), linear damage calculation by "Palmgren-Miner" summation is appropriate.

Table 1 summarizes the allowable amplitudes of the stresses (fatigue strength) in accordance with the National Standard [2] and IIW Recommendations [3] in the different zones of the side frame welded structure, taking into account the specified [2] stress spectrum of load at 10^7 cycles during long-term operation. Comparison shows insufficient conservatism of the Standard in assessment of fatigue strength of welded joints.

Table 1. Fatigue strength (stress amplitudes) in accordance with the current Standard and IIW Recommendations, and calculated static stresses in different zones of the welded side frame

	Zone of the side frame structure	Standard [2]		IIW Recommendations [3]	Calculation
		Reduction factor of endurance limit $\overline{K_\sigma}$	Fatigue strength (in amplitude) max(σa), MPa	Fatigue strength (in amplitude) [σa], MPa $\gamma_M = 1/1{,}4$	Maximum principal stresses σa, MPa
1	Base material of bottom belt in R55 of axle box pedestal	1,5	150	–	81
2	Longitudinal fillet weld joint of sidewall and coil springs support plate	3,0	78	40/50	56
3	Transverse fillet weld joint of axle box support	3,0	78	40/50	37
4	Longitudinal fillet weld joint of sidewall and bottom belt in R55 of axle box pedestal	3,0	78	40/50	53
5	Transverse butt weld joint of sidewall	4,7	51	44/56	33

Also in Table 1 the calculated static stresses in the different zones of the side frame due to acting of maximum design operational load *23,5 tf* are presented. In the most loaded areas of the side frame welded structure to provide the enough level of fatigue resistance it is advisable to apply additional improvement of residual stress conditions (stress relieving thermal treatment, hammer-peening) in the transition zone of the longitudinal welds (pos.2 and 4).

The prototypes of welded side frames were manufactured and accelerated fatigue testing were carried out (Fig. 2) with a cycle stress amplitude increased in 2 times. After $N = 8.8$ million cycles of vertical loading (amplitude $P_{ai} = 245\ kN = 25\ tf$, constant average cycle load $P_m = 363\ kN = 37\ tf$), a macro crack formed on the side wall of the frame between the lower belt and the support plate of the springs. The prototypes of the welded side frame showed a high value of durability, which is ten times greater than the design life of the cast structure of the side frame with an axial load of 23.5 tons. After $N = 13$ million cycles the failure along the transverse weld joint of the sidewall occurred (Fig. 2).

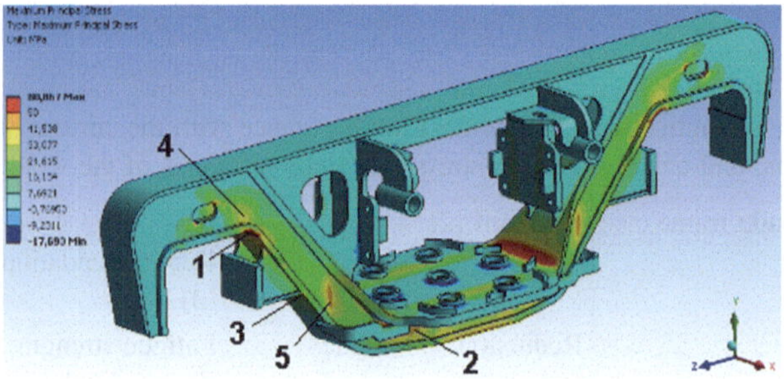

Fig. 2. Distribution of maximum principal stresses in the structure of welded side frame under the action of maximum design operational load *23,5 tf*

A prototype of the welded side frame demonstrated a high value of the relative survivability *(Ni-Ni_cr)/Ni* = 0,32, and in absolute value survivability of welded structure exceeds several times the service life of the cast structure of the side frame (Fig. 3).

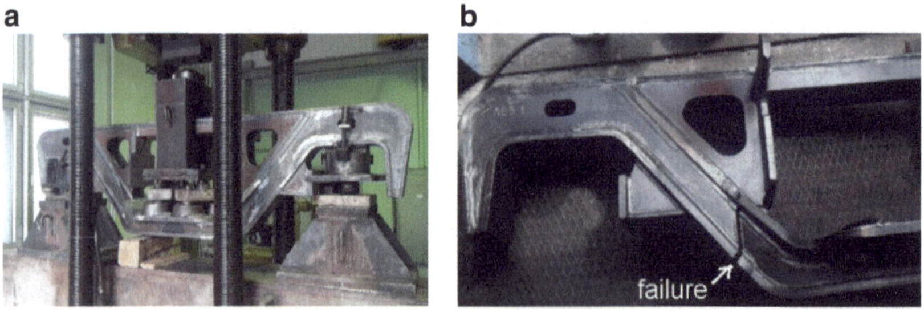

Fig. 3. Prototype of side frame during accelerated fatigue test (a) and failure in the weld zone (b)

1 Conclusions

1. Welded structures of load-bearing elements of freight rail cars bogie designed within the requirements of the National Standard may not have sufficient durability, because the exist Standard are not sufficiently conservative. The design of new welded structures must be carried out in accordance with the current Recommendations of the International Welding Institute for fatigue design of welded joints and components.
2. The results of the accelerated tests of the prototypes of the developed side frame welded structure of a three-element rail freight car bogie with an axial load of 23,5 ts showed a significant increase in service life and survivability, higher in tens time in comparison with a side frame of cast design at operational cyclic loads.

References

1. Makhnenko, O.V., Saprykina, G.J., Mirzov, I.V., Pustovoi, A.D.: Prospects for the creation of welded structures of load-bearing elements of a freight car bogie. J. Autom. Weld. **3**(3), 36–42 (2014). (in Russian)
2. GosNIIV – VNIIGT, M.: Standards for the calculation and design of railroad cars of the Ministry of Railways of 1520 mm gauge (non-self-propelled), p. 317 (1995). (in Russian)
3. Recommendations for Fatigue Design of Welded Joints and Components. IIW Doc XIII-1965к14-03/XV-1127к14-03, p. 147, 10 July 2006

Electron Beam Additive Technology Optimization Using Mathematical Modeling

Oleh Makhnenko[1(✉)], Nikita Ananchenko[1], Stepan Kandala[1], and Andriy Babenko[2]

[1] E.O. Paton Electric Welding Institute of NASU, Kyiv 03680, Ukraine
makhnenko@paton.kiev.ua
[2] Igor Sikorsky Kyiv Polytechnic Institute, NTUU, Kyiv 02056, Ukraine

Abstract. Additive manufacturing technology are developing fast these days. Economical profit of manufacturing parts with complex geometry which have either hollow section or complex three-dimensional curvature, using 3D printers are higher than using standard manufacturing methods. Which means that 3D printer could be an alternative to traditional ways of manufacturing. This technology used in aerospace and medicine right now, and have a future in other industries. Currently the technology of wire electron *xBeam 3D Metal Printer* of parts with complex geometry from titanium alloy has been developing in Ukraine. Using of mathematical modeling for optimization of the technology parameters is very promising.

Keywords: Additive technology · Mathematical modelling · Microstructure · Residual stresses and deformations · Thermal fields

1 Introduction

Additive manufacturing is becoming an alternative to traditional manufacturing methods for titanium parts and structural components these days, in aerospace and medicine industries first of all. Additive technology makes manufacturing of parts with complex three-dimensional surface curvature or limited edition parts with nonstandard dimensions much cheaper and easier. Using of filler wire makes additive manufacturing process much faster, electron beam heating sources are efficient in vacuum condition that also is advisable for preventing titanium alloy from oxidation. But using of wire in the manufacturing process leads to a high level of residual stresses and distortions in structure element. Samples, made on 3D printer, could be seen on Fig. 1 [1]. Optimization of additive manufacturing process, using mathematical modeling and FEM analysis, makes sense these days.

2 Microstructure

Mechanical properties of titanium alloy Ti-6Al-4V depends on its microstructure: grain size [2, 3] and percentage of α'' phase in $\alpha + \beta$ structure [4]. Grain size defined by cooling rate during polymorphic transformation [5]: the higher cooling rate, the less

E. Gdoutos and M. Konsta-Gdoutos (Eds.): ICTAEM 2020, STIN 16, pp. 280–286, 2020.
https://doi.org/10.1007/978-3-030-47883-4_50

(a)

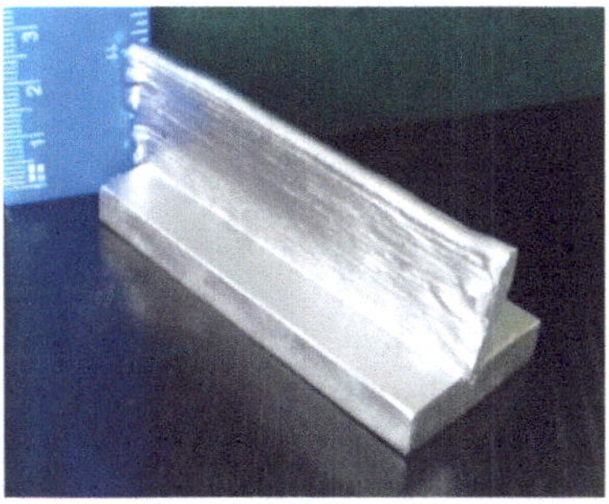
(b)

Fig. 1. Samples, formed by xBeam 3D metal printer technology, with small (a) and large (b) substrates.

grain size (Eq. 1) [2]. Percentage of martensitic α″ phase also defined by cooling rate: when cooling rate is 410 °C/s and higher – fully α″ structure formed; in interval from 20°C/s to 410°C/s – α″ + α+β structure formed, and percentage of α″ phase linearly decreases, with decreasing of cooling rate from 410 °C/s to 20 °C/s; when cooling rate is 20 °C/s and less – α + β structure formed [4].

Mechanical properties of material, such as yield strength and elongation, defined by grain size and percentage of α″ phase in titanium alloy Ti-6Al-4V. Yield strength of material defined, mostly, by grain size: increasing of cooling rates, which means decreasing of grain size, leads to growth of yield strength [3, 6]; but, as it mentioned above, too high cooling rates leads to α″ phase formation, which reduces yield strength a little bit (Eq. 2) [4]. Though, impact of α″ phase percentage, on strength characteristics much less, than grain size. Elongation of material mostly defined by quantity of martensitic α″ phase in titanium alloy: increasing of α″ phase percentage in alloy from 0% to 100% leads to linear decreasing of elongation from 11% to 2% (Eq. 3) [4].

$$d_{grain} = \begin{cases} 8220 \cdot V_{cooling}^{-0,58},\ \mu m\ where\ V_{cooling} \geq 6\ {}^{\circ}C/s \\ 2500,\ \mu m\ where\ V_{cooling} < 6\ {}^{\circ}C/s \end{cases} \tag{1}$$

$$\sigma_y = 905 + 8,757 \cdot 10^6 \cdot d_{grain}^{-1,724} - 100 \cdot v_{\alpha'}, MPa \tag{2}$$

$$\delta = -9 \cdot v_{\alpha'} + 11, \% \tag{3}$$

3 Mathematical Model

3.1 Mathematical Model for Temperature Fields

Finite element model of printed samples with small and large substrate was created, to calculate thermal fields during layer surfacing of a sample, using heat equation (Eq. 4) and heat sink boundary conditions (Eq. 5).

$$c\gamma(T) \cdot \frac{\partial T}{\partial t} = \nabla[\lambda(T) \cdot \nabla T] \tag{4}$$

$$-\lambda(T) \cdot \frac{\partial T}{\partial n} = \begin{cases} \alpha_T \cdot (T - T_{out}), \textit{heat transfer with machining attachment} \\ \varepsilon \cdot \sigma_{SF} \cdot \left(T^4 - T_{out}^4\right) - q, \textit{thermal radiation from surfaces} \end{cases} \tag{5}$$

Also, was created the thermo elastic plastic FE analysis model to calculate residual stresses and distortions after finish of forming and complete cooling of a sample.

3.2 Comparing of 2D and 3D Thermal Field Models

Two dimensional thermal field model can be used for assumption of the fast moving heat source, which allows to reduce time of calculation sufficiently. Calculation's accuracy is, from one side, increasing because of smaller mesh size, but from other side decreasing, because 2D thermal field model don't consider terminal effects on the beginning and end of surfaced layer. Results of 2D and 3D solution for typical example of titanium alloy Ti-6Al-4V sample forming (electron beam power 4,5 kW, wire diameter 1,6 mm, cladding speed 14 mm/s, dimensions of substrate 8 × 30 × 70 mm, wall width 3,2 mm), are compared on Fig. 2.

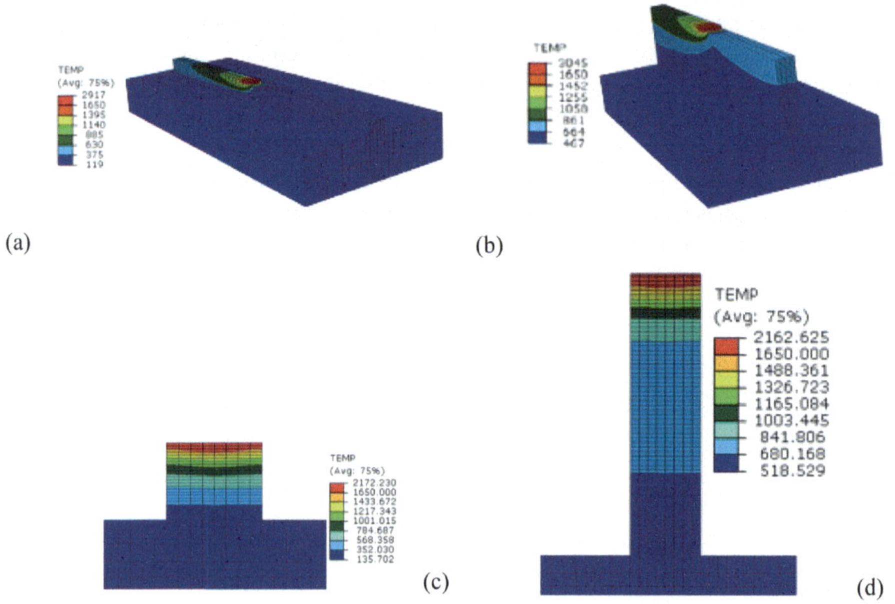

Fig. 2. Comparing of 3D (a, b) and 2D (c, d) thermal field models on example of 5th (a, c), and 25th (b, d) layer surfacing.

3.3 Results for Residual Stresses and Distortions

Residual stresses and displacements in printed structure elements could limit usage of additive technology. After surfacing of 25 layers and complete cooling of a sample, residual deflection on a 70 mm long sample with large substrate are less than 0,4 mm, when its height 21 mm; and for sample with small substrate - are more than 3 mm, when its height 16 mm (Fig. 3a, b). Longitudinal component has the highest level of residual stresses. For large substrate sample, maximum residual tensile stress reach's 480 MPa, and for small substrate sample – 280 MPa (Fig. 3c, d). Because of small substrate sample has less transverse stiffness than the large substrate one, residual stresses in it is less almost in two times, but residual deflection almost in ten times more, which can lead to culling this kind of samples.

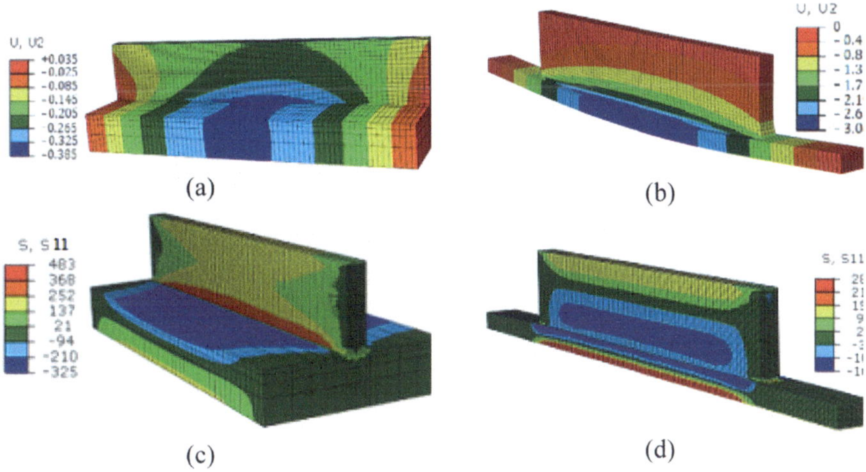

Fig. 3. Residual deflections (a, b) and longitudinal stresses (c, d) of samples with large (a, c) and small (b, d) substrate.

4 Optimization of Technology, Varying Time Δt and Temperature

Because of irregular heat sink from each formed layer, which related to cladding first layers on a cold substrate, and last ones – on heated metal, high cooling rates received on first layers, and as a result – small grain size with big percentage of α'' phase; and on the last layers received low cooling rates and large grain size (about 2500 μm). Which causes low plasticity of material on first layers, and low strength characteristics on last layers.

Delay time Δt between cladding of each layer is one of factors to change for getting efficient technology parameters. Heat sink on last layers, comes on stationary mode with increasing of time Δt, on example of sample with large substrate. So, grain size and yield strength distribution becomes more regular on the last layers, with increasing of Δt (Fig. 4a, b). Considering, that Δt = 24 s mode will be optimal, because Δt = 60 s mode leads to higher level of residual stresses and deformations.

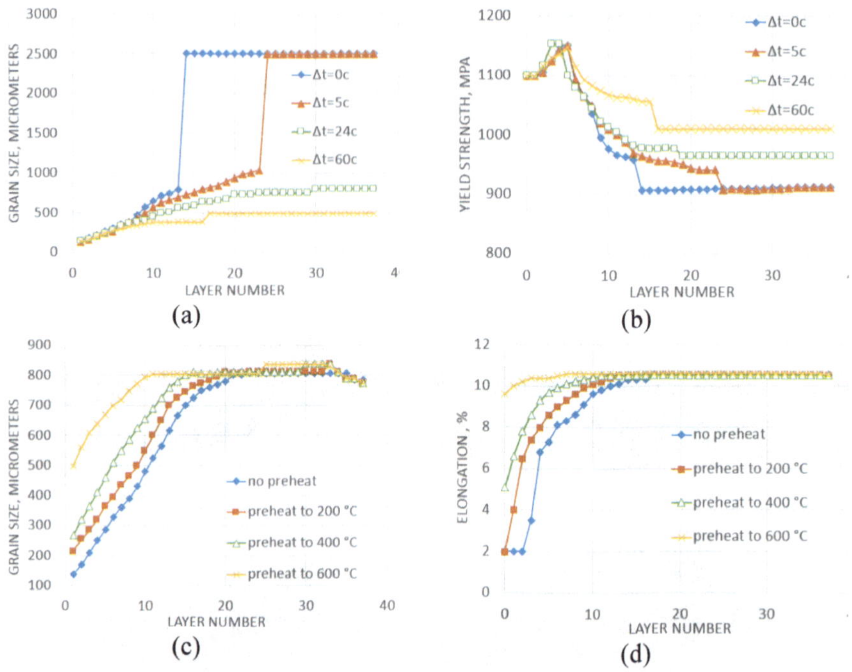

Fig. 4. Relation between grain size (a, c), yield strength (b), elongation (d) and layer number for different delay time Δt (a, b) and preheat temperature (c, d).

Substrate preheat is also one of process parameters to change, for getting optimal microstructure. Increasing of preheat temperature until 600 °C, with combination of delay time Δt = 24 s, leads to stationary heat sink mode on first layers, which means more regularly distribution of α'' phase, grain size, and elongation of material (Fig. 4c, d).

Comparing of Residual Stresses and Distortions for Different Technology Parameters. In Table 1 shown modeled residual stresses and deflections of samples with large and small substrates, with using different delay time Δt between layer cladding. Results shows that too long delay time Δt affects on residual deflection negatively.

Table 1. Residual stresses and deflections of samples, made with different delay time Δt.

	Delay time Δt, s	Residual deflection, mm	Residual stresses, MPa			
			Mises	Longitudinal	Transverse	On height
Small substrate	5	3.07	302	258/−166	83/−115	87/−102
	24	3.81	622	472/−245	98/−142	91/−94
	60	4.11	844	642/−410	149/−222	117/−98
Large substrate	5	0.39	778	483/−325	350/−189	175/−180
	24	0.43	875	505/−364	360/−286	193/−183
	60	0.48	872	423/−411	420/−288	190/−160

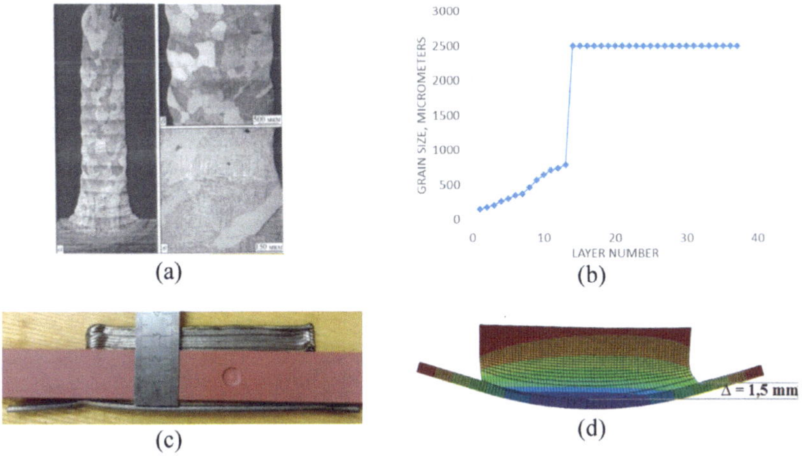

Fig. 5. Macrosection of manufactured sample (a), modeled grain size (b), residual deflection of manufactured sample (c) and modeled residual deflection (d).

5 Comparing with Experiment

Grain size was compared on a samples with large substrate, using initial technology parameters, i.e. without extra delay time (Δt = 0 s) and without substrate preheat. In result of experiment, macrosection analysis gives grain size 200–300 μm on first layers, and 400–600 μm on middle layers (Fig. 5a) [1]. Modeled grain size on first layers is 150–250 μm and 400–700 μm on middle layers (Fig. 5b).

Also, residual deflection of manufactured sample with small substrate was compared with modeled deflection. Modeled residual deflection of sample with small substrate equals to 1,5 mm (Fig. 5d) and shows adequately correlation with deflection of manufactured sample – about 2 mm (Fig. 5c).

References

1. Kovalchuk, D., Melnyk, V., Melnyk, I.: New capabilities of additive manufacturing with xBeam 3D Metal Printing technology. In: College of Works of Eight International Conference on Beam Technology in Welding and Material Handling, Kiev, Ukraine, p. 51 (2017)
2. Polkin, I.: Progressive technologies and new processes. In: Progressive Technologies of Light and Special Alloys, Fizmatlit, Moscow, pp. 66–73 (2006)
3. Lutjering, G., Williams, J.C.: Microstructure and properties of materials V.2. In: Li, J.C.M.: Microstructure and Mechanical Properties of Titanium Alloys, pp. 26–29 (2000)
4. Liu, S.: Additive manufacturing of Ti6Al4V alloy: a review. Mater. Des. **164**, 1–23 (2018)
5. Sieniawski, J., Ziaja, W., Kubiak, K.: Microstructure and mechanical properties of high strength two-phase titanium alloys. In: Sieniawski, J.: Titanium Alloys, pp. 69–79 (2013)
6. Grabin, V.: Structure and Properties of Welded Joints from Titanium Alloys, p. 59. Naukova Dumka, Kiev (1964)

Non-stationary Contact Problems for Thin Shells and Solids

Grigory Fedotenkov[1,2(✉)] and Dmitry Tarlakovskii[2]

[1] Moscow Aviation Institute, National Research University, Moscow, Russia
greghome@mail.ru
[2] Institute of Mechanics, Lomonosov Moscow State University, Moscow, Russia

Abstract. A spatial non-stationary contact problem with moving boundaries of the interaction region for a thin elastic cylindrical shell and an absolutely rigid indenter bounded by a smooth convex surface is considered. A closed mathematical formulation is given and a system of resolving equations is constructed. The system of resolving equations is based on the spatial-temporal integral equation resulting from the principle of superposition and contact conditions. The core of this equation is the transient function for the cylindrical shell. To a closed system of resolving equations, it is supplemented by a kinematic relation for determining the moving boundary of the contact area and the equation of motion of the indenter as an absolutely rigid body. An algorithm for solving the spatial non-stationary contact problem for an infinitely long cylindrical shell and absolutely rigid indenter in the case of a normal impact on the side surface of the shell is constructed and implemented. Examples of calculations are given.

Keywords: Spatial non-stationary contact problems · Cylindrical shell · Superposition method · Fourier series · Integral transforms · Transient functions · Numerical-analytical algorithms

1 Problem Formulation

At the initial instant of time, an absolutely rigid indenter, moving with a given initial velocity V_0, comes into contact with the lateral surface of the infinitely long thin elastic circular cylindrical shell. The indenter is bounded with a smooth convex surface, the vector of its initial velocity is directed normal to the side surface of the shell (Fig. 1).

The equations of motion of the model S.P. Timoshenko in displacements recorded in the main orthogonal coordinates associated with the directions of the main curvatures of the middle surface are used to describe the motion of the shell [1, 2]:

$$\begin{gathered}\frac{\partial^2 W}{\partial \tau^2} = \mathbf{L}W + P,\\ \mathbf{L} = \left(L_{ij}\right)_{5\times 5},\ W = \left(u, v, w, \chi_\alpha, \chi_z\right)^{\mathrm{T}},\ P = (0, 0, p, 0, 0)^{\mathrm{T}},\end{gathered} \tag{1}$$

E. Gdoutos and M. Konsta-Gdoutos (Eds.): ICTAEM 2020, STIN 16, pp. 287–292, 2020.
https://doi.org/10.1007/978-3-030-47883-4_51

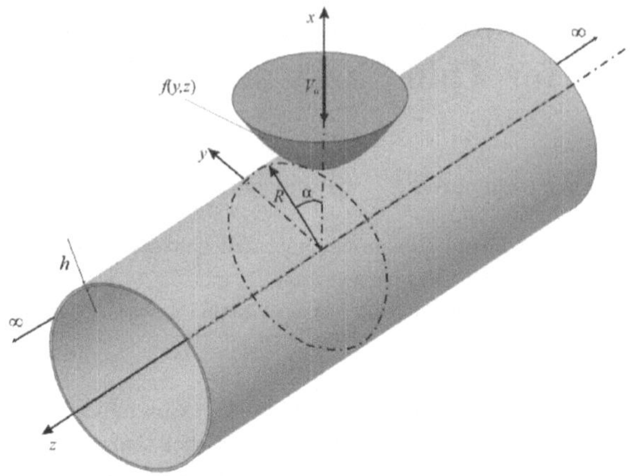

Fig. 1. Problem formulation.

$$L_{11} = \frac{\partial^2}{\partial \alpha^2} + \eta^2\left(\frac{\partial^2}{\partial z^2} - k^2\right),\ L_{12} = \left(1 - \eta^2\right)\frac{\partial^2}{\partial \alpha \partial z},\ L_{13} = \left(1 + k\eta^2\right)\frac{\partial}{\partial \alpha},$$

$$L_{14} = -\gamma^2\frac{\partial^2}{\partial \alpha^2} + \eta^2 k^2,\ L_{15} = -\gamma^2\left(1 - 2\eta^2\right)\frac{\partial^2}{\partial \alpha \partial z},\ L_{21} = L_{12},\ L_{22} = \frac{\partial^2}{\partial z^2} + \eta^2\frac{\partial^2}{\partial \alpha^2},$$

$$L_{23} = \left(1 - 2\eta^2\right)\frac{\partial}{\partial z},\ L_{24} = -\gamma^2\eta^2\frac{\partial^2}{\partial \alpha \partial z},\ L_{25} = -\gamma^2\eta^2\frac{\partial^2}{\partial \alpha^2},\ L_{31} = -L_{13},$$

$$L_{32} = -L_{23},\ L_{33} = \eta^2 k^2\left(\frac{\partial^2}{\partial \alpha^2} + \frac{\partial^2}{\partial z^2}\right) - 1,\ L_{34} = \eta^2 k^2\frac{\partial}{\partial \alpha},\ L_{35} = \eta^2 k^2\frac{\partial}{\partial z},$$

$$L_{41} = \gamma^{-2}L_{14},\ L_{42} = \gamma^{-2}L_{24},\ L_{43} = -\gamma^{-2}L_{34},\ L_{44} = \frac{\partial^2}{\partial \alpha^2} + \eta^2\left(\frac{\partial^2}{\partial z^2} - k^2\gamma^{-2}\right),$$

$$L_{45} = L_{12},\ L_{51} = \gamma^{-2}L_{15},\ L_{52} = \gamma^{-2}L_{25},\ L_{53} = -\gamma^{-2}L_{35},$$

$$L_{54} = L_{12},\ L_{55} = \frac{\partial^2}{\partial z^2} + \eta^2\left(\frac{\partial^2}{\partial \alpha^2} - k^2\gamma^{-2}\right).$$

u, v – tangential displacements of the shell in the direction of the coordinate lines α (angular coordinate) and z (longitudinal coordinate) respectively, w – normal displacement of the shell χ_α, χ_z – angles of rotation of the normal fiber, p – normal pressure, τ – dimensionless time, $\mathbf{L}$ – matrix-operator, $k^2 = 5/6$ – shear coefficient.

Here and further all functions, variables, and parameters are reduced to a dimensionless form using a system of dimensionless quantities (primed variables correspond to dimensional ones):

$$u = \frac{u'}{R}, v = \frac{v'}{R}, w = \frac{w'}{R}, u_c = \frac{u'_c}{R}, z = \frac{z'}{R}, \tau = \frac{c_1 t}{R}, \eta^2 = \frac{c_2^2}{c_1^2}, c_1^2 = \frac{\lambda + 2\mu}{\rho},$$

$$c_2^2 = \frac{\mu}{\rho}, \gamma^2 = \frac{h^2}{12R^2}, p = \frac{p'R}{h(\lambda + 2\mu)}, V_0 = \frac{V'_0}{c_1}, m_u = \frac{m'_u c_1^2}{R^2 h(\lambda + 2\mu)}.$$

ρ – density of the shell material, λ, μ – Lame parameters of the shell material, R and h – radius and thickness of the shell, u_c – displacement of the center of mass of the indenter, c_1 and c_2 – speed of the tension-compression waves and shear waves in the shell material, m_u – the mass of the indenter, t – dimensional time.

The equation of motion of the striker as an absolutely rigid body has the following form:

$$m_u \ddot{u}_c(\tau) = P(\tau) = \iint\limits_{\Omega(\tau)} p(\xi, \zeta, \tau) d\xi d\zeta, \ u_c(0) = 0, \ \dot{u}_c(0) = V_0. \tag{2}$$

P – the resulting force of contact pressure p, $\Omega(\tau)$ – the contact area, the boundary of which depends on time.

We assume that the contact between the shell and the striker occurs under conditions of free slippage.

Let Π_1 and Π_2 be the surfaces bounding the shell and the striker at the moment of time $\tau > 0$. With respect to an inertial Cartesian rectangular coordinate system whose axis coincides with the axis of the shell, they are given by the equations

$$\Pi_k : x = f_k(y, z, \tau), (y, z) \in \Omega_k, k = 1, 2,$$

where Ω_k – projections of surfaces Π_1 and Π_2 on a plane Oyz.

Then the true contact area is defined by the following conditions:

$$\Pi_* : \Delta(y, z, \tau) = f_1(y, z, \tau) - f_2(y, z, \tau) = 0, \ p(y, z, \tau) < 0, (y, z) \in \Omega_k, \tag{3}$$

where Δ – is the gap between the boundary surfaces Π_1 and Π_2.

The equations that implicitly define the boundary $\partial\Pi_*$ of the contact area follow from (3):

$$\partial\Pi_* : \Delta(y, z, \tau) = 0, p(y, z, \tau) = 0$$

In the linearized formulation of the problem, the true contact area Π_* is replaced by a fictitious area $\Omega(\tau)$ belonging to the plane that is a common tangent to the surfaces Π_1 and Π_2 at the initial time of contact interaction (Fig. 2).

Suppose that the position of the surface bounding the striker at the current time τ is given by an explicit equation in the coordinate system $Oxyz$:

$$x = f(y, z) + u_c(\tau). \tag{4}$$

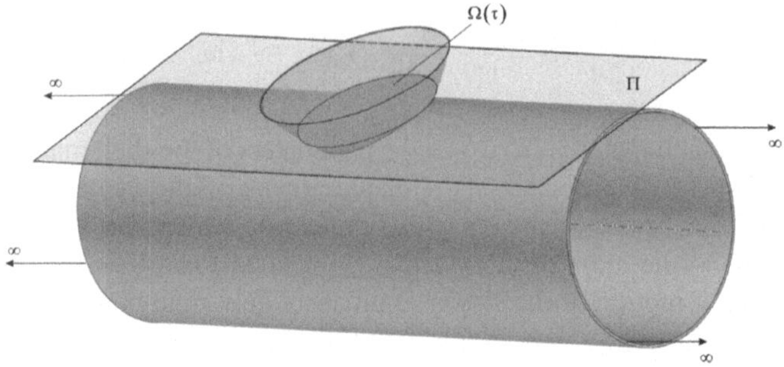

Fig. 2. Contact area.

The boundary $\partial\Omega(\tau)$ of the area $\Omega(\tau)$ is determined from the condition of intersection of the surface bounding the striker with the plane Π:

$$\partial\Omega(\tau) : f(y,z) + u_c(\tau) = 1. \tag{5}$$

Taking into account (4), the contact conditions of the indenter and the shell have the form

$$f(y,z) + u_c(\tau) = w(\alpha,z,\tau),\ p(\alpha,z,\tau) < 0,\ y,z \in \Omega(\tau),\ y = \sin\alpha. \tag{6}$$

At the initial time, the shell is in an unreformed state, which corresponds to zero initial conditions

$$u|_{\tau=0} = v|_{\tau=0} = w|_{\tau=0} = \chi_\alpha|_{\tau=0} = \chi_z|_{\tau=0} = \dot{u}|_{\tau=0} = \dot{v}|_{\tau=0} = \dot{w}|_{\tau=0} = \dot{\chi}_\alpha|_{\tau=0} = \dot{\chi}_z|_{\tau=0} = 0. \tag{7}$$

Equations and relations (1), (2), (4)–(7) are a closed mathematical formulation of the non-stationary contact problem with movable boundaries for a cylindrical shell and an absolutely rigid indenter.

2 Calculation Examples

To solve the system of resolving equations, we use a numerical-analytical algorithm based on the method of mechanical quadratures [3–6].

As an example, consider the non-stationary contact interaction of a circular cylindrical shell ($R = 1$, $h = 0.025$), with an absolutely rigid indenter bounded by a surface that has the shape of rotating paraboloid: $f(y,z) = y^2 + z^2 + 1$. As the shell material, we use steel with the following parameters: $\eta = 0.53$, $\gamma = 1.3 \times 10^{-6}$.

In Fig. 3a–3c shows the dependence of the movement of the center of mass of the striker on time (Fig. 3a), the radius of the contact area from time (Fig. 3b) and the

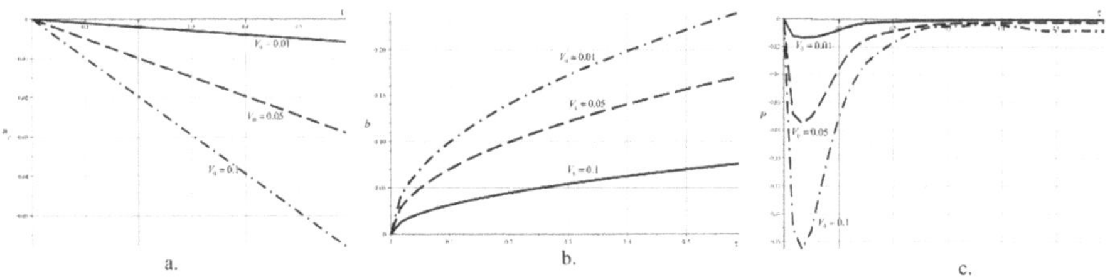

Fig. 3. Dependences of the movement of the center of mass of the indenter, the radius of the contact area and the resulting contact force on time.

resulting contact force from time (Fig. 3c). Here, the solid curves correspond to $V_0 = 0.01$, dashed $V_0 = 0.05$, dash-dotted – $V_0 = 0.1$.

In Fig. 4a–4c shows the distribution of contact pressure in the angular coordinate α at the time $\tau = 0.3$. Figure 4a corresponds to the coordinate value $z = 0$, 4b – $z = 0.15$, 4c – $z = 0.3$. Solid curves correspond to $V_0 = 0.01$, dashed $V_0 = 0.05$, dash-dotted – $V_0 = 0.1$.

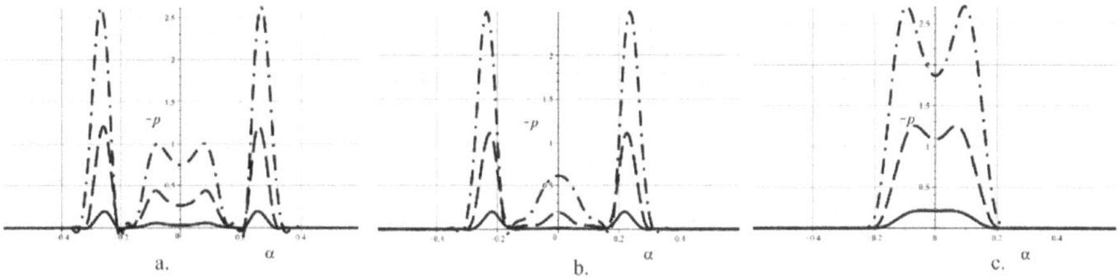

Fig. 4. Distribution of contact pressure by angular coordinate at a time.

Figure 5a and 5b represent distributions of normal displacements along the angular coordinate (Fig. 5a) and along the longitudinal coordinate (Fig. 5b). Here, the results correspond to the time $\tau = 0.6$ and the initial speed $V_0 = 0.1$. In Fig. 5a, the solid curves correspond to $z = 0$, dashed $z = 0.15$, dash-dotted – $z = 0.3$. Solid curves in Fig. 5b correspond to $\alpha = 0$, dashed $\alpha = 0.15$, dash-dotted – $\alpha = 0.3$.

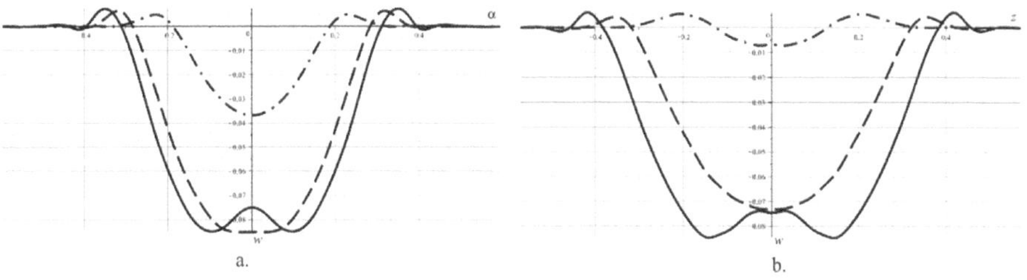

Fig. 5. The distribution of normal displacements along the angular and longitudinal coordinates.

Acknowledgments. This work was financially supported by the Russian Foundation for Basic Research (RFBR), project 19-08-00438 A.

References

1. Fedotenkov, G.V., Kalinchuk, V.V., Mitin, A.Y.: Three-dimensional non-stationary motion of Timoshenko-type. Lobachevskii J. Math. **40**(3), 311–320 (2019)
2. Fedotenkov G.V., Tarlakovskii D.V., Mitin A.Y.: Transient spatial motion of cylindrical shell under influence of non-stationary pressure. In: Gdoutos E. (eds) Proceedings of the Second International Conference on Theoretical, Applied and Experimental Mechanics. ICTAEM 2019. Structural Integrity, vol. 8, pp. 264–269. Springer, Cham (2019)
3. Tarlakovskii, D.V., Fedotenkov, G.V.: Two-dimensional nonstationary contact of elastic cylindrical or spherical shells. J. Mach. Manuf. Reliab. **43**(2), 145–152 (2014)
4. Tarlakovskii, D.V., Fedotenkov, G.V.: Fedotenkov nonstationary 3D motion of an elastic spherical shell. Mech. Solids **50**(2), 208–217 (2015)
5. Fedotenkov, G.V., Mikhailova, E.Y., Kuznetsova, E.L., Rabinskiy, L.N.: Modeling the unsteady contact of spherical shell made with applying the additive technologies with the perfectly rigid stamp. Int. J. Pure Appl. Math. **111**(2), 331–342 (2016)
6. Mikhailova, E.Yu., Tarlakovskii, D.V., Fedotenkov, G.V.: The unsteady contact interaction problem of an absolutely rigid body and a membrane. In: Gdoutos E. (eds) Proceedings of the Second International Conference on Theoretical, Applied and Experimental Mechanics. ICTAEM 2019. Structural Integrity, vol. 8, pp. 289–293. Springer, Cham (2019)

Shear Buckling Mode and Failure of Sandwich Specimen Facing Layer Under Four-Point Bending

V. N. Paimushin[1,2(✉)], M. V. Makarov[1,2], S. A. Kholmogorov[1], and M. A. Shishov[1]

[1] Kazan National Research Technical University named after A.N. Tupolev, K. Marks Str. 10, 420111 Kazan, Russia
vpajmushin@mail.ru

[2] Kazan Federal University, Kremlyovskay Str. 18, 420008 Kazan, Russia

Abstract. The geometrically and physically nonlinear problem of four-point bending of a sandwich specimen with a transversally flexible core and facing layers from fiber reinforced plastic, characterized by a nonlinear relationship between transverse shear stresses and the corresponding shear strains, is formulated. The statement of the problem is given taking into account the contact interaction of the facing layers with the support and loading rollers, a numerical method for solving it is developed based on the finite sum method (method of integrating matrices). Equations based on the refined kinematic model of S. P. Tymoshenko taking into account the transverse compression of the facing layers and the equations of elasticity theory for the core, simplified in the framework of the model of the transversally flexible layuer. The latter allow integration along the transverse coordinate when introducing as unknown transverse shear stresses in the core, constant in thickness. An investigation of the prebuckling and postbuckling behavior of the specimen was carried out, based on the method of continuing the solution with respect to the parameter, when the transverse shear strain in the loaded facing layer was chosen as the parameter. It was shown that during tests for four-point bending of specimens of the class under consideration, their failure can be due to the implementation of transverse-shear buckling modes of the facing layer in the vicinity of the loading roller.

Keywords: Sandwich specimen · Transversely soft core · Facing layer · Fiber reinforced plastic · Four-point bending · Integration matrices method · Iterative process · Geometrically and physically nonlinearity · Parameter continuing method · Shear buckling mode · Postbuckling behaviour

1 Introduction

The most rational use of fiber reinforced plastics (FRP) is possible in structural elements of a sandwich structure. Their facing (loaded) layers made of FRP have, as a rule, low transverse shear stiffness. If compressive stresses are formed in them during loading, then one of the mechanisms of their failure is the transverse-shear buckling

E. Gdoutos and M. Konsta-Gdoutos (Eds.): ICTAEM 2020, STIN 16, pp. 293–300, 2020.
https://doi.org/10.1007/978-3-030-47883-4_52

mode. In the mechanics of composites, theoretical studies of this buckling mode, which is realized in standard specimens under compression, were carried out in different approximations in many works [1, 2, etc.]. In accordance with the results of these works, in order to identify the described buckling mode, which can also be realized in sandwich structures with facing layers of fibrous composites, the construction of the corresponding theory requires taking into account the transverse shear strains in the facing layers. Such refined geometrically and physically nonlinear equations of the theory of sandwich plates and shells with facing hard layers of fibrous composites and a medium transversely flexible layer were constructed in [3].

Tests of specimens for four-point bending according to the scheme shown in Fig. 1 are one of the types of experimental studies of sandwich structural elements with facing rigid layers of fibrous composites.

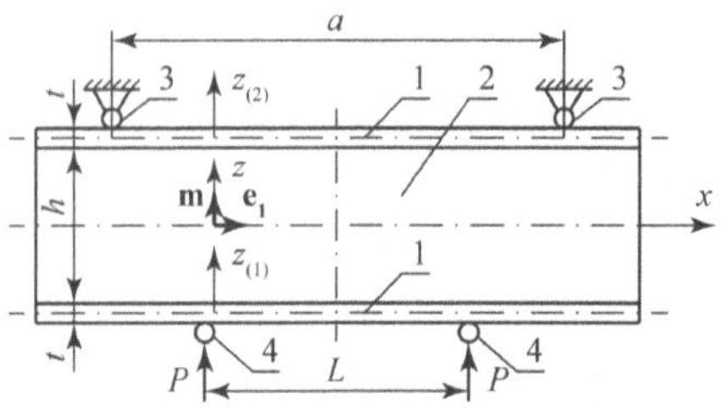

Fig. 1. Installing diagram of sandwich specimen on four-point bending test

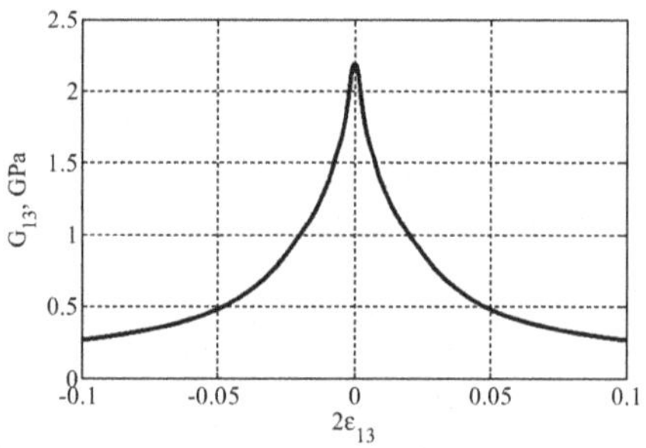

Fig. 2. Relationship of shear modulus from shear strain

These tests are carried out, as a rule, in order to determine the value of the ultimate compressive stress that is formed in the bottom facing layer within the area L of the working length a of the specimen, upon reaching which the specified layer is failure. It can be assumed that one of the causes of such failure is the buckling of the compressed facing layer in one of the possible modes, the classification of which is given in [4, 5]. As will be shown below, using the equations constructed in [3], it is possible to identify both bending and bending-shear and purely shear buckling modes of the lower facing layer of the specimen under compression under its four-point bending according to the diagram in Fig. 1.

2 Statement of the Problem

In accordance with [3], a refined kinematic model of S.P. Tymoshenko $\mathbf{U}^{z(k)} = u^{(k)}\mathbf{e}_1 + w^{(k)}\mathbf{m} + z_{(k)}\left(\gamma^{(k)}\mathbf{e}_1 + \varphi^{(k)}\mathbf{m}\right)$ was adopted to describe the mechanics of the deformation of facing layers at small and medium bends of the plate, and for the core a model of a transverse flexible layer was used. After integrating the simplified equations of theory of elasticity, written for the core along the transverse coordinate z, and introducing an unknown function q into consideration, the equilibrium equations corresponding to the accepted model and the kinematic condition for the interface of the facing layers with the core were derived in [3]. With four-point bending of the plate in a cylindrical shape, they have the form ($\delta_{(1)} = -\delta_{(2)} = 1$)

$$\begin{aligned}
&T^{11}_{(k),x} + \delta_{(k)} q = 0, \quad \left(T^{13}_{(k)} - T^{11}_{(k)} w^{(k)}_{,x} + hq/2\right)_{,x} + \delta_{(k)}\sigma^0_{33} + X^3_{(k)} = 0, \\
&M^{11}_{(k),x} - T^{13}_{(k)} + tq/2 = 0, \quad \left(M^{13}_{(k)} + \delta_{(k)} htq/4\right)_{,x} - T^{33}_{(k)} + \sigma^0_{33} t/2 - \delta_{(k)} X^3_{(k)} t/2 = 0, \\
&u^{(2)} - u^{(1)} - (\gamma^{(1)} + \gamma^{(2)})t/2 + \left(w^{(1)}_{,x} + w^{(2)}_{,x}\right)h/2 + \left(\varphi^{(1)}_{,x} - \varphi^{(2)}_{,x}\right)ht/4 \\
&- hq/G^0_{13} + q_{,xx} h^3/(12E^0_3) = 0,
\end{aligned} \tag{1}$$

where, with the linearly elastic behavior of the material of the facing layer in the x, z directions and the physically nonlinear dependence between the shear stresses $\sigma^{(k)}_{13}$ and the corresponding shear strain $2\varepsilon^{(k)}_{13}$, the internal forces and moments introduced into the analysis and the desired functions $u^{(k)}, w^{(k)}, \gamma^{(k)}, \varphi^{(k)}$ are related by the dependences

$$\begin{aligned}
&T^{11}_{(k)} = B_1\left(u^{(k)}_{,x} + (w^{(k)}_{,x})^2/2 + \nu_{31}\varphi^{(k)}\right), \quad T^{33}_{(k)} = B_3\left(\nu_{13}(u^{(k)}_{,x} + (w^{(k)}_{,x})^2/2) + \varphi^{(k)}\right), \\
&T^{13}_{(k)} = B_{13}(w^{(k)}_{,x} + \gamma^{(k)}), \quad \sigma^0_{33} = \left(w^{(2)} - w^{(1)} - t(\varphi^{(1)} + \varphi^{(2)})/2\right)E^0_3/h, \\
&M^{11}_{(k)} = D_1\gamma^{(k)}_{,x}, \quad M^{13}_{(k)} = D_{13}\varphi^{(k)}_{,x}, \quad B_1 = E_1 t/(1 - \nu_{13}\nu_{31}), \quad B_3 = E_3 t/(1 - \nu_{13}\nu_{31}), \\
&B_{13} = G_{13} t, \quad D_1 = B_1 t^2/12, \quad D_{13} = B_{13} t^2/12.
\end{aligned}$$

For Eqs. (1) for $x = -l, l$ we formulate the boundary conditions for free edges ($2l$ – plate length)

$$\begin{aligned}
&T^{11}_{(k)}(x) = 0, \ M^{11}_{(k)}(x) = 0, \ \left(M^{13}_{(k)} + \delta_{(k)} qht/4\right)(x) = 0, \ q(x) = 0, \\
&\left(T^{13}_{(k)} - T^{11}_{(k)} w^{(k)}_{,x} + hq/2\right)(x) = 0,
\end{aligned} \tag{2}$$

for $x = 0$ – symmetry conditions with respect to the central section, and for $x = a/2$ – condition in the form of equal to zero of transverse displacement of the upper facing layer in the center of the support roller:

$$\begin{aligned}
&u^{(k)}(0) = 0, \ \gamma^{(k)}(0) = 0, \ \left(M^{13}_{(k)} + \delta_{(k)} htq/4\right)(0) = 0, \ q(0) = 0, \\
&\left(T^{13}_{(k)} - T^{11}_{(k)} w^{(k)}_{,x} + hq/2\right)(0) = 0, \ w^{(2)}(a/2) + \varphi^{(2)}(a/2)t/2 = 0.
\end{aligned} \tag{3}$$

To simulate external loading (see Fig. 1), the distributed load $X^3_{(1)}$ applied to the bottom layer at the contact area of the loading roller with the facing layer and it is specified by a cosine function with an amplitude value p. To simulate the contact interaction between the support and the top facings layer in the support section $x = a/2$, the condition was set in the form of the last equation of system (3), and the unknown reaction $X^3_{(2)}$ from the support roller was approximated by the cosine function with an amplitude value p_z, that can be determined from the solution of the problem.

For an approximate solution of the formulated boundary value problem (1)–(3), the finite sum method (the method of integrating matrices) was used [6, 7]. In accordance with this method, we reduce the initial differential equations to Volterra-type integral equations of the second kind with additional relations for determining unknown integration constants. Such a reduction is carried out by integrating Eqs. (1) and satisfying conditions (2)–(3). Note that the initial boundary-value problem has the form of a second-order differential problem, while integral equations will contain only derivatives of the desired functions. In [7], a method of collocation by Gaussian nodes and a method for constructing integrating matrices were proposed, and in [8, 9] a constructive application of the method of integrating matrices in version [7] for problems of the theory of plates and shells was described in detail. Therefore, we restrict ourselves to the final discrete statement of the problem in the operator form, which is obtained by replacing the integral operators in the problem with integrating matrices: $A(U) = A_1 U + A_2(U) = F,$ where $U = (w^{(1)}_{,x}, w^{(2)}_{,x}, u^{(1)}_{,x}, u^{(2)}_{,x}, q_{,x}, w^{(1)}_l, w^{(2)}_l, \varphi^{(1)}_l, \varphi^{(2)}_l, q_{a,1}, p_z) \in H_h^{\times 5} \times R^{\times 6}$, A_1 – linear, A_2 – nonlinear operators; the vector function F of the right-hand sides depends only on the transverse load function $X^3_{(1)}(p)$ applied to the bottom facing layer.

Experimental investigations were carried out in [10, 11] to determine the shear modulus of fibrous composite materials used for facing layers of a sandwich plate. As a result of the studies, averaged experimental dependences $\sigma_{13} = G_{13}(2\varepsilon^{(k)}_{13})2\varepsilon^{(k)}_{13}$ of shear stresses on shear strains were obtained, on the basis of which the dependences of the secant shear modulus on the corresponding shear strains were constructed (see Fig. 2). The last dependences in the used mathematical model of deformation are included as terms, which are taken into account in the form of a physically nonlinear dependence $G_{13}(2\varepsilon^{(k)}_{13}) = G_{13}(w^{(k)}_{,x} + \gamma^{(k)})$ during piecewise linear interpolation of the experimental dependence $G_{13}(w^{(k)}_{,x} + \gamma^{(k)})$ obtained in [11].

The described numerical method for solving the formulated nonlinear problem is implemented on the basis of a modified Newton's iterative process

$$A'_{(n)}(U^{(n)})\,(U^{(n+1)} - U^{(n)}) + (A_1^{(n)} + A_2^{(n)})\,U^{(n)} = F. \tag{4}$$

The nonlinear dependence of the transverse shear modulus G_{13} on the shear strain $2\varepsilon^{(k)}_{13} = w^{(k)}_{,x} + \gamma^{(k)}$, obtained by the experimental method in [11], is included in the problem as the dependence of the operators $A'_{(n)}, A_1^{(n)}, A_2^{(n)}$ on $U^{(n)}$, where $U^{(0)}$ – is the given initial approximation, i.e. at each n-th iterative step, it is necessary to re-calculate the stiffness matrix depending on the n-th approximation of the solution $(U^{(n)},\ (2\varepsilon_{13}^{(k)})^{(n)})$; $A'_{(n)}(U^{(n)})\Delta U$ is the differential of the operator $A^{(n)}$ at the point $U^{(n)}$. Calculations according to the iteration process (4) were carried out until the norms of the residual and the difference between the iterations remained greater than the specified accuracy.

In the calculations, the elastic characteristics of the facing layers are taken to be $E_1 = 103$ GPa, $E_3 = 5,9$ GPa, $\nu_{13} = 0,3$; transversely flexible core – $E_3^0 = 80$ MPa, $G_{13}^0 = 30$ MPa, and geometric parameters equal to $t = 0,3$ cm, $h = 1$ cm, $L = 7$ cm, $a = 25$ cm, $2l = 27$ cm. To describe the desired functions having large gradients on very short regions (in particular, in the loading roller zone), the region was decomposed using piecewise Lagrangian interpolation to construct integrating matrices. To this end, we divide the region $[0, l]$ into three segments $e_1 = [0, L/2]$, $e_2 = [L/2, L/2 + R]$, $e_3 = [L/2 + R, l]$, introduce a global grid $\omega_h = \{x_i : 0 < x_1 < x_2 < \ldots x_N < l\}$ on the segment $[0, l]$ as follows: on each segment e_1, e_2, e_3 we introduce independent Gaussian nodes associated with the roots of the Legendre polynomial; the number of nodes on these segments were taken equal $N_1 = 40$, $N_2 = 81$, $N_3 = 130$ respectively.

3 Results of Numerical Experiments

By tabulating the parameter of the amplitude value p of the transverse load $X_3^{(1)}$, an analysis was made of the sequence of equilibrium states (solutions of the problem) at various loading levels, including the maximum values of deflections and shear strains of the bottom loaded facing layer. It is worth noting that the "load-deflection" relationship is linear in nature, therefore it is not of particular interest. Figure 3 shows the dependence of the load on the maximum modulus of the shear strain, where, $P = b \int_{-l}^{l} X_{(1)}^3$, $b = 2.5$ cm, $\left\| \varepsilon_{13}^{(1)} \right\| = \max_{i=1,\ldots N} |\varepsilon_{13}^{(1)}(x_i)|$.

Note that in Fig. 3 dashed line represents the solution of a geometrically nonlinear problem without taking into account physical nonlinearity, the solid line represents the solution of a geometrically and physically nonlinear problem.

In the vicinity of some load value $P = P^*$ indicated in Fig. 3 by triangle, the iterative process ceases to converge. To search for the unstable equilibrium positions indicated in Fig. 3 by a dashed line, the parameter continuation method was implemented (see, for example, [8, 9], which describes in detail the constructive implementation of the parameter continuation method [12]), which is one of the versions of the modified globally incremental Lagrange theory (the problem regarding increments, in contrast to the classical incremental Lagrange theory, is solved in a nonlinear formulation).

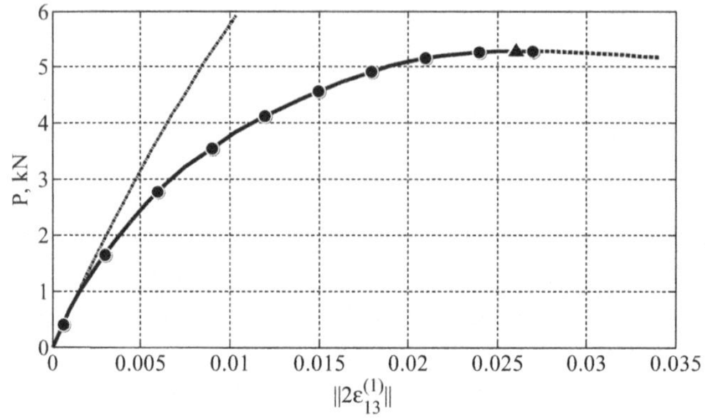

Fig. 3. "Load-shear transversal strain" curve

The shear strains at the point ξ was used as loading parameter in the method, where ξ – the point at which the shear strains reach the maximum value in absolute value. The Fig. 4a shows the distribution along the length of the specimen of the transverse shear strains of the bottom facing layer at various load values P; Fig. 4b shows a plot of Fig. 4a on an enlarged scale, indicating the corresponding values P of the applied force for each of the solutions shown in the form of shear strains. The load values P in Fig. 4b are also marked with black round markers in the diagram "load – transverse shear strain" (see Fig. 3).

Figure 5 shows the distribution of the transverse shear modulus along the length of the bottom compressed facing layer, corresponding to the shear strain at the point indicated in Fig. 3 by triangle marker, at $P = P^*$.

It should be noted that the maximum modulus values of the transverse shear strains and shear stresses, as well as the minimum absolute values of the transverse shear moduli, are reached at the edge of the loading roller (see Fig. 4, 5, 6). It can be seen that in its vicinity localized zones of concentration of shear stresses and corresponding

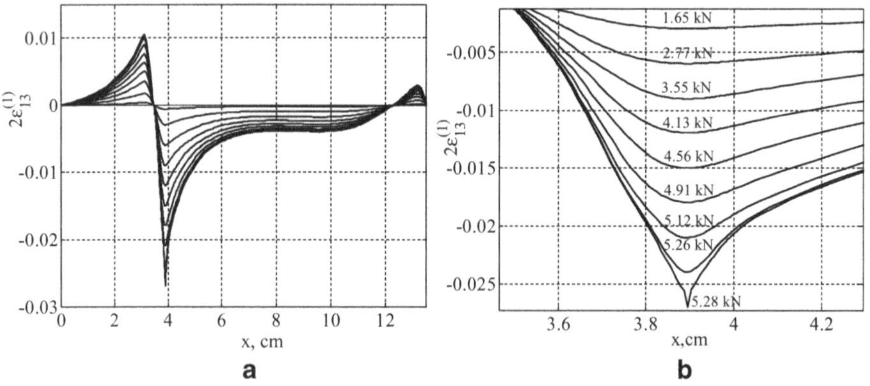

Fig. 4. Shear transversal strain for different load values

shear strains are formed, and, consequently, a sharp drop in the values of secant shear moduli. Figure 6 shows the increments of the transverse shear strains of the bottom facing layer in the postbuckling region, which were obtained as a solution to the problem with respect to the increment functions using the parameter continuation method. The analysis showed that they characterize the buckling by the transverse-shear mode, which is realized when the transverse shear stresses in the facing layer reach a certain limiting value. This buckling mode is similar to the transverse-shear mode [10], which is realized in the specimen of fiber reinforced plastic at three-point bending and is the cause of their failure during testing.

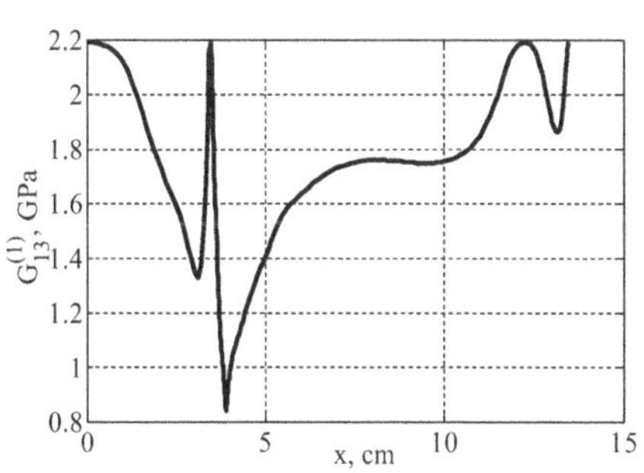

Fig. 5. The distribution of transverse shear modulus along the length

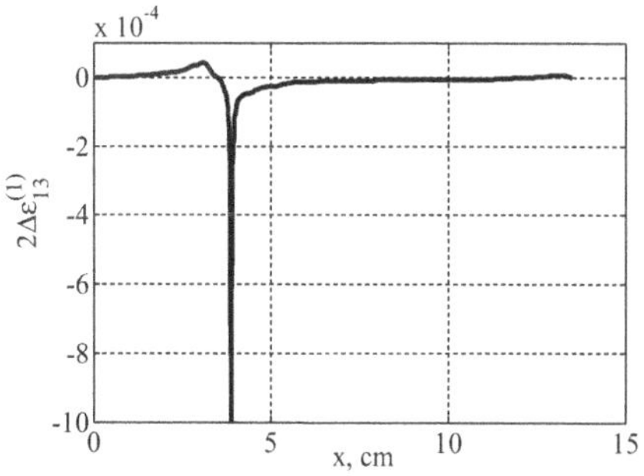

Fig. 6. The distribution of increments of transverse shear strains along the length

Acknowledgments. The research results were obtained in the framework of the fulfillment of the state task of the Ministry Science and High Education of Russia (evaluation of mechanical characteristics of CFRP), supported by Russian Science Foundation (project No. 19-79-10018, Section 2 and 3) and by Russian Foundation for Basic Research (project No. 19-08-00073, development of numerical method).

References

1. Naik, N.K., Kumar, R.S.: Compressive strength of unidirectional composites: evaluation and comparison of prediction models. Compos. Struct. **46**, 299–308 (1999)
2. Budiansky, B., Fleck, N.A.: Compressive failure of fibre composites. J. Mech. Phys. Solids **41**(1), 183–211 (1993)
3. Badriev, I.B., Paimushin, V.N., Shihov, M.A.: Refined equations of the sandwich shells theory with composite external layers and a transverse soft core at average bending. Lobachevskii J. Math. **40**(11), 1094–1914 (2019)
4. Paimushin, V.N.: A stability theory of sandwich structural elements analysis of the current state and a refined classification of buckling forms. Mech. Compos. Mater. **35**(6), 465–470 (1999)
5. Paimushin, V.N.: Theory of stability for three-layer plates and shells: stages of development, state-of-the-art, and prospects. Mech. Solids **36**(2), 127–137 (2001)

6. Vakhitov, M.B.: Integration matrices: apparatus for numerical solving of differential equations of structural mechanics. Russ. Aeronaut. **3**, 50–61 (1966). (In Russian)
7. Dautov, R.Z., Paimushin, V.N.: On the method of integrating matrices for the solution of boundary value problems for fourth-order ordinary equations. Russ. Math. **40**(10), 11–23 (1996)
8. Badriev, I.B., Makarov, M.V., Paimushin, V.N.: Solvability of physically and geometrically nonlinear problem of the theory of sandwich plates with transversally-soft core. Russ. Math. **59**(10), 57–60 (2015)
9. Badriev, I.B., Makarov, M.V., Paimushin, V.N.: Geometrically nonlinear problem of longitudinal and transverse bending of a sandwich plate with transversally soft core. Lobachevskii J. Math. **39**(3), 448–457 (2018)
10. Paimushin, V.N., Kholmogorov, S.A., Makarov, M.V., Tarlakovskii, D.V., Lukaszewicz, A.: Mechanics of fiber composites: forms of loss of stability and fracture of test specimens resulting from three-point bending tests. Zeitschrift fur Angewandte Mathematik und Mechanik **99**(1), 1–25 (2019)
11. Paimushin, V.N., Kholmogorov, S.A.: Physical-mechanical properties of a fiber-reinforced composite based on an ELUR-P carbon tape and XT-118 binder. Mech. Compos. Mater. **54** (1), 2–12 (2018)
12. Grugoluk, E.I., Shalashilin, V.I.: Nonlinear Deformation Problems: Method of Parameter Continuation in Nonlinear Problems of Mechanics of Rigid Bodies. Nauka, Moscow (1988)

Model of Low-Cycle Deformation of Stainless Steel Under Soft Loading

Vasilii Gorokhov(✉), Sergei Kapustin, Yuriy Churilov, Dmitriy Kazakov, and Dmitriy Zhegalov

Research Institute for Mechanics, National Research Lobachevsky State University of Nizhny Novgorod, Gagarin Avenue 23, Nizhny Novgorod 603950, Russia
vas-gor@rambler.ru

Abstract. A variant of the plasticity model is proposed that allows describing the kinetics of plastic deformation of materials under different modes of low-cycle loads. In the model, the radius of the flow surface is represented as a function depending on temperature and accumulated plastic deformation, and the coordinates of the center are described by two Armstrong-Frederick-Kadashevich-type evolutionary equations, the first of which additionally introduces corrective material functions. Changes in plastic deformations in the model are determined based on the associated flow law. Receiving material functions of the model and testing it was carried out on the basis of the results of experimental study of the deformation of a cylindrical sample under soft cyclic loading with maximum control and minimum effort in the cycle. The experiments were carried out on the Amsler HA 100 servo-hydraulic fatigue testing machine. Comparative analysis of the results of numerical simulation and experiment shows that the proposed model allows us to describe the main effects of plastic deformation of the material under consideration under soft asymmetric loads (displacement of the plastic hysteresis loop and reduction of its width) quite well.

Keywords: Plasticity · Soft loading · Low-cycle deformation · Yield surface

1 Introduction

Currently, there are a number of mathematical models proposed to describe the processes of plastic deformation under low-cycle loads. In [1], it was found that the variants of the models proposed in [2, 3] do not allow us to correctly describe the quantitative dependence of the displacement and change in the loop width on the number of cycles under soft cyclic loads. In [1], it is also shown that for these types of loads, the models [4–6] satisfactorily describe the laws of displacement of the plastic deformation loop, but there is practically no information about the possibility of describing the kinetics of the plastic hysteresis loop width within these models, which actually determine the cyclic component of the fracture energy.

A variant of the plasticity model with combined hardening is considered below to describe the main effects of deformation under soft, hard, symmetric and asymmetric low-cycle loads, which is a development of the model variant [3], supplemented with refined relations for describing kinematic hardening [1, 4, 6].

E. Gdoutos and M. Konsta-Gdoutos (Eds.): ICTAEM 2020, STIN 16, pp. 301–304, 2020.
https://doi.org/10.1007/978-3-030-47883-4_53

2 Constitutive Equations of the Plasticity Model

In the proposed version of the model, the radius of the yield surface C_p is represented as a function depending on the temperature T and the accumulated plastic deformation characterized by parameter k_p - plastic strain path length:

$$C_p = C_p(T, k_p), \tag{1}$$

$$k_p = \bar{k}_p + \Delta k_p,\ \Delta k_p = \sqrt{\frac{2}{3}\left(\Delta e_{ij}^p \Delta e_{ij}^p\right)}. \tag{2}$$

The components of the ρ_{ij} tensor are the functional of the plastic deformation process:

$$\rho_{ij} = \rho_{ij} + \Delta\bar{\rho}_{ij}, \tag{3}$$

where $\Delta\rho_{ij}$ is the change in ρ_{ij} at the elementary step of changing external influences, which is the sum of two types of back-stress described by evolutionary equations of the Armstrong-Frederic-Kadashevich type [7, 8]:

$$\Delta\rho_{ij} = \sum_{m=1}^{2} \Delta\rho_{ij}^m, \tag{4}$$

$$\Delta\rho_{ij}^1 = G_1^1 \cdot \Delta e_{ij}^p - G_2^1 \cdot \rho_{ij}^1 \cdot \Delta k_p + g_T^p \cdot \rho_{ij}^1 \langle \Delta T \rangle, \tag{5}$$

$$\Delta\rho_{ij}^2 = G_1^2 \cdot \Delta e_{ij}^p - G_2^2 \cdot \rho_{ij}^2 \cdot \Delta k_p + g_T^p \cdot \rho_{ij}^2 \langle \Delta T \rangle. \tag{6}$$

In ratios (3):

$$\langle \Delta T \rangle = \Delta T \quad \text{at} \quad \Delta T \geq 0, \quad \langle \Delta T \rangle = 0 \quad \text{at} \quad \Delta T < 0, \tag{7}$$

$$G_1^1 = G_1^0 \cdot (1 - n_1(T, k_p)),\ G_2^1 = G_2^0 \cdot (1 - n_2(T, e_u^p)). \tag{8}$$

In the above ratios (5), (6) and (8): $G_1^0(T),\ G_2^0(T),\ G_1^2(T),\ G_2^2(T),\ g_T^p(T)$ – temperature – dependent T material functions; $n_1(T, k_p),\ n_2(T, e_u^p)$ – corrective material functions (e_u^p – intensity of plastic deformations).

3 Results of the Experiment and Numerical Simulation

Obtaining the material parameters of the model and checking its operability was carried out on the basis of the results of an experimental study of the deformation of a sample made of material of the 12Kh18N10T type with a cylindrical working part, under soft cyclic loading with control of the maximum and minimum values of forces in the cycle.

In Fig. 1 and Fig. 2 shows graphs of the dependence of the displacement of the loop $\Delta l(n) = l(n) - l(50)$ (Fig. 1) and the magnitude of the total deformations of the loop in the $h(n)$ cycle (Fig. 2) versus the number of cycles n. The results of numerical simulation are shown by solid lines in the figure, and the experimental results by dots.

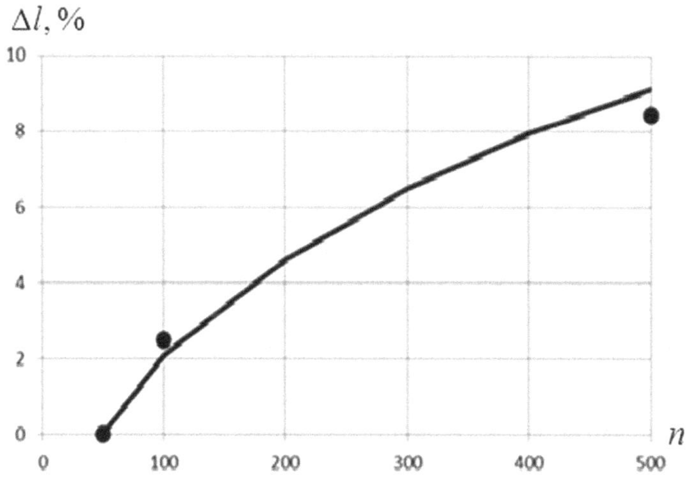

Fig. 1. Loop offset depending on the number of loading cycles. Solid line - numerical modeling, points – experiment.

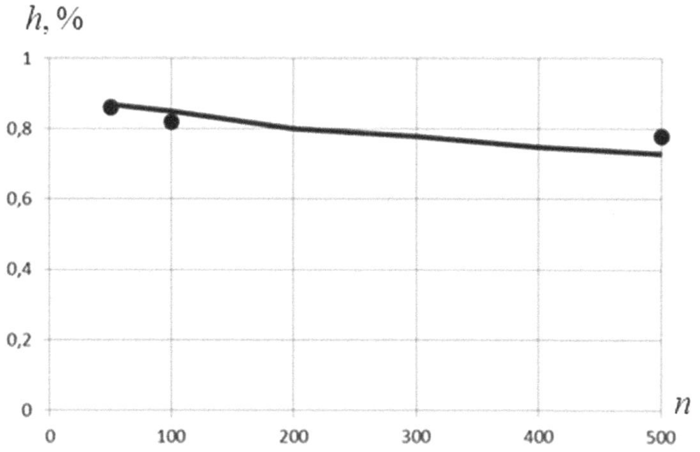

Fig. 2. The range of total deformations in the cycle depending on the number of loading cycles. Solid line - numerical modeling, points – experiment.

4 Conclusion

A variant of the plasticity model is proposed that allows describing the kinetics of plastic deformation of materials under soft modes of low-cycle loading. A good agreement between the results of numerical simulation of plastic deformation of stainless steel and experimental data is demonstrated.

Acknowledgements. This work was supported by the Ministry of Science and Higher Education of the Russian Federation (Agreement No. 075-11-2019-050).

References

1. Bondar, V.S., Abashev, D.R., Petrov, V.K.: Comparative analysis of variants of plasticity theories under cyclic loading. PNRPU Mech. Bull. **2**, 23–44 (2017)
2. Kazakov, D.A., Kapustin, S.A., Korotkikh, Y.: Modeling of Deformation and Fracture Processes of Materials and Structures. Publishing house of Nizhny Novgorod state University, Nizhny Novgorod (1999)
3. Kapustin, S.A., Churilov, Y., Gorokhov, V.A.: Modeling of Nonlinear Deformation and Fracture of Structures Under Multi-Factor Influences on the Basis of FEM. Publishing house of Nizhny Novgorod state University, Nizhny Novgorod (2015)
4. Bondar, V.S.: Inelasticity: Variants of the Theory. Fizmatlit, Moscow (2004)
5. Chaboche, J.-L.: A review of some plasticity and viscoplasticity constitutive theories. Int. J. Plast. **24**, 1642–1692 (2008)
6. Bondar', V.S., Danshin, V.V., Makarov, D.A.: Mathematical modeling of deformation processes and damage accumulation under cyclic loading. PNRPU Mech. Bull. **2**, 125–152 (2014)
7. Amstrong, P.J., Frederick, C.O.: A mathematical representation of the multiaxial Bauscinger effect. CEGB Report No. RD/B/N/731 (1966)
8. Kadashevich, Y.: On various tensor-linear relations in the theory of plasticity. Stud. Elast. Plast. **6**, 39–45 (1967)

Creep and Long-Term Strength of Structures

Ivan Volkov, Leonid Igumnov(✉), and Svetlana Litvinchuk

Research Institute for Mechanics, National Research Lobachevsky State University of Nizhny Novgorod, Gagarin Avenue 23, Nizhny Novgorod 603950, Russia
litvinchuk@mech.unn.ru

Abstract. The issue of evaluating strength and service life is discussed as applied to structures performance of which is characterized by multi-parametric nonstationary thermal-mechanical effects. The major degradation mechanisms of structural materials are considered. In the framework of mechanics of damaged media, a mathematical model describing the processes of viscoplastic deformation and damage accumulation due to creep is developed. The results of numerical simulation of the carrying capacity of a power plant reactor vessel in the event of a hypothetical emergency are presented. A number of characteristic features accompanying the deformation and failure processes connected with the time and place of macrocrack nucleation, the stress-strain state history and damage degree are given. By comparing numerical and experimental results we can conclude that the proposed constitutive relations of MDM adequately describe degradation of the initial strength properties of the material in terms of creep and long-term strength.

Keywords: Long-term strength · Modeling · Constitutive relations · Mechanics of damaged media

1 Introduction

To guarantee safe exploitation of critical engineering facilities and to prolong their standard service life, it is necessary to monitor damage growth rates in the most hazardous zones of structural elements and predict the development of such processes up to the limit state [1].

The currently available methods for evaluating service life of structural elements do not account for actual processes taking place in materials. Elastic analysis used in the standard approach does not enable us to consider actual characteristics of viscoplastic deformation, on which the service life of structural elements depends [1].

In this connection, it becomes vital to develop new methods for evaluating service life of structural elements based on equations of thermal viscoplasticity, equations of damage accumulation and failure criteria with their comprehensive substantiation by conducting the related full-scale and numerical experiments on laboratory specimens and numerically analyzing the deformation and failure processes of structural elements under service conditions [1–6].

The most effective tool for these purposes is mathematical modeling of degradation processes of structural materials, using modern models and methods of mechanics of

E. Gdoutos and M. Konsta-Gdoutos (Eds.): ICTAEM 2020, STIN 16, pp. 305–308, 2020.
https://doi.org/10.1007/978-3-030-47883-4_54

damaged media and fracture mechanics. Obtained by modeling, the values of damage degree of the material in accessible zones must be updated with the help of modern physical methods using periodic non-destructive control [7].

In the framework of MDM, a mathematical model has been developed that describes the processes of damage accumulation in structural materials due to degradation mechanisms, determined by growing creep strains under multiaxial stress and arbitrarily complex regimes of thermal-mechanical loading.

2 Technique and Numerical Results

The main assumptions of the introduced version of MDM are as follows:

- the material is initially isotropic and free of damage (only the anisotropy caused by deformation processes is accounted for, whereas damage-induced anisotropy of material elastic properties is not taken into account);
- the components of strain tensors and strain rates are the sum of momentary and time components;
- the evolution of equipotential creep surfaces is described by the change of its radius and displacement of its center;
- the volume change of the body element is elastic;
- small deformation processes are considered;
- the only structural parameter characterizing the material damage degree at the macro-level is scalar parameter – damage degree;
- the accumulated damage degree affecting the deformation process of the material is taken into account by introducing effective stresses.

The relation between tensor components of stresses and elastic strains is established on the basis of the equations of thermal elasticity.

To describe creep processes, a family of equipotential creep surfaces (surfaces with a constant creep strain rate) is introduced in the stress space, which has a common center and different radii defined by a current stressed state [8–10].

Tensor components of creep strain rates are defined on the basis of the associated flow law and gradientality principle.

At the stage of the development of defects scattered over the volume, the damage degree effect on the material physical mechanical properties is observed. In the first approximation, this effect can be taken into account based on the concept of a degrading continuum, by introducing effective stresses [2, 7, 11].

The rate of damage accumulation during creep is determined by the evolutionary equation accounting for the volumetric stressed state, the level of accumulated damage, accumulated relative damage energy, spent on the nucleation of defects and the rate of change of damage energy [3, 5, 6, 12].

The condition when damage degree reaches its critical value is taken as the criterion of termination of the growth phase of scattered macrodefects.

The results of numerically analyzing the carrying capacity of a NPP reactor vessel made of 15X2NMFA steel in the event of a hypothetical emergency for the long-term strength mechanism are presented.

The calculation scheme is represented by an axisymmetric structure of a reactor vessel, consisting of a cylindrical wall and an elliptic bottom. An emergency was simulated by applying internal hydrostatic pressure, varying from zero at a height of $h = 1{,}5$ m from the bottom lowest point and modeling the force effect of meltdown, internal pressure and temperature, varying in the limits of the considered part of the reactor vessel within 184° to 1510 °C.

The reliability of the developed MDM model was numerically assessed prior to emergency simulation and material parameters of 15X2NMFA steel were determined within the temperature range of 20 °C to 1200 °C.

The problem of evaluating long-term strength of a NPP reactor vessel under thermal dynamical loading was numerically solved in two steps.

At the first step, the stage when the pressure and temperature increase up to their maximal values during a short period was analyzed. At the second step, the pressure and temperature were kept constant.

A number of calculations were carried out which differ in the value of internal pressure.

The numerical analyses proved that, for all the versions of the computations from the viewpoint of long-term strength, the second stage (with the constant pressure and temperature) accompanied by the intensive development of creep strains and growth of defects is the most important.

The maximum allowable pressure that doesn't give rise to macrocrack nucleation has been defined.

The obtained numerical results on determining the maximum allowable pressure agrees with the results presented in [13].

3 Conclusion

The numerical computations conducted and their comparison with the available experimental data make it possible to make the conclusion about the adequacy of the constitutive relations of MDM in modeling degradation of materials according to the long-term strength mechanism and the possibility of effectively using the developed constitutive relations of MDM for evaluating long-term strength of materials and structures.

The mathematical model of MDM has been developed, that describes the processes of inelastic deformation and damage accumulation in structural materials (metals and their alloys) during the degradation of initial strength properties of materials according to the long-term strength mechanism.

Using the numerical modeling method and comparing the obtained results with experimental data, the reliability of the constitutive relations of MDM for creep has been assessed. This made it possible to make the conclusions about the reliability of the developed modeling representations and the accuracy of the developed method for determining the material parameters included into these relations.

The results of numerically analyzing the carrying capacity of a NPP reactor vessel in the event of a hypothetical accident are presented proving the validity of the carrying capacity failure of such facilities.

Acknowledgements. This work was supported by the Ministry of Science and Higher Education of the Russian Federation (Agreement No. 075-11-2019-050).

References

1. Mitenkov, F.M., Kaidalov, V.F., Korotkikh, Y., et al.: Methods of Justifying the Resource of Nuclear Power Plants. Mashinostroenie, Moscow (2007)
2. Volkov, I.A., Korotkikh, Y.: The Equations of State of Viscoelastoplastic Media with Damage. Fizmatlit, Moscow (2008)
3. Lemaitre, J.: Damage modelling for prediction of plastic or creep fatigue failure in structures. In: Trans. 5th International Conference on SMRiT, Paper no. L5/1b, North Holland (1979)
4. Muracami, S., Imaizumi, T.: Mechanical description of creep damage and its experimental verification. J. Mech. Theor. Appl. **1**, 743–761 (1982)
5. Lemaitre, J.: Continual damage model used to calculate plastic damage. In: Proceedings of the American Society of Mechanical Engineers, vol. 107, pp. 90–98 (1985)
6. Chaboche, J.L.: Constitutive equation for cyclic plasticity and cyclic viscoplasticity. Int. J. Plast **5**(3), 247–302 (1989)
7. Volkov, I.A., Igumnov, L.A.: Introduction to the Continuum Mechanics of the Damaged Environment. Fizmatlit, Moscow (2017)
8. Nikitenko, A.F.: Experimental substantiation of the hypothesis of existence of a creep surface under complex loading. Probl. Strength **8**, 3–11 (1984). Report 1, 2
9. Volkov, I.A., Igumnov, L.A., Korotkih, Y.: Applied Theory of Viscoplasticity. Publishing House of Nizhny Novgorod State University, Nizhny Novgorod (2015)
10. Volkov, I.A., Igumnov, L.A., Kazakov, D.A., Shishulin, D.N., Smetanin, I.V.: Constitutive relations of non-stationary creep under a complex stress state. Probl. Strength Plast. **78**(4), 436–451 (2016)
11. Volkov, I.A., Igumnov, L.A., Kazakov, D.A., Mironov, A.A., Tarasov, I.S., Shishulin, D.N., Smetanin, I.V.: The damage media model for describing service life of structural materials. Probl. Strength Plast. **79**(3), 285–300 (2017)
12. Bodner, S.R., Lindholm, U.S.: An incremental criterion for time-dependent failure of materials. J. Eng. Mater. Technol. ASME **100**(2), 51–58 (1976)
13. Semishkin, V. P., et al.: Thermo-mechanical behavior of the VVR corps in a severe accident, In: Proceedings, 5th International Scientific and Technical Conference, Podolsk, Russia (2007)

Mathematical Modeling of Nonlinear Dynamic Deformation and Failure of Metal-Plastic Shells of Revolution

Leonid Igumnov(✉), Nikolai Abrosimov, Nadezhda Novoseltseva, and Vasilii Gorokhov

Research Institute for Mechanics, National Research Lobachevsky State University of Nizhni Novgorod, 23, bldg. 6, Prospekt Gagarina (Gagarin Avenue), Nizhny Novgorod 603950, Russia
abrosimov@mech.unn.ru

Abstract. A technique for numerical analysis of nonlinear dynamic deformation and progressive failure of multi-layered metal-plastic shells of revolution is developed with account for their strain-rate dependent strength characteristics. The geometric dependencies are formulated on the basis of quadratic version of the nonlinear theory of elasticity. The relationship between stress and strain tensors in a composite macrolayer is based of Hooke's law for an orthotropic body combined with the theory of effective modules. The process of progressive layer-by-layer failure is described in the framework of the degradation model of stiffness characteristics. The strain rate dependent stiffness and strength characteristics are accounted for. An energetically consistent system of equations of motion is constructed using the principle of possible displacements. A numerical method for solving the problem is based on an explicit variational-difference scheme. The proposed technique was verified on the problem of unsteady deformation of a cylindrical shell subjected to pulse pressure.

Keywords: Composite materials · Shells of revolution · Strength · Failure · Numerical methods · Explosive loading

Due to their effective energy absorption, increased crack-resistance and non-splintering character of possible failure, composite materials are widely used in constructing protective structures subjected to intense pulse loadings. In this connection, experimental and theoretical studies of strain rate effect of composite materials on their strength and stiffness characteristics appear to be highly topical since the material of structural elements subjected to pulse loading exhibits stress-strain behavior with a high degree of variability over time. In a number of experimental and theoretical works [1–3], the strain rate dependence of elastic and strength characteristics of composite materials is noted. However, the determination of the parameters characterizing the properties of composite materials at high strain rates is associated with certain difficulties due to the necessity to measure the effect of dynamic pressure on a specimen as a function of time.

The paper presents analytical results of the strain rate effect on dynamic behavior and progressive failure of shells of revolution made of hybrid metal-plastic materials.

E. Gdoutos and M. Konsta-Gdoutos (Eds.): ICTAEM 2020, STIN 16, pp. 309–311, 2020.
https://doi.org/10.1007/978-3-030-47883-4_55

We considered both homogeneous glass-fiber-reinforced plastic shells formed by double alternating winding of spiral and annular layers with a thickness ratio of 1:1, and inhomogeneous ones fabricated by spiral cross winding of unidirectional glass-fiber-reinforced plastic (according to the scheme of reinforced homogeneous shells) on a steel cylindrical mandrel made of mild steel.

Since the shells of revolution made from composite materials are inhomogeneous, have low shear stiffness and, in some cases, are rather thick, thus, in order to describe their stress-strain state, it is necessary to use nonclassical shell theories [4].

Geometrical dependencies are constructed using the relations of the simplest quadratic version of the nonlinear elasticity theory [4].

The stress-strain tensors in a homogeneous composite macrolayer are related by Hooke's law for an orthotropic body in combination with the theory of effective moduli [4]. The process of progressive layer-by-layer damage of layered shells of revolution is described in the framework of the model of degradation of their stiffness characteristics [5]. And furthermore, the strain-rate dependent strength characteristics of composite materials are taken into account. In particular, for unidirectional fiberglass, the material stiffness and strength characteristics can be described by the regression function [3]

$$F(\dot{e}) = \alpha + \beta \dot{e}^{\gamma} \quad (1)$$

where F and $\dot{e}$ are strength characteristics and strain rate; α, β, γ are experimentally determined material constants. The constitutive relations in the isotropic steel layer of the shell are formulated on the basis of the differential theory of plasticity with linear hardening [4].

In order to derive the equations of motion of an inhomogeneous shell of revolution, the principle of virtual displacements is used [6].

The equations derived are universal enough, because they allow one to describe nonlinear nonstationary deformation processes and to estimate the limiting deformability and dynamic strength of two-layered metal-plastic shells of revolution, and their geometrical and structural parameters changing over a wide range.

The numerical method for solving the problem formulated was based on the explicit variational-difference scheme [4, 7]. The calculations were carried out on a "Lobachevsky" supercomputer.

A numerical analysis of the effect of the strain rate on the dynamic strength of two-layered metal-plastic cylindrical shells was considered on the problem of their deformation under the action of a pressure pulse caused by a blast in a center of a shell of an explosion charge (EC), which describes the pressure profile in the incident shock wave by an empirical relation [8]. The physicomechanical characteristics of unidirectional glass-fiber-reinforced plastic shells were determined from the results of quasistatic tests on ring and plane unidirectional specimens by standard techniques for identifying the corresponding stiffness and strength characteristics [9]. The strain rate dependence of the strength characteristics of glass-fiber-reinforced plastics is described by function (1).

The results obtained attest that the account of strain rate dependence of strength characteristics for all the reinforcement schemes considered lead to a qualitative difference in the character and size of failure zones of the binder and fibers and to a significant increase in the load-carrying capacity of the shells compared with the calculations of constant strength characteristics.

Calculation models allowing one to analyze the processes of progressive failure of pulse-loaded metal-plastic cylindrical shells of revolution both with account of the strain rate dependence of their strength characteristics (dynamic model) and with constant characteristics (static model) have been proposed.

A comparative analysis of calculation results with experimental data testifies their better agreement in dynamic model. For various reinforcement schemes of composite macrolayer, the qualitative differences in character and size of failure zones calculated by both the methods were revealed.

The results obtained can be used in the design and evaluation of the dynamic strength of load-carrying elements.

This work was supported by a grant from the Government of the Russian Federation (contract No. 14.Y26.31.0031).

References

1. Aseev, A.V., Makarov, G.E., Stepanenko, S.V.: An experimental study of the dynamic behavior of tubular samples of fiber composite materials at the limit of their bearing capacity. Prikl. Mekh. Tekhn. Phys. **3**, 140–147 (1992)
2. Aseev, A.V., Gorshkov, N.N., Demeshkin, A.G., Makarov, G.E., Plastinin, A.V., Sil'vestrov, V.V., Stepanenko, S.: Experimental investigation of the deformability of fiberglass and organic plastics as a function of strain rate. Mech. Compos. Mater. **2**, 188–195 (1992). (in Russian)
3. Shokrieh, M.M., Karamnejad, A.A.: Investigation of strain rate effects on the dynamic response of a glass/epoxy composite plate under blast loading by using the finite-difference method. Mech. Compos. Mater. **50**(3), 295–310 (2014)
4. Abrosimov, N.A., Bazhenov, V.G.: Nonlinear Problems of Dynamics of Composite Structures, 400 p. N. Novgorod: Publishing House NNGU (2002). (in Russian)
5. Abrosimov, N.A., Novosel'tseva, N.A.: Numerical simulation of the level-by-level destruction of cylindrical shells under explosive loading. Mech. Compos. Mater. **51**(4), 579–594 (2015)
6. Vasidzu, K.: Variational Methods in the Theory of Elasticity and Plasticity. M., Mir, 512 p. (1987). (Russian translation)
7. Abrosimov, N.A., Elesin, A.V., Lazarev, L.N., Novosel'tseva, N.A.: Numerical analysis of the strength of glassfiber-reinforced plastic cylindrical shells of different structure at a pulse loading. Probl. Prochn. Plastychn. Mezhvuz. Sborn. **4**(75), 288–295 (2013). N. Novgorod: Publishing House NNGU
8. Adishchev, V.V., Kornev, V.M., Talzi, L.A.: Estimates of maximum stresses in closed cylindrical vessels under axisymmetric explosive loading. Institute of Hydrodynamics SB AS Russia, Novosibirsk. Department in VINITI, № 6588-83 Dep. (1983)
9. Rusak, V.N., Fedorenko, A.G., Syrunin, M.A., et al.: Ultimate strain and strength of basalt-fiber-reinforced shells under internal explosive loading. Prikl. Mekh. Tekhn. Phys. **43**(1), 186–195 (2002)

Impact of Transient Pressure on a Half-Space with Membrane Type Coating

Elena Mikhailova[1(✉)], Grigory Fedotenkov[1,2], and Dmitry Tarlakovskii[1,2]

[1] Moscow Aviation Institute (National Research University), Moscow, Russia
mihel6@yandex.ru

[2] Institute of Mechanics, Lomonosov Moscow State University, Moscow, Russia

Abstract. The unsteady influence of an external load on the base, which is an elastic half-space with a membrane-type coating is investigated. The load acting on the "half-space-membrane" system is directed normally to the surface of the membrane and is a function that depends on the coordinate and time.

A system of resolving equations is built. Its basic equation follows from the boundary condition, which corresponds to the equality of normal displacements on the surface of the half-space to the deflections of the membrane. The normal displacements of half-space and deflections of the membrane are connected with contact pressure and stresses by integral relations based on the principle of superposition. The basic equation is supplemented by initial conditions to a closed system of resolving equations.

To solve the system of resolving equations, a numerical-analytical algorithm based on the method of quadratures is used.

The results of calculations for the external load specified by various functions are presented.

Keywords: Membrane · Elastic half-space · Transient load · Numerical-analytical algorithm

1 Problem Formulation

The transient impact of normal pressure $p(\tau, x)$ on the elastic half-space with a membrane-type coating is considered.

We assume that the half-space is occupied by a homogeneous isotropic linearly elastic medium with constants λ, μ and density. A rectangular Cartesian coordinate system $Oxyz$ is introduced. In it, the coordinate plane $z = 0$ coincides with the undeformable free surface of the half-space, and the axis Oz is directed deep into the half-space.

Mathematical model of the problem contains [1–4]

E. Gdoutos and M. Konsta-Gdoutos (Eds.): ICTAEM 2020, STIN 16, pp. 312–315, 2020.
https://doi.org/10.1007/978-3-030-47883-4_56

- equations of motion of an elastic medium in displacement potentials φ and ψ

$$\ddot{\varphi} = \Delta\varphi,\ \eta^2\ddot{\psi} = \Delta\psi,\ \Delta = \frac{\partial^2}{\partial x^2} + \frac{\partial^2}{\partial z^2} \tag{1}$$

- relation between elastic displacement potentials and nonzero components of the displacement vectors u (along the axis Ox), w (along the axis Oz)

$$u = \frac{\partial\phi}{\partial x} - \frac{\partial\psi}{\partial z},\ w = \frac{\partial\phi}{\partial z} + \frac{\partial\psi}{\partial x}, \tag{2}$$

- relation between the components of the displacement vectors and the stress tensor σ_{ij}

$$\begin{gathered}\sigma_{11} = \frac{\partial u}{\partial x} + \kappa\frac{\partial w}{\partial z},\ \sigma_{22} = \kappa\left(\frac{\partial u}{\partial x} + \frac{\partial w}{\partial z}\right),\ \sigma_{33} = \kappa\frac{\partial u}{\partial x} + \frac{\partial w}{\partial z}, \\ \sigma_{13} = \frac{1-\kappa}{2}\left(\frac{\partial u}{\partial z} + \frac{\partial w}{\partial x}\right)\end{gathered} \tag{3}$$

- membrane motion equation (v – deflection, a – the speed of the waves of tension-compression in the membrane)

$$\ddot{v}(\tau, x) = a^2\frac{\partial^2 v}{\partial x^2} + p + \sigma_{330},\ \sigma_{330} = \sigma_{33}|_{z=0} \tag{4}$$

- initial conditions

$$v|_{\tau=0} = \dot{v}|_{\tau=0} = \psi|_{\tau=0} = \dot{\psi}|_{\tau=0} = \varphi|_{\tau=0} = \dot{\varphi}|_{\tau=0} = 0 \tag{5}$$

- boundary adhesion-free conditions between the membrane and the surface of the half-space

$$\sigma_{13}|_{z=0} = 0,\ v = w|_{z=0} \tag{6}$$

All variables and parameters are presented in a dimensionless form (the prime indicates the dimensional values):

$$\tau = \frac{c_1 t}{L}, v = \frac{v'}{L}, u = \frac{u'}{L}, w = \frac{w'}{L}, x = \frac{x'}{L}, z = \frac{z'}{L}, p = \frac{p' L}{c_1^2 \rho_m}, a = \frac{a'}{c_1}, c_1^2 = \frac{\lambda + 2\mu}{\rho_p}$$
$$c_2^2 = \frac{\mu}{\rho_p}, \eta^2 = \frac{c_1^2}{c_2^2}, \varphi = \frac{\varphi'}{L^2}, \psi = \frac{\psi'}{L^2}, \sigma_{ij} = \frac{\sigma'_{ij}}{\lambda + 2\mu}, \kappa = \frac{\lambda}{\lambda + 2\mu} \tag{7}$$

Here t - dimensional time, ρ_m, ρ_p - density of the membrane and the half-space material, c_1, c_2 - speed of the tension-compression waves and shear waves in half-space material, L - the characteristic linear parameter.

2 Method of Solution of Problem

The resolving equation is based on the principle of superposition and follows from the boundary condition (6)

$$G_m * p = \sigma_{330} * (G_p - G_m). \tag{8}$$

Here G_m, G_p - transient function for a for a membrane and half-space, respectively [2–4].

The left side of Eq. (8) can be represented as follows:

$$G_m * p = \frac{1}{2} \int_0^{\tau} dt \int_{x-\tau+t}^{x+\tau-t} p d\xi \tag{9}$$

This integral for a given function is calculated analytically.

To solve Eq. (8), a numerical-analytical algorithm based on the quadrature method is used. For the hyperbolic type of equations of motion of the half-space and the membrane an explicit integration scheme is used. The space-time domain is covered with a grid with a constant step in time and coordinate. The grid functions correspond to the required functions. A discrete analogue is constructed for the equation. Moreover, the kernel of convolution has the form

$$G_p(x, \tau) = [G_r(x, \tau) + G_s(x, \tau)] H(\tau - |x|). \tag{10}$$

Here G_r, G_s - the regular and singular components of the transient function for half-space, $H(\tau, x)$ - Heaviside function. The integrals from the first component are calculated using Gauss method, the integrals from the second are calculated using the weighting method and are understood in the meaning of a regularized value.

Acknowledgment. This work was financially supported by the Russian Foundation for Basic Research (RFBR), project 20-08-01099 A.

References

1. Gorshkov, A.G., Medvedsky, A.L., Rabinsky, L.N., Tarlakovsky, D.V.: Waves in Continuum Media. Fizmatlit, Moscow (2004). (in Russian)
2. Gorshkov, A.G., Tarlakovsky, D.V.: Dynamic Contact Problems with Moving Boundaries. Fizmatlit, Moscow (1995). (in Russian)
3. Mikhailova, E.Yu., Fedotenkov, G.V.: Nonstationary axisymmetric problem of the impact of a spherical shell on an elastic half-space (initial stage of interaction). Mech. Solids **46**(2), 239–247 (2011)
4. Mikhailova, E.Yu, Tarlakovskii, D.V., Fedotenkov, G.V.: The unsteady contact interaction problem of an absolutely rigid body and a membrane. In: Proceedings of the Second International Conference on Theoretical, Applied and Experimental Mechanics, ICTAEM 2019. Structural Integrity, vol. 8, pp. 289–293. Springer, Cham (2019)

Fracture Nanomechanics, Fatigue and Fracture at Small Scales (experiments and simulations) by Pasquale Gallo

Some Considerations on Stress Intensity Factor at Atomic Scale

Pasquale Gallo(✉)

Aalto University, P.O. Box 14100, 00076 Espoo, Finland
pasquale.gallo@aalto.fi

Abstract. This work reviews recent molecular statistics (MS) numerical experiments of cracked samples, and discusses the crack-tip region stress field of ideal brittle materials. Continuum-based linear elastic fracture mechanics, indeed, breaks down at extremely small scale, where the discrete nature of atoms is considered. Surprisingly, recent results have shown that the concept of stress intensity factor (SIF) is still valid. In this work, by means of MS simulations on single-edge cracked samples of ideal brittle silicon, it is shown that the stress intensity factor derived from the virial stress may be useful to describe the fracture at extremely small dimensions and to quantify the breakdown of continuum-based linear elastic fracture mechanics. However, it is still debated whether a continuum-based concept such as the "stress" should be applied to a system made of atoms.

Keywords: Virial stress · Atom · Crack · Stress intensity factor

1 Introduction

With recent developments in miniaturization of electronics devices such as nano and micro-electromechanical systems (NEMS/MEMS), issues usually addressed at the macroscale, i.e. fatigue and fracture, have been brought into a completely new "scale" [1, 2]. At the same time, developments in nanotechnology give nowadays a completely new way at which the fatigue and fracture can be studied with a remarkable potential impact in several fields of engineering [3–10]. At such small scales, where discrete nature of atoms can't be ignored, the continuum concepts largely used at the macro-scale become questionable [11–13].

While methods based on energy have shown good potential to be extended from continuum to discrete system [12, 14–18], it is still debated whether the concept of "stress" should be applied to a system made of atoms [19, 20]. Stress is, indeed, a continuum concept, originated from the study of strength and failure of solid, and commonly defined as the quantity that represents the internal forces on a defined plane of a continuous material. Thus, questions arise on the definition of "atomic stress". When investigating fracture at the atomic scale, the virial stress tensor is commonly employed to derive mechanical stresses acting on atoms [20–22]. Setting aside the debate on the validity of the virial stress as a representation of mechanical stress at atomic scale, in the present work the focus is on the applicability of classic continuum concepts in the presence of defects, such as the stress intensity factor (SIF). By

E. Gdoutos and M. Konsta-Gdoutos (Eds.): ICTAEM 2020, STIN 16, pp. 319–324, 2020.
https://doi.org/10.1007/978-3-030-47883-4_57

reviewing recent molecular statistics (MS) analyses on single-edged cracked samples loaded under mode I [23], and representative of ideal brittle fracture, it is demonstrated that the virial stress shows the trend of inverse square root singularity and that computation of the SIF according to Irwin's concept is possible. Furthermore, the breakdown of continuum linear elastic fracture mechanics, recently defined by means of energy concepts [11–13], is here quantified by using merely the stress fields.

It is concluded that the SIF from atomic stress may be useful in characterizing fracture at atomic scale, provided that the virial stress is accepted as representative of atomic mechanical stress. On the other hand, energy concepts should be anyway preferred for future developments, since they can provide a direct equivalence between continuum, discrete systems and among different scales.

2 Review of Recent Molecular Statistics Simulations on Cracked Samples

Recently, fracture tests by means of MS simulations were conducted by using open-source code LAMMPS [24]. While details can be found in [23], important aspects are presented hereafter. The modified Stillinger-Weber (SW) interatomic potential [25] was employed. The SW potential is representative of ideal brittle fracture, and often used when studying single crystal silicon. The focus was on single-edge cracked sample, loaded under mode I. Figure 1 depicts an example of the samples and orientation.

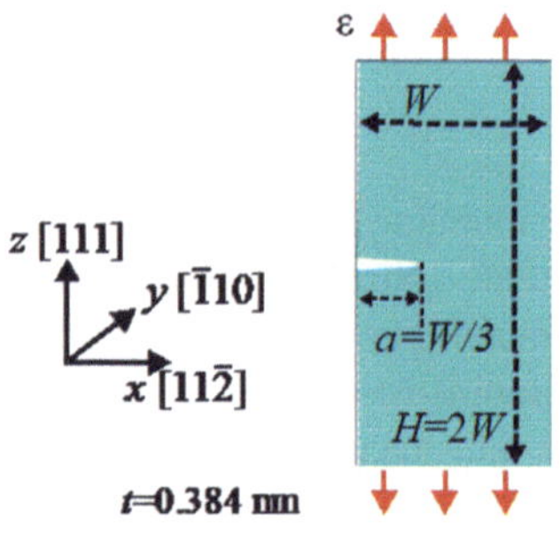

Fig. 1. Cracked samples employed in the molecular statistics analyses and orientation [23]. The thickness of the simulation cell is indicated as *t*.

Several specimens where considered, scaling their size until few nanometers: the width of the sample *W* varied from 198 nm to 9.8 nm, while the crack length *a* was kept equal to *W*/3. The mechanical properties were given by the following material constants: C_{11} = 201 GPa, C_{12} = 51.4 GPa and C_{44} = 90.5 GPa. Along the direction [111], an ideal material strength σ_{IS} of 35 GPa at critical strain ε_C = 0.3 was obtained. A stepwise increment of strain ε is applied at the upper and lower layers of atoms according to Fig. 1, and the strain is increased until final fracture. At fracture, the critical displacement d_C (maximum displacement before failure) is obtained. The mechanical stress is obtained dividing the virial stress (*stress* × *volume* quantity) by the per-atom volume at $\varepsilon_C = 0.3$, i.e. 24.44 Å^3. This value is, indeed, a more reliable estimation of the atomic volume in regions that are highly deformed, such as close to the crack tip at failure conditions. Analyses were conducted at 0 K and fully included the lattice trapping.

3 Results and Discussion

The crack tip stress fields of some selected considered geometries are presented in Fig. 2a for the sake of clarity. The values of the atoms at the crack tip are plotted at $r = 0.1$ Å for convenience, since as well known the log-log scale does not allow the value of 0. The depicted stress is the mechanical atomic stress, i.e. virial stress divided by the per-atom volume as explained in the previous section.

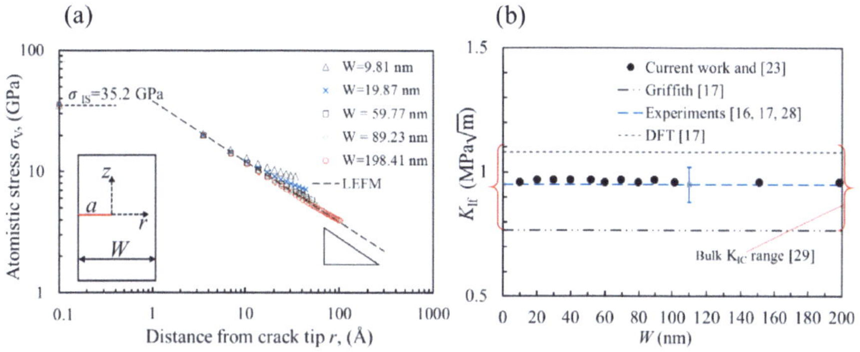

Fig. 2. (a) Near crack-tip atomic stress distribution for selected geometries; atomistic stress is derived from *virial stress/per-atom volume*. The stress values of the atoms at the crack tip are plotted for $r = 0.1$ Å rather than 0 Å in order to summarize in a single log-log scale graph both stress distribution and crack-tip values. (b) Critical stress intensity factors versus the variation of the specimen width W.

The figure shows two very important results:

- The MS analyses do not predict the infinite stress at the crack-tip but rather a finite value, i.e. the ideal material strength σ_{IS}, regardless of the specimen size. This indicates that the fracture, in the case of ideal brittle materials, is ultimately governed by atoms at crack tip.
- Even if the stress singularity is missing, near the crack tip region the stress still varies with $1/r^{0.5}$ as expected from continuum LEFM. Figure 2a depicts only selected geometries for the sake of clarity, but same results are obtained for all the considered samples.

The results allow the quantification of the SIF at failure K_{If} (i.e. at critical displacement d_C), according to the classic Irwin's definition [26, 27]:

$$K_{\mathrm{If}} = \sigma_{\mathrm{V}}\sqrt{2\pi r}\,, \tag{1}$$

where σ_V is the atomistic mechanical stress, perpendicular to the crack plane; r is the distance from the crack-tip along the crack plane. The results are presented in Fig. 2b and compared with other experimental and numerical works by other authors [16, 17, 28]. $K_{\mathrm{If}} \approx 0.97$ MPa·m$^{0.5}$ is constant for all the geometries, and it agrees well with the fracture toughness of single crystal silicon, including bulk samples [29].

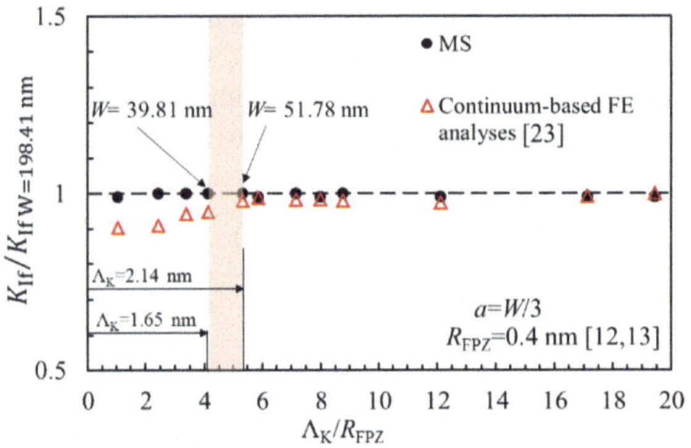

Fig. 3. Normalized SIF at failure versus the ratio of singular stress field length Λ_K and fracture process zone length R_{FPZ}. Numerical values are reported in [23].

Finally, the comparison with continuum-based linear elastic solutions conducted in [23] is summarized in Fig. 3. The stress intensity factors at failure are normalized vs the K_{If} of the largest sample $W = 198.41$ nm, and plotted versus the length of the singular stress field Λ_K normalized by the fracture process zone R_{FPZ}. Λ_K is the distance from the crack-tip at which the stress deviates more than 5% from the expected $1/r^{0.5}$, while the R_{FPZ} is a constant value taken from the literature [12, 13]. When continuum and atomistic simulations are overlapped, the continuum-based formulation breaks down when the ratio between Λ_K and R_{FPZ} is approximately 4-5, in agreement with [12, 13]. The atomistic simulations, instead, show a clearly scale-independence, confirming that ideal brittle fracture is ultimately governed by atomic bond breaking [7, 17].

Concluding, the concept of SIF is still surprisingly valid if the atoms are modeled, and static crack and ideal brittle material containing no other defects are considered. This result also agrees well with conclusions made by other authors [30]. Provided that the virial stress is affectively accepted as representative of atomic stress, crack tip region stress field of molecular system may be described by continuum-based SIF concept.

References

1. Kitamura, T., Sumigawa, T., Shimada, T., Van Lich, L.: Challenge toward nanometer scale fracture mechanics. Eng. Fract. Mech. **187**, 33–44 (2018)
2. Kitamura, T., Sumigawa, T., Hirakata, H., Shimada, T.: Fracture Nanomechanics, 2nd edn. Pan Stanford Publishing, Singapore (2016)
3. Xu, K., Li, B., Li, S., Luo, M., Gao, X., Jiang, C., Song, L.: In situ observation for the fatigue crack growth mechanism of 316L stainless steel fabricated by laser engineered net shaping. Int. J. Fatigue **130**, 105272 (2019). https://doi.org/10.1016/j.ijfatigue.2019.105272
4. Andric, P., Curtin, W.A.: Atomistic modeling of fracture. Model. Simul. Mater. Sci. Eng. **27** (1), 013001 (2018). https://doi.org/10.1088/1361-651x/aae40c
5. Ghayesh, M.H., Farajpour, A.: A review on the mechanics of functionally graded nanoscale and microscale structures. Int. J. Eng. Sci. **137**, 8–36 (2019). https://doi.org/10.1016/j.ijengsci.2018.12.001
6. Taloni, A., Vodret, M., Costantini, G., Zapperi, S.: Size effects on the fracture of microscale and nanoscale materials. Nat. Rev. Mater. **3**, 211–224 (2018). https://doi.org/10.1038/s41578-018-0029-4
7. Shimada, T., Kitamura, T.: Fracture mechanics at atomic scales. In: Altenbach, H., Matsuda, T., Okumura, D. (eds.) Advanced Structured Materials, pp. 379–396. Springer International Publishing, Cham (2015)
8. Kumar, S., Curtin, W.A.: Crack interaction with microstructure. Mater. Today **10**, 34–44 (2007). https://doi.org/10.1016/S1369-7021(07)70207-9
9. Wang, Y.Q., Fritz, R., Kiener, D., Zhang, J.Y., Liu, G., Kolednik, O., Pippan, R., Sun, J.: Fracture behavior and deformation mechanisms in nanolaminated crystalline/amorphous micro-cantilevers. Acta Mater. **180**, 73–83 (2019). https://doi.org/10.1016/j.actamat.2019.09.002
10. Pippan, R., Wurster, S., Kiener, D.: Fracture mechanics of micro samples: fundamental considerations. Mater. Des. **159**, 252–267 (2019). https://doi.org/10.1016/j.matdes.2018.09.004
11. Gallo, P., Sumigawa, T., Shimada, T., Yan, Y., Kitamura, T.: Investigation into the breakdown of continuum fracture mechanics at the nanoscale: synthesis of recent results on silicon. In: Gdoutos, E.E. (ed.) Proceedings of the First International Conference on Theoretical, Applied and Experimental Mechanics, pp. 205–210. Springer International Publishing, Cham (2019)
12. Gallo, P., Hagiwara, Y., Shimada, T., Kitamura, T.: Strain energy density approach for brittle fracture from nano to macroscale and breakdown of continuum theory. Theor. Appl. Fract. Mech. **103**, 102300 (2019). https://doi.org/10.1016/j.tafmec.2019.102300
13. Shimada, T., Ouchi, K., Chihara, Y., Kitamura, T.: Breakdown of continuum fracture mechanics at the nanoscale. Sci. Rep. **5**, 8596 (2015). https://doi.org/10.1038/srep08596
14. Gallo, P., Sumigawa, T., Kitamura, T., Berto, F.: Static assessment of nanoscale notched silicon beams using the averaged strain energy density method. Theor. Appl. Fract. Mech. **95**, 261–269 (2018). https://doi.org/10.1016/j.tafmec.2018.03.007
15. Gallo, P., Sumigawa, T., Kitamura, T.: Experimental characterization at nanoscale of single crystal silicon fracture toughness. Frat ed Integrità Strutt **13**, 408–415 (2019). https://doi.org/10.3221/IGF-ESIS.47.31
16. Gallo, P., Yan, Y., Sumigawa, T., Kitamura, T.: Fracture behavior of nanoscale notched silicon beams investigated by the theory of critical distances. Adv. Theory Simul. **1**, 1700006 (2018). https://doi.org/10.1002/adts.201700006

17. Sumigawa, T., Shimada, T., Tanaka, S., Unno, H., Ozaki, N., Ashida, S., Kitamura, T.: Griffith criterion for nanoscale stress singularity in brittle silicon. ACS Nano **11**, 6271–6276 (2017). https://doi.org/10.1021/acsnano.7b02493
18. Huang, K., Shimada, T., Ozaki, N., Hagiwara, Y., Sumigawa, T., Guo, L., Kitamura, T.: A unified and universal Griffith-based criterion for brittle fracture. Int. J. Solids Struct. **128**, 67–72 (2017). https://doi.org/10.1016/j.ijsolstr.2017.08.018
19. Zhou, M.: A new look at the atomic level virial stress: on continuum-molecular system equivalence. Proc. R. Soc. Math. Phys. Eng. Sci. **459**, 2347–2392 (2003). https://doi.org/10.1098/rspa.2003.1127
20. Cheng, S.H., Sun, C.T.: Convergence of local atomistic stress based on periodic lattice. Int. J. Solids Struct. **51**, 2027–2035 (2014). https://doi.org/10.1016/j.ijsolstr.2014.02.023
21. Subramaniyan, A.K., Sun, C.T.: Continuum interpretation of virial stress in molecular simulations. Int. J. Solids Struct. **45**, 4340–4346 (2008). https://doi.org/10.1016/j.ijsolstr.2008.03.016
22. Zimmerman, J.A., WebbIII, E.B., Hoyt, J.J., Jones, R.E., Klein, P.A., Bammann, D.J.: Calculation of stress in atomistic simulation. Modell. Simul. Mater. Sci. Eng. **12**(4), S319–S332 (2004). https://doi.org/10.1088/0965-0393/12/4/S03
23. Gallo, P.: On the crack-tip region stress field in molecular systems: the case of ideal brittle fracture. Adv. Theory Simul. **2**, 1900146 (2019). https://doi.org/10.1002/adts.201900146
24. Plimpton, S.: Fast parallel algorithms for short-range molecular dynamics. J. Comput. Phys. **117**, 1–19 (1995). https://doi.org/10.1006/jcph.1995.1039
25. Sillinger, F.H., Weber, T.A.: Computer simulation of local order in condensed phases of silicon. Phys. Rev. B **31**, 5262–5271 (1985). https://doi.org/10.1103/PhysRevB.31.5262
26. Irwin, G.R.: Linear fracure mechanics, fracture transition, and fracture control. Eng. Fract. Mech. **1**(2), 241–257 (1968)
27. Irwin, G.R.: Analysis of stresses and strains near the end of a crack traversing a plate. J. Appl. Mech. **24**, 361–364 (1957)
28. Sumigawa, T., Ashida, S., Tanaka, S., Sanada, K., Kitamura, T.: Fracture toughness of silicon in nanometer-scale singular stress field. Eng. Fract. Mech. **150**, 161–167 (2015). https://doi.org/10.1016/j.engfracmech.2015.05.054
29. Stan, G., Krylyuk, S., Davydov, A.V., Cook, R.F.: Bending manipulation and measurements of fracture strength of silicon and oxidized silicon nanowires by atomic force microscopy. J. Mater. Res. **27**, 562–570 (2011). https://doi.org/10.1557/jmr.2011.354
30. Le, M.Q., Batra, R.C.: Mode-I stress intensity factor in single layer graphene sheets. Comput. Mater. Sci. **118**, 251–258 (2016). https://doi.org/10.1016/j.commatsci.2016.03.027

Initial Intergranular Cracking of Ni-Base Superalloys Due to the Degradation of the Crystallinity of Grain Boundaries Under Creep-Fatigue Loading

Wataru Suzuki[1], Yifan Luo[1], Kenta Ishihara[1], Kens Suzuki[2], and Hideo Miura[2(✉)]

[1] Department of Finemechanics, Graduate School of Engineering, Tohoku University, Sendai, Miyagi 9808579, Japan
[2] Fracture and Reliability Research Institute, Graduate School of Engineering, Tohoku University, Sendai, Miyagi 9808579, Japan
hmiura@rift.mech.tohoku.ac.jp

Abstract. The mechanism of the drastic decrease in the lifetime of Ni-base superalloy, Alloy 617 under creep-fatigue loading at elevated temperatures was clarified by using EBSD (Electron Back-Scatter Diffraction) analysis. The degradation process was monitored by using an intermittent creep-fatigue test and EBSD analysis. The change of the crystallinity of grains and grain boundaries was quantitatively analyzed by using the image quality (IQ) value obtained from the EBSD analysis. The IQ value indicated the density of defects such as vacancies, dislocations, local strain, and so on. The decrease in the IQ value corresponded to the decrease in the crystallinity of the observed area. The accumulation of fine voids was found to be accelerated under the creep-fatigue loading, and it caused the drastic decrease of not only the IQ value, but also the strength of the grain boundaries. Intergranular cracking started to occur when the crystallinity of grain boundaries decreased to the critical value due to the degradation of the crystallinity caused by the local accumulation of dislocations and voids around the grain boundaries.

Keywords: Creep-fatigue damage · Ni-base superalloy · Intergranular cracking. EBSD analysis · EBSD analysis

1 Introduction

Ni-base superalloys are strong candidate for the various components used in advanced thermal and chemical plants and jet engines, such as pressure vessels, pipes, turbine blades, and so on, because of their high strength at elevated temperatures higher than 700 °C. In order to minimize the emission of CO_2 during their operations, it is very important to improve their thermal efficiency by elevating their operating temperatures. In the case of thermal power plants, the frequent random change of output is required to compensate the unexpected changes of the outputs of renewable energies, comparing with the constant output of the conventional plants. Thus, the creep-fatigue loads are applied to the components used in the advanced plants at the elevated temperatures.

E. Gdoutos and M. Konsta-Gdoutos (Eds.): ICTAEM 2020, STIN 16, pp. 325–331, 2020.
https://doi.org/10.1007/978-3-030-47883-4_58

It was reported that the effective lifetime of Ni-base superalloys dropped down drastically under the creep-fatigue loading at elevated temperatures [1–3]. The main reason for the drastic decrease was attributed to the change of the crack propagation path from conventional transgranular to intergranular. The mechanism of the change, however, is not clarified yet. Therefore, in this study, the degradation process of the strength of grain boundaries in Alloy 617, one of the representatives of Ni-base superalloy, was investigated in detail by using an intermittent creep-fatigue test and EBSD analysis. The degradation of the local crystallinity of grains and grain boundaries was continuously monitored in the same area.

2 Experimental Methods

Table 1 shows the chemical composition of Alloy 617 used in this study. This alloy consists of two phases, γ phase (Ni) and γ' phase (Ni_3Al). The alloy was heat-treated by solution annealing at 1150 °C for 1.12 h after hot rolling. After that a small specimen was cut into the dumbbell shape as shown in Fig. 1. This size was good for the continuous observation of the change of its crystallinity by EBSD in SEM. The intermittent creep-fatigue test was applied to the specimen at 800 °C in an inert gas of Ar. The test conditions were as follows: The maximum applied stress was 100 MPa, and the stress ration was 0.03. The loading rate was 100 N/s and the holding time at the maximum stress was 10 min. The test was frequently stopped at certain plastic strains because all the fractured specimen showed the almost the same normalized-time dependency of plastic strain.

Table 1. Chemical composition of Alloy 617 (wt%).

C	S	Cr	Ni	Mn	Si	Mo
0.06	<0.002	21.98	54.71	0.03	0.06	8.8
Ti	Cu	Fe	P	Al	Co	B
0.4	0.02	0.8	<0.002	1.14	11.69	0.001

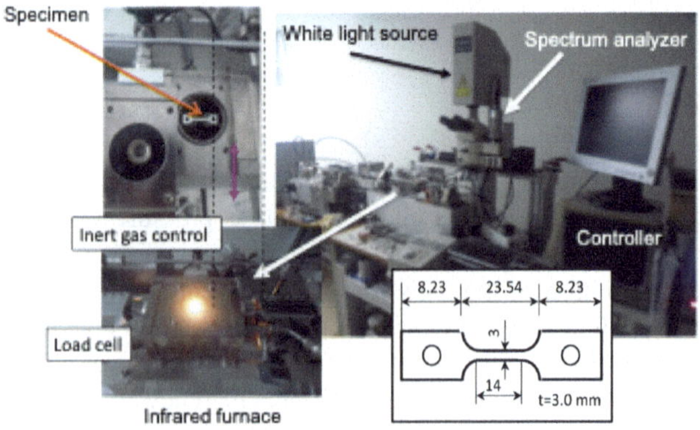

Fig. 1. Outlook of the test equipment and the shape of a test specimen.

The crystallinity of a grain and a grain boundary was quantitatively measured by using IQ value obtained from EBSD analysis. This value indicates the average sharpness of Kikuchi pattern obtained from the observed area. Since the clarity of the pattern is determined by the uniformity of the diffracted beams based on Bragg's law, it indicates the width of the fluctuation of the measured local atomic plane distance, and therefore, the density of various defects such as vacancies, dislocations, impurities, local strain, and so on which cause the variation of local lattice constant. It was confirmed that this value is effective for evaluating the quality of grains and grain boundaries in the observed area [4, 5]. In addition, it shows the strong relationship with the strength of a grain and a grain boundary [6].

The strength of a grain and a grain boundary was measured by using a micro tensile test method [7, 8], which consisted of a scanning electron microscope (Hitachi, SU-70), with focused ion beam (Hitachi, FB2200) as shown in Fig. 2. A bicrystal sample was cut from the tested specimen by considering the local IQ value of the two grains and the grain boundary in the sample.

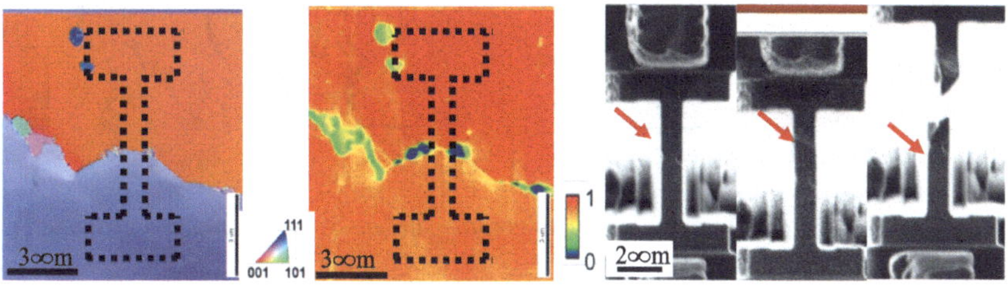

(a) IPF map (b) IQ map (c) Micro tensile test

Fig. 2. Micro tensile test of a bicrystal specimen.

3 Degradation of the Crystallinity of Alloy 617

The change of the surface morphology of the specimen observed in the same area during the creep-fatigue test is shown in Fig. 3. Initially, its surface was flat and no specific defects were observed on the surface. When the testing time t reached $t/t_f = 12\%$, where t_f is the rapture time, fine white dots started to appear along grain boundaries as shown in Fig. 3(b). These white dots were confirmed to be fine voids. When the normalized time t/t_f reached 48%, some grain boundaries perpendicular to the uniaxial loading direction started to crack as shown in Fig. 3(c). Therefore, the main degradation mechanism of Alloy 617 during the creep-fatigue test was the accumulation of vacancies along grain boundaries.

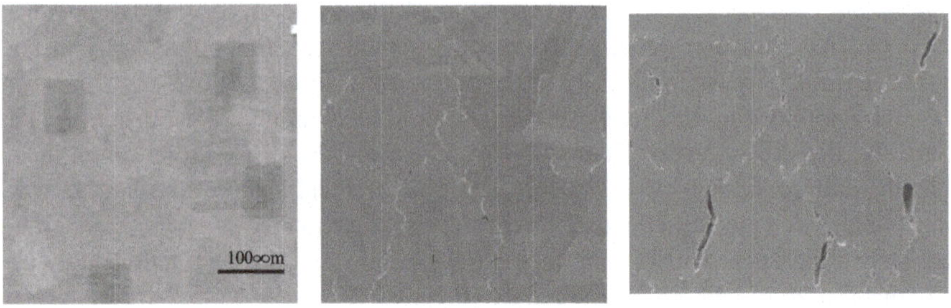

(a) $t/t_f = 0\%$ (b) $t/t_f = 12\%$ (c) $t/t_f = 48\%$

Fig. 3. Change of the surface morphology of the specimen during the creep-fatigue test.

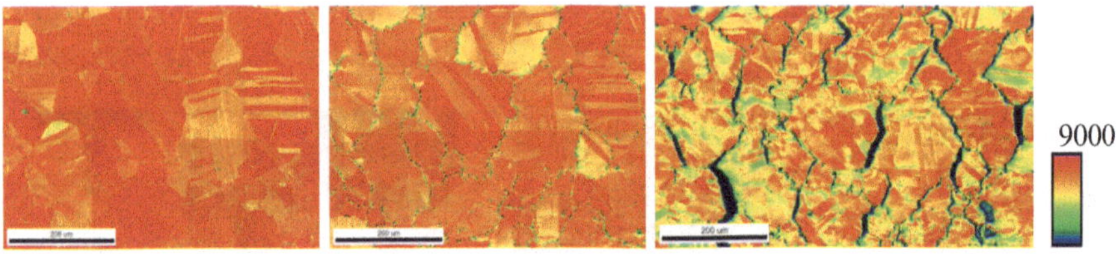

(a) $t/t_f = 0\%$ (b) $t/t_f = 12\%$ (c) $t/t_f = 48\%$

Fig. 4. Change of the distribution of IQ value on the surface of the specimen during the creep-fatigue test.

The degradation process was observed by using IQ value as shown in Fig. 4. The red color area is the area with high IQ value, in other word, high crystallinity, and the blue color indicates the low crystallinity. Initially, all the grains showed high crystallinity as shown in Fig. 4(a). The crsytallinity near grain boundaries started to degrade at $t/t_f = 12\%$ as shown in Fig. 4(b). The reason for the degradation was the accumulation of vacancies (fine voids). In addition, the accumulation of dislocations around grain boundaries was confirmed by KAM (Kernel Average Misorientation) value obtained from EBSD analysis. Thus, the vacancies were emitted from the heavily accumulated dislocations around the grain boundaries. Figure 4(c) shows the distribution of IQ value in the same area. Since plastic deformation proceeded, the shape of grains was deformed heavily comparing with that observed in Fig. 4(b). The crystallinity around grain boundaries were further degraded and a lot of intergranular cracks appeared. Most cracked grain boundaries were perpendicular to the uniaxial loading direction. Thus, the strength of these cracked grain boundaries was lower than that of uncracked grains.

Figure 5 shows the degradation of the stress-strain curve of the typical bicrystal structures cut from the specimen in Fig. 4(c). One bicrystal structure consisted of a grain boundary with relatively high IQ value of about 5080. This sample was fractured in the lower grain in the bicrystal structure. Clear plastic deformation of this sample was observed as shown in the figure (orange dotted line) and the tensile strength of this sample was about 650 MPa. Thus, the strength of the grain boundary of this sample was higher than this value. The other bicrystal sample consisted of the grain boundary with IQ value of about 3100. Intergranular fracture occurred in this sample clearly, and the strength of the grain boundary was about 400 MPa (black dots in Fig. 5). The lower IQ value of a grain boundary indicates the higher density of defects, in this sample, vacancies (fine voids). Therefore, this result clearly validated that there was a strong relationship between the crsytallinity and the strength of a grain boundary. The accumulation of vacancies in and around a grain boundary degrades the strength of the grain boundary. Therefore, the accumulation of fine voids during the creep-fatigue voids around grain boundaries degraded their strength drastically, and thus, caused intergranular cracking.

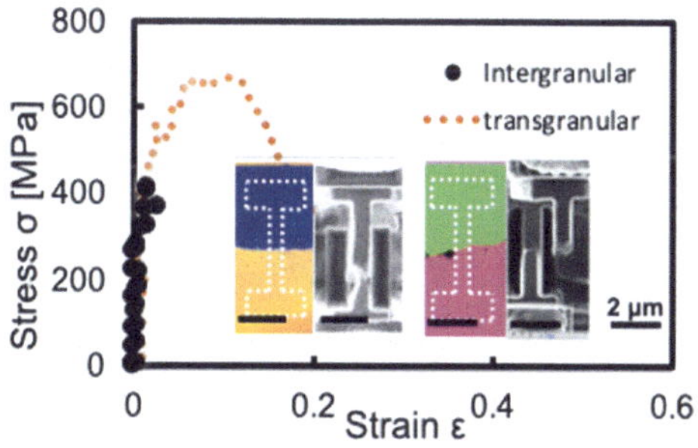

Fig. 5. Change of the distribution of IQ value on the surface of the specimen during the creep-fatigue test.

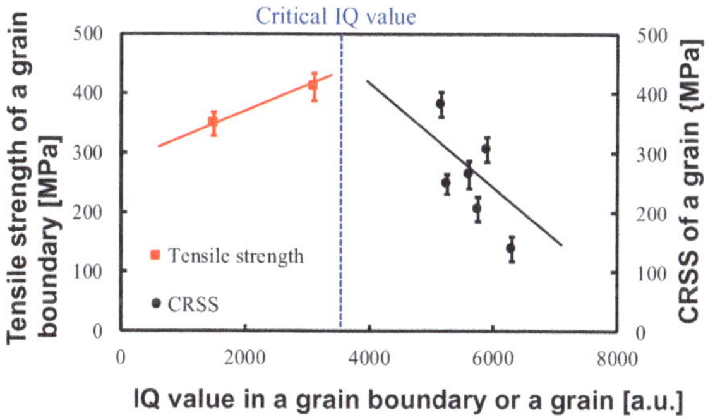

Fig. 6. Change of the distribution of IQ value on the surface of the specimen during the creep-fatigue test.

Figure 6 summarizes the IQ-value dependence of the strength of a grain boundary and the critical resolved stress of a grain measured by the micro tensile test. There was clear relationship between the IQ value and each strength. The strength of a grain boundary decreased monotonically with IQ value because of the decrease in the density of atomic bonding in the grain boundary due to the accumulation of fine voids. On the other hand, the critical resolved shear stress of a grain decreased monotonically with IQ value because a grain with higher IQ value consists of lower density of defects, in other words, pinning centers for the movement of dislocations. As a result, there should be a critical IQ value at which the fracture mode of a bicrystal sample change from ductile transgranular cracking to brittle intergranular one. Thus, intergranular cracking in Alloy 617 was dominated by the degradation of IQ value of grain boundaries.

Summary

The initial creep-fatigue damage of Ni-base superalloy, Alloy 617, at elevated temperature was analyzed continuously by suing an intermittent creep-fatigue test and EBSD analysis of a small dumbbell-shape specimen. It was found that the accumulation of fine voids started to appear in the early stage of the testing time of about 12%. The accumulation degraded the effective strength if grain boundaries. Once the crystallinity of a grain boundary reached a critical value, intergranular cracking started to occur. Thus, the initial intergranular cracking of the alloy was dominated by the acceleration of the accumulation of vacancies around grain boundaries under creep-fatigue loading.

Acknowledgments. This research was supported partly by JSPS KAKENHI Grant Number JP16H06357.

This research activity has been also supported partially by Japanese special coordination funds for promoting science and technology, Japanese Grants-in-aid for Scientific Research, and Tohoku University.

References

1. Carroll, L.: Progress Report on Long Hold-Time Creep-Fatigue of Alloy 617 at 850 °C. Idaho National Laboratory INL ART Program, INL/EXT-15-35132, 18 p (2015)
2. Murakoshi, T., Suzuki, K, Nonaka, I., Miura. H.: Microscopic analysis of the initiation of high-temperature damage of Ni-based heat-resistant alloy. In: Proceedings on ASME2016 International Mechanical Engineering Congress And Exposition (IMECE2016), Phoenix, AZ, USA, no. 67599, 6 p (2016)
3. Ishihara, K., Suzuki, W., Luo, Y., Suzuki K., Miura, H.: Initial grain-boundary cracking of Ni-base superalloys under creep-fatigue loadings at 750 °C. In: Proceedings on 13th International Conference on the Mechanical Behavior of Materials, Melbourne, pp. 342–345 (2019)
4. Murata, N., Suzuki, K., Miura, H.: Quantitative evaluation of the crystallinity of grain boundaries in polycrystalline materials. In: Proceedings on ASME2012 International Mechanical Engineering Congress and Exposition. (IMECE2012), Houston, TX, USA, no. 7426, 7 p (2012)
5. Murata, N., Saito, N., Tamakawa, K., Suzuki, K., Miura, H.: Effect of crystallographic quality of grain boundaries on both mechanical and electrical properties of electroplated copper thin film interconnections. J. Electronic Packaging ASME **137**(3), 031001 (2015)
6. ASME 2017 International Mechanical Engineering Congress and Exposition (2018). https://doi.org/10.1115/IMECE 2017-70302

7. Nakanishi, T., Suzuki, K., Miura, H.: Improvement of the long-term reliability of interconnections by controlling the crystallinity of grain boundaries. In: Proceedings on ASME2015 International Mechanical Engineering Congress and Exposition (IMECE2015), San Francisco, CA, USA, no. 48200, 6 p (2015)
8. Suzuki, W., Ishihara, K., Kikuchi, R., Suzuki, K., Miura, H.: Change of the effective strength of grain boundaries in Alloy 617 under creep-fatigue loadings at 800 °C. In: Proceedings on ASME2015 Int. Mechanical Engineering Congress and Exposition, (IMECE2015), Salt Lake City, UT, USA, no. 11210, 6 p (2019)

Characterization of Microstructurally Small Fatigue Crack Behavior

Evgenii Malitckii(✉), Heikki Remes, Pauli Lehto, and Sven Bossuyt

Department of Mechanical Engineering, Aalto University School of Engineering, Espoo, Finland
evgeny.malitckiy@aalto.fi

Keywords: Microstructurally small fatigue crack · Digital image correlation · Shear strain localization

1 Extended Abstract

Aiming to understand the fracture mechanism accompanying the small fatigue crack propagation in polycrystalline metallic materials like steels the full-field strain measurement using digital image correlation (DIC) method was performed. The developed experimental approach was applied to study the small fatigue crack behavior in body-centered Fe-Cr ferritic stainless steel with the relatively large grain size of about 350 μm. The high spatial measurement accuracy of the strain field analysis was performed using the unique patterning technique with a characteristic speckle size of approximately 10 μm. The experimental procedure is described in detail by Malitckii et al. [1]. The developed experimental procedure combines the full-field strain measurement, electron backscatter diffraction (EBSD) and fracture surface analysis in order to reveal the crack growth rate, path and mechanics for microstructurally small crack in high-cycle fatigue range. Figure 1 shows the schematic representation of the small fatigue crack growth mechanism.

DIC analysis evidences the periodic appearance of the pairs of highly tensioned and compressed strain localization areas within the grains forming simultaneously with the fatigue crack propagation (see insert of Fig. 1). Such a behavior manifests the presence of shear deformation that was found to be highly localized, while the macroscopic deformation is small. Cumulative analysis of the strain field and crack growth rate reveals that the observed shear strain localization areas are formed subsequently ahead of the crack tip and control apparently the small fatigue crack growth rate. The fatigue crack growth rate is affected by the accumulated shear strain deformation areas rather than by microstructural features such as grain boundaries.

The relationship between the accumulation of the shear deformation areas and grain orientation i.e. Schmid factor was studied in the first three grains, where the intermittent appearance of the shear localization zones ahead of the crack tip was observed. Schmid factor of the grains was found to be inversely proportional to the strain intensity factor calculated as quotient between cumulative maximum shear strain and deformation area. The shear localization at the grain boundary was found to be smaller compared to that within the grains. Worth to note, however, the small fatigue crack propagates across the grain boundary of low misorientation angle of about 15°–20°.

E. Gdoutos and M. Konsta-Gdoutos (Eds.): ICTAEM 2020, STIN 16, pp. 332–333, 2020.
https://doi.org/10.1007/978-3-030-47883-4_59

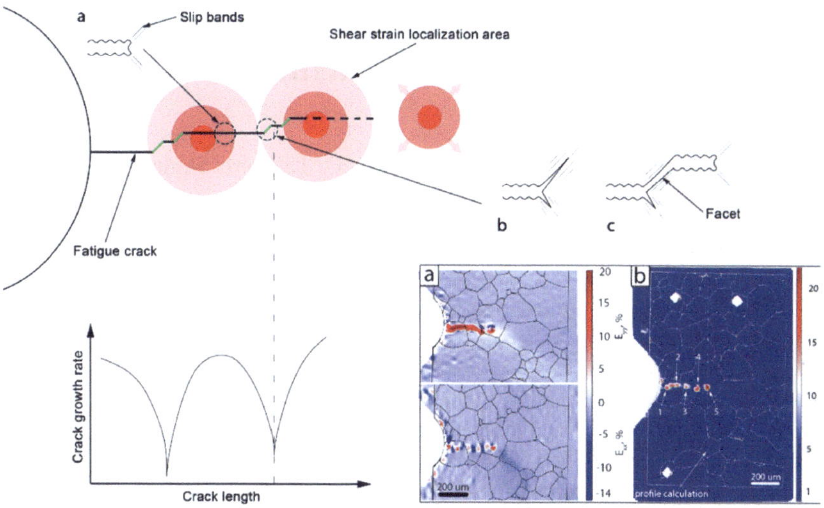

Fig. 1. The behavior of a small fatigue crack growth in a bcc ferritic steel (ASTM UNS S43940 ferritic stainless steel). Fatigue crack growth by Stage II multiple-slip mechanism is shown by the black line and fatigue crack growth by Stage I single-slip mechanism by the green line, respectively. After crack tip approaches the shear strain localization area the Stage II fatigue crack growth mechanism (a) changes intermittently to Stage I (b, c) forming a faceted profile. The intermittent change of the crack growth mechanism accompanied by fatigue crack growth rate retardation is shown schematically on the crack growth rate diagram. The insert at the right bottom corner shows strain field of linear deformations E_{xx} and E_{yy} (a) and maximum shear deformation (b) calculated by DIC method during a fatigue test of the ferritic stainless steel specimen [2].

The small fatigue crack growth rate decreases shortly while the crack tip approaches the localized deformation area. The crack retardation is accompanied by the change of the crack growth mechanism from multiple-slip (Stage II) to single-slip (Stage I). The cumulative maximum shear strain increases slightly with crack traversing from the first to the following grain, but the increase is controlled, apparently, by the change in stress conditions due to the reduction of the cross-section area of the specimen rather than by the orientation of the grains.

References

1. Malitckii, E., Remes, H., Lehto, P., Bossuyt, S.: Full-field strain measurement for microstructurally small fatigue crack propagation using digital image correlation method. J. Vis. Exp. (143), e59134 (2019)
2. Malitckii, E., Remes, H., Lehto, P., Yagodzinskyy, Y., Bossuyt, S., Hänninen, H.: Strain accumulation during microstructurally small fatigue crack propagation in bcc Fe-Cr ferritic stainless steel. Acta Mater. **144**, 51–59 (2018)

An Atomic-Level Unified Criterion for Brittle Fracture

Kai Huang[1,2(✉)], Takashi Sumigawa[1], Takahiro Shimada[1], and Takayuki Kitamura[1]

[1] Kyoto University, Kyoto 615-8540, Japan
huangkai@hit.edu.cn
[2] Harbin Institute of Technology, Harbin 150001, China

Abstract. This paper aims to propose a unified criterion for brittle fracture of general (blunt) notches and ideal (sharp) cracks. A new fracture parameter, i.e., atomic energy release rate (ERR), for which the atomic discrete nature is fully taken into account, is defined based on the concept of atomic fracture mechanics (AFM). The results show that the proposed atomic ERR criterion works well for a nano-sized notch even when the notched sample becomes extremely small. Compared with an ideal crack, it is found that the magnitude of the critical atomic ERR at fracture is in good agreement with that of an ideal crack. Hence, brittle fracture of both a notch and a crack can be described by using a unique fracture parameter, providing a unified description at the atomic level.

Keywords: Brittle fracture · Criterion · Atomic level

1 Introduction

Brittle fracture, a catastrophic failure pattern, has been always happening in a wide range of engineering materials. Generally, there are mainly two kinds of defects which can easily bring about fracture, i.e., sharp cracks and blunt notches. Due to the absence of singular stress field at the notch root, brittle fracture of a blunt notch is generally distinguished from a crack case on the basis of the continuum fracture mechanics [1].

In this paper, fracture experiments *in silico* are carried out by means of Molecular Statics (MS) simulations. The atomic ERR G_{atomic} is proposed by considering the discrete nature of brittle fracture at the atomic level. It is found that G_{atomic} criterion is able to describe brittle fracture of a notch. Finally, a unification of brittle fracture for different kinds of defects is provided.

2 Results and Discussion

Figure 1a illustrates the geometry and loading condition for the nano-sized notched sample. As shown in Fig. 1a, the notch is orientated along the (111) cleavage plane with the $[\bar{1}10]$ front. The sample height H is relatively large enough compared with width W to eliminate the boundary effect. The notch length a is one third of W. A series

E. Gdoutos and M. Konsta-Gdoutos (Eds.): ICTAEM 2020, STIN 16, pp. 334–336, 2020.
https://doi.org/10.1007/978-3-030-47883-4_60

of W (= 3.86, 19.85, 33.15, 46.46, 66.42 and 79.73 nm) are adopted. Tension tests are carried out by means of MS simulations with bond-order potential of the modified Stillinger-Weber (SW) form [2, 3].

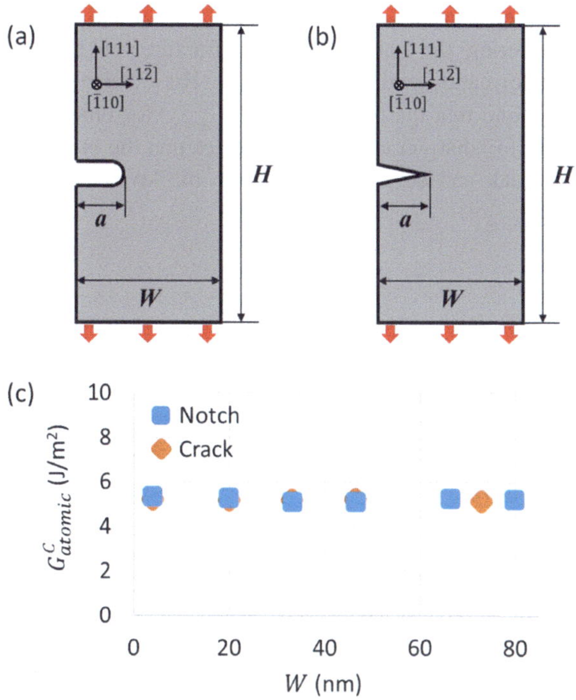

Fig. 1. Geometry and loading condition of the nano-sized sample with a blunt notch (a) and a sharp crack (b), respectively; (c) Critical atomic ERR G^C_{atomic} as a function of sample size W for notched and cracked samples.

The conventional ERR criterion has been one of the most widely accepted criteria for brittle fracture on the basis of Griffth theory, which is expressed as [4, 5]

$$G = -\frac{d\Pi(A)}{dA} = -\lim_{\Delta A \to 0} \frac{\Pi(A + \Delta A) - \Pi(A)}{\Delta A} \tag{1}$$

where $\Pi(A)$ is the strain energy of a continuum system with a crack cross-section A. However, G criterion cannot apply to a blunt notch, because the conventional G criterion requires the existence of a finite crack.

On the basis of AFM, in this study we propose a new of definition for the atomic ERR G_{atomic} with consideration of atomic discreteness, i.e., [6, 7]

$$G_{atomic} = -\frac{\Delta\Pi_{atomic}(A)}{\Delta A} = -\frac{\Pi_{atomic}(A + \Delta A) - \Pi_{atomic}(A)}{\Delta A} \tag{2}$$

where ΔA is the finite change of the new surface by the single bond breaking at the notch root. Figure 1c shows the critical atomic ERR G^C_{atomic}, when fracture occurs, as a function of sample size W. As the sample shrinks down, G^C_{atomic} remains at a constant level. Therefore, G_{atomic} criterion is effective to describe brittle fracture of a blunt notch despite the sample size.

Moreover, considering the nature of brittle fracture, an atomic event of bond breaking, the results for nano-sized cracks (see Fig. 1b) are also provided in Fig. 1c as a comparison. It is found that the magnitudes of G^C_{atomic} for blunt notches are in good agreement with those for sharp cracks. This indicates that the brittle fracture of a blunt notch and a sharp crack can be unified at the atomic level by the proposed G_{atomic} criterion.

3 Conclusions

In this study, we have proposed a new atomic ERR by considering the atomic discreteness of brittle fracture. The atomic ERR criterion works well for describing the brittle fracture of a blunt notch despite the extremely small sample size. A unification for notches and cracks is finally provided with an in-depth understanding of brittle fracture at the atomic level.

Acknowledgments. This study is supported by the Grant-in-Aid for Specially Promoted Research (Grant number 25000012) and Scientific Research(s) (Grant number 18H05241) from the Japan Society of Promotion of Science (JSPS), and China Postdoctoral Science Foundation (Grant number 2019M661267).

References

1. Ritchie, R.O.: The conflicts between strength and toughness. Nat. Mater. **10**, 817–822 (2011)
2. Stillinger, F.H., Weber, T.A.: Computer simulation of local order in condensed phases of silicon. Phys. Rev. B **31**, 5262–5271 (1985)
3. Holland, D., Marder, M.: Ideal brittle fracture of silicon studied with molecular dynamics. Phys. Rev. Lett. **80**, 746–749 (1998)
4. Griffith, A.A.: The phenomena of rupture and flow in solids. Phil. Trans. Roy. Soc. Lond. A **221**, 163–198 (1920)
5. Orowan, E.: Fracture and strength of solids. Rep. Prog. Phys. **12**, 185–232 (1949)
6. Shimada, T., Ouchi, K., Chihara, Y., Kitamura, T.: Breakdown of continuum fracture mechanics at the nanoscale. Sci. Rep. **5**, 8596 (2015)
7. Huang, K., Shimada, T., Ozaki, N., Hagiwara, Y., Sumigawa, T., Guo, L., Kitamura, T.: A unified and universal Griffith-based criterion for brittle fracture. Int. J. Solids Struct. **128**, 67–72 (2017)

Fatigue of Micro/Nano Metals

Takayuki Kitamura[1(✉)], Takashi Sumigawa[1], and Kai Huang[1,2]

[1] Kyoto University, Kyoto 615-8540, Japan
Kitamura@kues.kyoto-u.ac.jp
[2] Harbin Institute of Technology, Harbin 150001, China

Abstract. As materials with nanoscale dimensions are used in various industrial applications such as semiconductors, sensors, and micro- or nano-electromechanical-systems, the investigations on their mechanical behavior have been attracting many engineers/researchers. Especially, the knowledge on fracture mechanics/mechanism in the nanometer scale becomes important in terms of the reliability of small devices. Of course, this contributes an essential understanding of the strength of macro-components as well. The fatigue behavior is one of the key issues in terms of reliability.

Extensive experimental works have been conducted on the crack initiation mechanism in the fatigue of bulk metals. These have reported that the dislocation structures induce specific slips on the surface known as persistent slip bands (PSBs), which brings about a crack on the surface. However, the size of under-structure is in micron-scale independent of specimen size. As small metals in the micron- or nanometer-scale cannot have enough space to form the understructure, the cracking mechanism must be different from the one of bulk counterparts. My team has investigated the fatigue behavior of micro- and nano-metals in this decade [1–11]. In this talk, we present the characteristic fatigue damage/failure process and mechanics of small metals. Especially, we introduce our novel experimental methodology of TEM/SEM *in situ* observations for the investigation.

Keywords: Fatigue · Micro/Nano-scale · Metals · *In situ*

Acknowledgments. This study is supported by the Grant-in-Aid for Scientific Researches (S) (Grant number 18H05241) and (A)(18H03753) from the Japan Society of Promotion of Science (JSPS), and China Postdoctoral Science Foundation (Grant number 2019M661267).

References

1. Sumigawa, T., Murakami, T., Shishido, T., Kitamura, T.: Cu/Si interface fracture due to fatigue of cooper film in nanometer scale. Mater. Sci. Eng. A **527**, 6518–6523 (2010)
2. Sumigawa, T., Murakami, T., Kitamura, T.: Fatigue strength of Cu/Si interface in nano-component. Mater. Sci. Eng. A **528**, 5158–5163 (2011)
3. Yan, Y., Sumigawa, T., Kitamura, T.: Effect of environment on fatigue strength of Cu/Si interface in nanoscale components. Mater. Sci. Eng. A **556**, 147–154 (2012)
4. Sumigawa, T., Mtatsumoto, K., Tsuchiya, T., Kitamura, T.: Fatigue of 1 μm-scale gold by vibration with reduced resonant frequency. Mater. Sci. Eng. A **556**, 429–436 (2012)

E. Gdoutos and M. Konsta-Gdoutos (Eds.): ICTAEM 2020, STIN 16, pp. 337–338, 2020.
https://doi.org/10.1007/978-3-030-47883-4_61

5. Sumigawa, T., Shiohara, R., Mtatsumoto, K., Kitamura, T.: Characteristic feature of slip bands in submicron single crystal gold in fatigue. Acta Mater. **61**, 2692–2700 (2013)
6. Sumigawa, T., Mtatsumoto, K., Fang, H., Kitamura, T.: Formation of slip bands in nano-polycrystalline copper under high-cycle fatigue of Si/Ti/Cu/SiN nano-scale material. Mater. Sci. Eng. A **608**, 221–228 (2014)
7. Fang, H., Shiobara, R., Sumigawa, T., Kitamura, T.: Size dependence of fatigue damage in sub-micrometer single crystal gold. Mater. Sci. Eng., A **618**, 416–423 (2014)
8. Sumigawa, T., Byungwoon, K., Mizuno, Y., Morimura, T., Kitamura, T.: *In situ* observation on formation process of nanoscale cracking during tension-compression fatigue of single crystal copper micron-scale specimen. Acta Mater. **153**, 270–278 (2018)
9. Sumigawa, T., Uegaki, S., Yukishita, T., Arai, S., Takahashi, Y., Kitamura, T.: FE-SEM in situ observation of damage evolution in tension-compression fatigue of micro-sized single-crystal copper. Mater. Sci. Eng., A **764**, 138218 (2019)
10. Huang, K., Sumigawa, T., Kitamura, T.: Load-dependency of damage process in tension-compression fatigue of microscale single-crystal copper. Int. J. Fatigue **133**, 105415 (2020)
11. Yan, Y., Sumigawa, T., Wang, X., Chen, W., Xuan, F., Kitamura, T.: Fatigue curve of microscale single-crystal copper: An *in situ* SEM tension-compression study. Int. J. Mech. Sci. **171**, 105361 (2020)

Author Index

A
Abrosimov, Nikolai, 309
Afanasieva, O. A., 181
Al-Abedy, H. K., 106
Alexandrov, Sergei, 151, 158
Ananchenko, Nikita, 280

B
Babenko, Andriy, 280
Barile, Claudia, 26
Bikakis, George S. E., 20
Bogdanov, Viacheslav, 228, 268
Bogdanova, Olga, 235
Borovyk, Y. V., 198
Bossuyt, Sven, 332
Boulay, C., 64

C
Cao, Jingjing, 86
Casavola, Caterina, 26
Cendón Franco, D. A., 166
Chen, Daolun, 14
Chen, Hao, 106
Chen, Tianyu, 3
Churilov, Yuriy, 301

D
Danoglidis, Panagiotis A., 58, 88
Dmytrakh, I. M., 32
Dovzhyk, Mykhailo, 268
Dvorak, Jiri, 93

F
Fedotenkov, Grigory, 287, 312
Fedotova, Daria, 117
Feng, Aihan, 14
Firsov, V. A., 254

G
Gallo, Pasquale, 319
Gazizullin, R. K., 248, 254
Gdoutos, Emmanuel E., 58
Godio-Raboutet, Y., 64
Gorokhov, Vasilii, 301, 309
Granichin, Nickolay, 100
Gusev, Boris, 162

H
Harvey, Christopher M., 3
Huang, C., 9
Huang, Kai, 334, 337
Hughes, James Martin, 133

I
Igumnov, Leonid, 305, 309
Ilchuk, Nataliia, 242
Ishihara, Kenta, 325

J
Jiang, Peicheng, 151

K
Kalfountzos, Costas D., 20
Kandala, S. M., 198
Kandala, Stepan, 280
Kapustin, Sergei, 301
Kazakov, Dmitriy, 301
Kholmogorov, S. A., 293
Kipnis, A. L., 216
Kitamura, Takayuki, 334, 337

E. Gdoutos and M. Konsta-Gdoutos (Eds.): ICTAEM 2020, STIN 16, pp. 339–341, 2020.
https://doi.org/10.1007/978-3-030-47883-4

Ko, Tsai-Wen, 168
Kong, Guofang, 86
Konsta-Gdoutos, Maria S., 58, 88
Kotousov, Andrei, 133, 145
Kral, Petr, 93
Kryvyi, Oleksandr, 204
Kucharova, Kveta, 93
Kulkarni, Rishikesh, 76
Kuo, Tzu-Yu, 82
Kushnir, Roman, 210
Kvapilova, Marie, 93
Kyrylova, O., 187

L
Lai, Hsiang-Wei, 168
Lang, Lihui, 151
Lehto, Pauli, 332
Leshchak, R. L., 32
Li, Hezong, 52, 106
Litvinchuk, Svetlana, 305
Liu, Jiahua, 38
Liu, Yang, 38, 70
Lobanov, L., 273
Lobanov, L. M., 198
Logachev, Andrey, 100
Luo, Yifan, 325

M
Maglogianni, Myrsini E., 88
Makarov, M. V., 293
Makhnenko, O., 273
Makhnenko, O. V., 198
Makhnenko, Oleh, 280
Malitckii, Evgenii, 332
Mikhailova, Elena, 312
Milenin, Alexey, 175
Miura, Hideo, 325
Morozov, Yurii, 204

N
Nazarenko, V. M., 216
Nazarenko, Vladimir, 228, 268
Ng, Ching-Tai, 133, 145
Novoseltseva, Nadezhda, 309

P
Paimushin, V. N., 248, 254, 293
Pappalettera, Giovanni, 26
Pappalettere, Carmine, 26
Pasternak, Iaroslav, 242, 261
Patil, Shanta, 76
Pérez Gallego, D., 166
Petrov, Yuri, 100
Pham, Thong D., 222
Pivtorak, Nina, 175
Poliakov, Vladimir, 164
Popov, V., 187
Pshenichnov, Sergey, 193
Pustovoy, A., 273

Q
Qin, Bo, 44
Qu, Shoujiang, 14

R
Rastogi, Pramod, 76
Remes, Heikki, 332
Rozynka, Galina, 175
Ruiz Hervías, J., 166

S
Saurin, Vasily, 162, 164
Scherbakov, Vasily, 140
Shimada, Takahiro, 334
Shishkin, V. M., 254
Shishov, M. A., 293
Shlyannikov, V. N., 112
Shlyannikov, Valery, 117
Silberschmidt, Vadim V., 3
Sklenicka, Vaclav, 93
Solovey, S., 273
Song, Xiu, 38, 70
Sulym, Heorhiy, 242, 261
Sumigawa, Takashi, 334, 337
Sun, Wei, 106
Suzuki, Kens, 325
Suzuki, Wataru, 325
Svoboda, Milan, 93
Syrotyuk, A. M., 32

T
Tailhan, J.-L., 64
Tarlakovskii, Dmitry, 287, 312
Tarlakovskii, Dmitry V., 222
Theotokoglou, Efstathios E., 20
Tokovyy, Yuriy, 210
Tumanov, A. V., 112

V
Vasylyshyn, Andrii, 261
Velikoivanenko, Elena, 175
Vestyak, Vladimir, 140
Vestyak, Vladimir A., 222
Vidler, James, 145
Vilotic, Dragisa, 158
Vilotic, Marko, 158
Vimalathithan, Paramsamy Kannan, 26

Volkov, Grigory, 100
Volkov, Ivan, 305
Volkov, V. S., 198

W
Wang, Lei, 38, 70
Wang, Simon, 3
Wang, Wei-Chung, 82
Wang, Xiaoyu, 86
Wang, Yanhui, 52
Wang, Yao, 70
Wang, Yun-Che, 168
Wang, Zhichong, 127
Wu, Ningxiang, 44

X
Xie, Liyang, 44

Y
Yang, Qian, 52
Yang, Zhinan, 52

Z
Zemskov, A. V., 181
Zhang, Fucheng, 52
Zhang, Yinling, 14
Zhao, Yifan, 86
Zhegalov, Dmitriy, 301
Zhelyazov, Todor, 170

The manufacturer's authorised representative in the EU is Springer Nature Customer Service Centre GmbH, Europaplatz 3, 69115 Heidelberg, Germany. If you have any concerns regarding our products, please contact ProductSafety@springernature.com

Printed and bound by CPI Group (UK) Ltd, Croydon, CR0 4YY
05/10/2026
02228124-0013